Sense and Avoid in UAS

Aerospace Series List

Sense and Avoid in UAS: Research and Applications
Angelov
April 2012

Morphing Aerospace Vehicles and Structures
Valasek
March 2012

Gas Turbine Propulsion Systems
MacIsaac and Langton
July 2011

Basic Helicopter Aerodynamics, Third Edition
Seddon and Newman
June 2011

Advanced Control of Aircraft, Rockets and Spacecraft
Tewari
July 2011

Cooperative Path Planning of Unmanned Aerial Vehicles
Tsourdos et al.
November 2010

Principles of Flight for Pilots
Swatton
October 2010

Air Travel and Health: A Systems Perspective
Seabridge et al.
September 2010

Design and Analysis of Composite Structures: With Applications to
Aerospace Structures
Kassapoglou
September 2010

Unmanned Aircraft Systems: UAVS Design, Development and Deployment
Austin
April 2010

Introduction to Antenna Placement and Installations
Macnamara
April 2010

Principles of Flight Simulation
Allerton
October 2009

Aircraft Fuel Systems
Langton et al.
May 2009

The Global Airline Industry
Belobaba
April 2009

Computational Modelling and Simulation of Aircraft and the Environment:
Volume 1 – Platform Kinematics and Synthetic Environment
Diston
April 2009

Sense and Avoid in UAS

Research and Applications

Edited by

Plamen Angelov

School of Computing and Communications, Lancaster University, UK

A John Wiley & Sons, Ltd., Publication

Library of Congress Cataloging-in-Publication Data

Sense and avoid in UAS : research and applications / edited by Plamen Angelov.
    p. cm.
  Includes bibliographical references and index.
  ISBN 978-0-470-97975-4 (hardback)
 1. Airplanes–Collision avoidance.   2. Drone aircraft–Control systems.   I. Angelov, Plamen P.
  TL696.C6S46 2012
  629.135'2–dc23   2011044007

A catalogue record for this book is available from the British Library.

ISBN: 978-0-470-97975-4

Typeset in 10/12pt Times by Aptara Inc., New Delhi, India
Printed and bound in Singapore by Markono Print Media Pte Ltd

Contents

Part IV   SAA Applications

Preface

This book is very special in several respects. On the one hand, it is the first of its kind where the reader can find in one place recent research results by leading academics from British, American, Australian and European universities as well as reports of implementation activities by leading industry-based researchers from giants like Boeing and authorities like MITRE. On the other hand, it combines topics such as human factors and regulation issues with technical aspects such as sensors, algorithms, methodologies and results. It is also unique because it reports the latest results from simulations, real experiments and implementation. Further, because the area of unmanned aircraft systems (UAS) is projected to grow exponentially in the next few decades in the most developed countries. Because of its nature (being closer to defence developments and thus being less open), publications (especially books, guides, instructions and reviews) are difficult to access. Indeed, the UAS market is forecast to grow from its present $5.9B to $11.3B annually during the next decade, totalling $94B for the period [1]. Moreover, it is envisaged that the F-35 Lightning II (Joint Strike Fighter) and respectively the Russian equivalent T-50 (PAK-FA) fifth generation jets will be the last major manned fighter aircraft types and the focus will shift to UAS. Large (multimillion) research and development programmes such as ASTRAEA, Taranis, SUAV[E], Mantis, etc. have taken place in the UK and similarly in the USA (two Grand Challenge competitions by DARPA; WASP III, Raven, Scan Eagle, MQ-9 and MQ-18, RQ-4 Blk and more recently, X47-B and RQ-170 (Sentinel) which was downed recently over Iran), leading European countries (France, Sweden, Germany, Czech Republic) and Israel during the last decade or so. UAS are critically important for future military capability in areas such as intelligence, surveillance, suppression of enemy air defence, close air support, situational awareness and missile defence. Their role in operations in Afghanistan and Libya cannot be underestimated. In 2009, the US Air Force started training more pilots to operate unmanned systems than to fly fighters and bombers [2]. The US Congress has mandated that, by 2015, one-third of ground combat vehicles will be unmanned [2].

There is also an embryonic, but very fast growing, civil market for UAS in areas as diverse and important for society as the police force, fire service, ambulance, coast guard, air sea rescue, fishing patrols, mountain rescue, utility companies, highway agencies, environmental protection, agriculture, nuclear industry, volcanoes research, postal services, communications, etc. It is reported [3] that currently there are some 300 UAS worldwide with over 100 (unsurprisingly) in the USA, followed by France and Russia and (somewhat surprisingly) the UK in 13th position with only 5, behind Switzerland, Norway, the Czech Republic, Japan and Israel.

Yet, the number of publications – and especially organised in books, guides and proceedings – on this specific topic of obvious interest is insignificant, if not non-existent. This book aims to fill the gap.

Before the reader is engulfed by technical details, it is worthwhile outlining the main topic, problem and terminology. First of all, it is important to clarify the meaning of the terms *autonomy* and *autonomous*. Broadly speaking, an autonomous system is one that can operate (including make decisions, plan actions, reach goals) without human intervention in any environmental conditions. In this sense, an autonomous system possesses a much higher level of automation and a higher level of complexity and intelligence than a (simply) automatic system, the theory (and industrial applications) of which was well developed half a century ago. In a more narrow sense, they distinguish different levels of autonomy, where the highest, sixth level is 'full autonomy' as described above. Below that there are five more levels starting from the lowest, first level of 'human operated' system, which often takes the form of a remotely operated vehicle (ROV). At this level, all the activities of the system are directly initiated by the human operator and the system has no control over the environment. The second, higher level, which can be called a 'human assisting' system, can perform actions if asked and authorised by the human. It can also be called 'advice only if requested' type of autonomy. The human asks the machine to propose actions and the human selects the actual action. At the higher, third level, which can be called 'human delegated', the machine suggests options to the human. The difference with the previous level is that it provides advice/suggestions even if not asked. Such a UAS can perform limited control activity on a delegated basis, for example automatic flight control, engine control. All of these, however, are being activated and deactivated by the human operator. A UAS of level four, which may be called 'human supervised' or 'advise and if authorised act', can suggest options and even propose one of them. This needs to be approved by the human operator, though, before being undertaken/activated! The penultimate level five, which can be called 'machine backed by human' or 'act unless revoked', includes UAS which can choose actions and perform them unless a human operator disapproves. This is, in fact, the highest level of autonomy of practical interest, because the highest level of 'full autonomy' is somewhat controversial (see, for example, Isaac Azimov's principles of robotics [4]).

In conclusion, there are several levels of *autonomy* and of practical interest are all levels but the last, the highest. Autonomous systems differ significantly from automatic systems known and used for over half a century. For the example of airborne systems, an automatic system would include vehicles that fly on a pre-programmed route through waypoint navigation, with landing controlled by the ground stations, payload switching on and off at predetermined points in the flight plan and capable of tracking a target. A UAS of interest (that is the subject of this book and offers huge potential for both military and civil applications) includes vehicle(s) that fly a mission based on tasks but has the ability to autonomously and adaptively react to threats and an evolving situation awareness capability, can adapt (evolve) the mission on the fly, where the payload can detect and manage the target and optimise performance, that can be activated and deactivated and where the interface between the ground and the vehicle is mission (task and information)-based, not control-based.

The topic of sense and avoid (SAA), which is also closely related to the term 'see and avoid' used in manned aircraft, is extremely important and was one of the main obstacles for wider application of UAS in non-segregated airspace related to the traffic safety and level of intelligence of the flying machines that are being produced and used both in military/defence and civilian domains. It has several aspects, including:

(a) Regulatory (traffic safety, rules of the air or rules of engagement, level of human involvement and autonomy, etc.).

(b) Technical (sensors, data processing, situation awareness and decision-making, aerodynamic limitations, etc.).

It has very intrinsic and strong links with a range of science and engineering subjects, such as:

- system engineering;

- automatic control;

- aerodynamics;

- image and video processing;

- machine learning and real-time data processing;

- decision-making;

- human–computer interaction, etc.

In this book, all of these issues are considered at some level of detail – including the implementation and experimental work which demonstrates ways to address or resolve them.

The book is composed of four parts, each one with a specific emphasis, namely Part I: Introduction (Chapters 1–3), Part II: Regulatory Issues and Human Factors (Chapters 4 and 5), Part III: Sense and Avoid Methodologies (Chapters 6–8) and, finally, Part IV: Sense and Avoid Applications (Chapters 9–11). The contributors are all experts in their field, and detailed biographies of each contributor can be found in About the Contributors at the start of the book.

An important goal of this book is to have a one-stop shop for engineers and researchers in this fast-moving and highly multi-disciplinary area, which covers many (if not all) aspects of the methodology and implementation of these new, exciting, yet challenging devices and complex artificial (yet very intelligent) systems which are bound to grow in number and complexity over the next decade and beyond. The aim was to combine the solid theoretical methodology based on a rigorous mathematical foundation, present a wide range of applications and, more importantly, provide illustrations that can be a useful guide for further research and development.

References

1. Teal Report, 2011. http://tealgroup.com/index.php?option=com_content&view=article&id=74 :teal-group-predicts-worldwide-uav-market-will-total-just-over-94-billion-&catid=3&Itemid=16. Accessed on 18 July 2011.

2. L. G. Weiss. 'Autonomous robots in the fog of war'. *IEEE Spectrum*, 8, 26–31, 2011.

3. UVS International. 2009/2010 UAS Yearbook, *UAS: The Global Perspective*, 7th edn, June 2009.

4. I. Azimov. 'The machine that won the war' (originally published in 1961), reprinted in I. Asimov, *Robot Dreams*. Victor Gollancz, London, pp. 191–197, 1989.

About the Editor

Plamen Angelov

Plamen Angelov is a Reader in Computational Intelligence and coordinator of the Intelligent Systems Research at Infolab21, Lancaster University, UK. He is a Senior Member of the Institute of Electrical and Electronics Engineers (IEEE) and Chair of two Technical Committees (TC): the TC on Standards, Computational Intelligence Society and the TC on Evolving Intelligent Systems, Systems, Man and Cybernetics Society. He is also a member of the UK Autonomous Systems National TC, of the Autonomous Systems Study Group, NorthWest Science Council, UK and of the Autonomous Systems Network of the Society of British Aerospace Companies. He is a very active academic and researcher who has authored or co-authored over 150 peer-reviewed publications in leading journals, 50+ peer-reviewed conference proceedings, a patent, a research monograph, a number of edited books, and has an active research portfolio in the area of computational intelligence and autonomous system modelling, identification and machine learning. He has internationally recognised pioneering results in online and evolving methodologies and algorithms for knowledge extraction in the form of human-intelligible fuzzy rule-based systems and autonomous machine learning. Angelov is also a very active researcher leading projects funded by EPSRC, ASHRAE-USA, EC FP6 and 7, The Royal Society, Nuffield Foundation, DTI/DBIS, MoD and other industry players (BAE Systems, 4S Information Systems, Sagem/SAFRAN, United Aircraft Corporation and Concern Avionica, NLR, etc.).

His research contributes to the competitiveness of the industry, defence and quality of life through projects such as ASTRAEA – a £32M (phase I and £30M phase II) programme, in which Angelov led projects on collision avoidance (£150K, 2006/08) and adaptive routeing (£75K, 2006/08). The work on this project was recognised by The Engineer Innovation and Technology 2008 Award in two categories: (i) Aerospace and Defence and (ii) The Special Award. Other examples of research that has direct impact on the competitiveness of UK industry and quality of life are the BAE Systems-funded project on sense and avoid (principal investigator, £66K, 2006/07), BAE-funded project on

UAS passive sense, detect and avoid algorithm development (£24K consultancy, a part of ASTRAEA-II, 2009), BAE Systems-funded project (co-investigator, £44K, 2008) on UAV safety support, EC-funded project (€1.3M, co-investigator) on safety (and maintenance) improvement through automated flight data analysis, Ministry of Defence-funded projects ('Multi-source Intelligence: STAKE: Real-time Spatio-Temporal Analysis and Knowledge Extraction through Evolving Clustering', £30K, principal investigator, 2011 and 'Assisted Carriage: Intelligent Leader–Follower Algorithms for Ground Platforms', £42K, 2009 which developed an unmanned ground-based vehicle prototype taken further by Boeing-UK in a demonstrator programme in 2009–11), the £9M project GAMMA: Growing Autonomous systems Mission Management, 2011–2014, in which PI of £480K work); funded by the Regional Growth Fund, UK Government; the £3M project CAST (Co-ordinated Airborne Studies in the Tropics) which envisages usage of the Global Hawk with NASA so-called 'innovation vouchers' by the North-West Development Agency-UK and Autonomous Vehicles International Ltd (£10K, 2010, principal investigator), MBDA-led project on algorithms for automatic feature extraction and object classification from aerial images (£56K, 2010) funded by the French and British defence ministries. Angelov is also the founding Editor-in-Chief of Springer's journal *Evolving Systems*, and serves as an Associate Editor of several other international journals. He chairs annual conferences organised by the IEEE, acts as Visiting Professor (2005, Brazil; 2007, Germany; 2010, Spain) and regularly gives invited and plenary talks at leading companies (Ford, Dow Chemical USA, QinetiQ, BAE Systems, Thales, etc.) and universities (Michigan, USA; Delft, the Netherlands; Leuven, Belgium; Linz, Austria; Campinas, Brazil; Wolfenbuettel, Germany; etc.).

More information can be found at www.lancs.ac.uk/staff/angelov.

About the Contributors

Chris Baber

Chris Baber is the Chair of Pervasive and Ubiquitous Computing at the University of Birmingham. His research interests focus on the many ways in which computing and communications technologies are becoming embedded in the environment around us and the things we use on a daily basis. Not only do we have significant computing power in the mobile phone in our pocket, but, increasingly, other domestic and personal products are gaining similar capabilities. Chris is interested in how such technologies will develop and how they will share the information they collect, and also in how these developments will affect human behaviour.

Cristina Barrado

Cristina Barrado was born in Barcelona in 1965 and is a computer science engineer from the Barcelona School of Informatics, which belongs to the Technical University of Catalonia (UPC). She also holds a PhD in Computer Architecture from the same university. Dr Barrado has been working with UPC since 1989 and is currently an associate professor at the School of Telecommunications and Aerospace Engineering of Castelldefels (Escola d'Enginyeria de Telecommunicació i Aeroespacial de Castelldefels, EETAC). Her current research interests are in the area of the UAS civil mission, including payload processing, avionics CNS capabilities and non-segregated airspace integration.

Richard Baumeister

Richard Baumeister from the Boeing Company has over 30 years' experience performing system engineering and management of complex missile and space programs. From 1979 to 1982 Rich was the lead mission planner and orbital/software analyst for the F-15 ASAT Program. In 1982–1986 Rich helped supervise the integration and operations of the Prototype Mission Operations Center into the NORAD Cheyenne Mountain Complex.

From 1987 to 1995 Rich was the Systems Engineering Manager for a classified complex national space system. During this period Rich oversaw the successful development of innovative techniques for the detection and resolution of system anomalies.

From 1996 to 2004 Rich was Director of Product Development for RESOURCE21 LLC, a Boeing-funded joint venture. Rich led the technical research and development of aerial and space-based remote sensing-based algorithms and associated information products for Production Agriculture, Commodities, Crop Insurance, and Forestry markets. He directed and participated in the creation of numerous proprietary research papers/presentations dealing with the detection of various crop stresses using multi-spectral imagery. Rich successfully managed the development of an atmospheric correction process and decision support tools in support of a commercial collection campaign.

From 2005 to the present Rich has been supporting automated air traffic control concepts and algorithms, and was the lead engineer for Boeing on the recently completed Smart Skies program.

Rich received his PhD in Mathematics/Physics from the University of Arizona in 1977 and was an Assistant Professor of Mathematics at Arizona State University prior to joining the Boeing company.

Marie Cahillane

Marie received her first degree, majoring in psychology, in 2003 from Bath Spa University and an MSc in research methods in psychology in 2005 from the University of Bristol. Marie was awarded her PhD in cognitive psychology in 2008, from the University of the West of England. Whilst conducting her doctoral research she lectured in psychology at Bath Spa University. Marie's research interests and expertise are in cognition and perception and her teaching specialisms include research methods in psychology, in particular quantitative methods and experimental design. Marie joined Cranfield Defence and Security as a Research Fellow in 2008 and is now a Lecturer in Applied Cognitive Psychology. At Cranfield Defence and Security, Marie leads several human factors research projects within the military domain. Research includes the acquisition and retention of skills required to operate systems and human interaction with complex systems.

Luis Delgado

Luis Delgado is an aeronautical engineer from the National School for Civil Aviation (École Nationale de l'Aviation Civile or ENAC) in Toulouse, France. He also holds a degree in Computer Science Engineering from the Barcelona School of Informatics (Facultat d'Informàtica de Barcelona, FIB) which belongs to the Technical University of Catalonia (Universitat Politècnica de Catalunya, UPC). He earned both degrees in 2007. His research interests include improving the performance and efficiency of the air traffic management (ATM) system and flexible, reliable and cost-efficient unmanned aircraft systems (UAS) operations in civil airspace.

He has been working with UPC since 2007 and currently is an assistant professor at EETAC. He is also a PhD student of the Aerospace Science and Technology doctorate program from UPC and expects to graduate in 2012.

Jason J. Ford

Jason J. Ford was born in Canberra, Australia in 1971. He received the BSc and BE degrees in 1995 and a PhD in 1998 from the Australian National University, Canberra. He was appointed a research scientist at the Australian Defence Science and Technology Organisation in 1998, and then promoted to a senior research scientist in 2000. He has held research fellow positions at the University of New South Wales, at the Australian Defence Force Academy in 2004 and at the Queensland University of Technology in 2005. He has held an academic appointment at the Queensland University of Technology since 2007. He has had academic visits to the Information Engineering Department at the Chinese University of Hong Kong in 2000 and to the University of New South Wales at the Australian Defence Force Academy from 2002 to 2004. He was awarded the 2011 Spitfire Memorial Defence Fellowship. His interests include signal processing and control for aerospace.

Štěpán Kopřiva

Štěpán Kopřiva is a researcher and PhD student at the Agent Technology Center of the Gerstner Laboratory, Department of Cybernetics, Czech Technical University. Štěpán graduated in 2009 from Imperial College London with an MSc degree in Advanced Computing. Prior to his current position, he worked as a programmer for the major POS systems manufacturer and researcher ATG.

Štěpán currently works on the AgentFly project – large-scale simulation and control in the air-traffic domain. His main research interests are logics and formal methods for multi-agent systems, classical planning, and large-scale simulations.

John Lai

John Lai was born in Taipei, Taiwan, in 1984. He received the BE (First Class Honours) degree in Aerospace Avionics in 2005 and a PhD in 2010, both from the Queensland University of Technology (QUT), Brisbane, Australia. Since obtaining his PhD, he has held a research fellow position at the Australian Research Centre for Aerospace Automation (ARCAA) – a joint research collaboration between the Commonwealth Scientific and Industrial Research Organisation (CSIRO) and QUT.

Juan Manuel Lema

Juan Manuel Lema was born in Montevideo, Uruguay in 1985 and is a technical telecommunications engineer from EETAC. He also holds a Master of Science in Telecommunications Engineering and Management. Mr Lema began his collaboration with the ICARUS group in January 2007, where he is a junior researcher. Currently he is a PhD student in the Computer Architecture doctoral program about UAS mission management.

George Limnaios

George Limnaios is a Major(Eng) at Hellenic Airforce. Since 1996 when he graduated from Hellenic Airforce Academy as an Avionics and Telecommunications Engineer he has been involved in the maintenance and support of A-7 and F-16 aircrafts serving the latter as a Technical Advisor and head of Quality Assurance Department. He is on educational leave seeking a post-graduate degree at the Technical University of Crete (Department of Electronic and Computer Engineering). His research interests include Renewable Energy Systems, Fault Tolerant Control, Fault Detection and Isolation and Unmanned Systems.

Luis Mejias

Luis Mejias received a degree in Electronic Engineering in November 1999 from UNEXPO (Venezuela), an MSc in Network and Telecommunication Systems from ETSIT-Universidad Politecnica de Madrid and a PhD from ETSII-Universidad Politecnica de Madrid. He has gained extensive experience with UAVs, investigating computer vision techniques for control and navigation. Currently, he is a lecturer in Aerospace Avionics at Queensland University of Technology, and a researcher at ARCAA.

Caroline Morin

Caroline obtained an M.A. and a Ph.D. in cognitive psychology from Laval University (Canada). She moved to the UK to take up a research fellowship at the University of Warwick where she was looking at the interaction between time and memory. In 2008, Caroline joined Cranfield University as a Research Fellow where she is leading a number of projects on Human Factors with a military population. Caroline's expertise is in human memory, categorization, time perception, decision making and human factors.

Peter O'Shea

Peter O'Shea is a Professor of Electrical Engineering at the Queensland University of Technology (QUT), Australia. He received the BE, DipEd and PhD from the University of Queensland, and then worked as an engineer at the Overseas Telecommunications Commission for three years. He has held academic appointments at RMIT's School of Electrical and Computer Systems Engineering for 7 years and at QUT's School of Engineering Systems for 10 years. He has won teaching awards from both the RMIT and QUT University Presidents, and has also won national teaching awards from Engineers Australia and the Australian Learning & Teaching Council. He was a co-recipient of the best technical paper award at the 2005 IEEE TENCON Conference. His interests are in (i) signal processing for communications, aerospace and power systems; (ii) reconfigurable computing; and (iii) engineering education.

Enric Pastor

Enric Pastor was born in Barcelona in 1968 and is a computer science engineer from the Barcelona School of Informatics, which belongs to the Technical University of Catalonia (UPC). He also holds a PhD in Computer Architecture from the same university. Dr Pastor has been working with UPC since 1992 and is currently an associate professor at EETAC. His research interests include new UAS architectures and the automation of mission processes in UAS civil applications.

Michal Pěchouček

Michal Pěchouček works as a Professor in Artificial Intelligence at the Department of Cybernetics, Czech Technical University, Prague. He graduated in Technical Cybernetics from FEE-CTU, obtained his MSc degree in IT: Knowledge Based Systems from the University of Edinburgh and completed his PhD in Artificial Intelligence and Biocybernetics at the CTU, Prague. He is Head of the Agent Technology Center at the Department of Cybernetics.

His research focuses on problems related to multi-agent systems, especially topics related to social knowledge, meta-reasoning, acting in communication inaccessibility, coalition formation, agent reflection, and multi-agent planning. Michal is an author or co-author of cited publications in proceedings of international conferences and journal papers. In addition, he is a member of the program committee of relevant conferences and workshops.

Xavier Prats

Xavier Prats is an aeronautical engineer from ENAC. He also holds a degree in Telecommunications Engineering from Telecom Barcelona (Escola Tècnica Superior d'Enginyeria de Telecomunicació de Barcelona, ETSETB) which belongs to the Technical University of Catalonia (Universitat Politècnica de Catalunya, UPC) in Barcelona (Spain). He earned both degrees in 2001. Furthermore, he received his PhD in Aerospace Science and Technology from UPC in 2010. His research interests include improving the performance and efficiency of the air traffic management (ATM) system and flexible, reliable and cost-efficient unmanned aircraft systems (UAS) operations in civil airspace.

He has been working with UPC since 2001 and currently is an assistant professor at EETAC. He co-founded the ICARUS research group and currently leads the group's air transportation research activities.

Jorge Ramirez

Jorge Ramirez is an aeronautical engineer from ENAC. He also holds a degree in Computer Science Engineering from the Barcelona School of Informatics (Facultat d'Informàtica de Barcelona, FIB) which belongs to the Technical University of Catalonia (Universitat Politècnica de Catalunya, UPC). He earned both degrees in 2000. His research interests include flexible, reliable and cost-efficient unmanned aircraft systems (UAS) operations in civil airspace and the use and optimization of communications navigation and surveillance (CNS) technologies for UAS.

He has been working with UPC since 2007 and currently is a lecturer at the Castelldefels School of Technology (Escola Politècnica Superior de Castelldefels or EPSC). He is also a PhD student of the Aerospace Science and Technology doctorate program from UPC and expects to graduate in 2012. Before joining UPC, Jorge was a software engineer at GMV during the 2000–2002 period and worked on the operational implementation of the European Geostationary Navigation Overlay Service (EGNOS). During the period 2002–2007 he worked as a system engineer at EADS–CASA, focusing on the interoperability assessment of tactical datalink systems in different projects such as the European airlifter A400M, the British tanker FSTA and the Australian MRTT.

Pablo Royo

Pablo Royo is a telecommunications engineer from EETAC. He earned his degree in 2004. Furthermore, he received his PhD in Computer Architecture from the same university in 2010. His research interests include improving the performance and efficiency of the air traffic management (ATM) system and flexible, reliable and cost-efficient unmanned aircraft systems (UAS) operations in civil airspace.

He has been working with UPC since 2002 and currently is a lecturer at the EETAC.

Eduard Santamaria

Eduard Santamaria was born in Sant Pere Pescador in 1974 and is an informatics engineer from the Barcelona School of Informatics, which belongs to the Technical University of Catalonia (UPC). He also holds a PhD in Computer Architecture from the same university. Dr Santamaria has been working with UPC since 2000 and is currently a lecturer at the School of Telecommunications and Aerospace Engineering of Castelldefels. His research is focused on mechanisms for mission specification and execution for UAS.

Hyo-Sang Shin

Hyo-Sang Shin is Lecturer on Guidance, Control and Navigation Systems in Centre for Autonomous Systems Group at Cranfield University, Defence College of Management and Technology. He gained an MSc on flight dynamics, guidance and control in Aerospace Engineering from KAIST and a PhD on cooperative missile guidance from Cranfield University. His experties include guidance, navigation, and control of UAVs, complex weapon systems, and spacecraft. He has published over 35 journal and conference papers and has been invited for many lectures both in Universities and industries mainly on path planning, cooperative control, collision avoidance and trajectory shaping guidance. His current research interests include cooperative guidance and control for multiple vehicles, optimal and adaptive nonlinear guidance, integrated guidance and control algorithm, coordinated heath monitoring and management, and air traffic management and sense-and-avoid for UAV.

David Šišlák

David Šišlák is a senior research scientist in the Agent Technology Center at the Department of Cybernetics, Czech Technical University, Prague. He is the chief system architect for the AgentFly and Aglobe systems. He participates in many research projects related to these systems, funded by Czech and also foreign research sponsors. His research interests are in technical cybernetics and multi-agent systems, focusing on decentralized collision avoidance algorithms in air-traffic domain, efficient communication, knowledge maintenance in inaccessible multi-agent environment, large-scale multi-agent simulations and agent frameworks.

David received a Master's degree in Technical Cybernetics and a PhD in Artificial Intelligence and Biocybernetics from the Czech Technical University, Prague. David is an author or co-author of many cited publications in proceedings of international conferences and journal papers. During his PhD studies, he obtained the IEEE/WIC/ACM WI-IAT Joint Conference 'Best Demo' Award, the international Cooperative Information Agents (CIA) workshop system innovation award for the Aglobe multi-agent platform and related simulations, and later he was a member of a team which won the main Engineering Academy prize of the Czech Republic. In 2011, David received the Antonin Svoboda prize for the best dissertation of 2010 awarded by the Czech Society for Cybernetics and Informatics.

Graham Spence

Graham Spence graduated from the University of Leeds (UK) in 1995 with a BSc in Computer Science with Artificial Intelligence. He continued as a postgraduate research student at Leeds and in 1999 was awarded his PhD on the subject of High Temperature Turbulent Diffusion Flame Modelling. For the next several years Graham worked in industry as a computer programmer, but was drawn back to a research post at the University of Sheffield (UK) in 2003, where he researched and developed a real-time model of aircraft interactions during wake vortex encounters. The project successfully integrated into a research flight simulator, large datasets resulting from large eddy simulations of the decay of aircraft wake vortices, enabling real-time simulations of the fly-through of computational fluid dynamics data. After completion of this project, Graham continued at the University of Sheffield where he researched and developed several automated airspace collision detection and avoidance algorithms. Recently, Graham has been involved in an international project that aimed to develop and demonstrate automation technologies that could assist with the challenge of UAS integration into non-segregated airspace. Graham currently works for Aerosoft Ltd in Sheffield (UK) and his research interests include aircraft separation algorithms, flight simulation, aircraft wake vortex interaction, data compression, computer networking and the application of recent smart phone and tablet technologies to airspace safety.

Antonios Tsourdos

Antonios Tsourdos is a Professor and Head of the Centre for Autonomous Systems at Cranfield University, Defence Academy of the United Kingdom. He was member of the Team Stellar, the winning team for the UK MoD Grand Challenge (2008) and the IET Innovation Award (Category Team, 2009). Antonios is an editorial board member of the Proceedings of the IMechE Part G Journal of Aerospace Engineering, the International Journal of Systems Science, the IEEE Transactions of Instrumentation and Measurement, the International Journal on Advances in Intelligent Systems, the Journal of Mathematics in Engineering, Science and Aerospace (MESA) and the International Journal of Aeronautical and Space Sciences. Professor Tsourdos is a member of the ADD KTN National Technical Committee on Autonomous Systems. Professor Tsourdos is co-author of the book *Cooperative Path Planning of Unmanned Aerial Vehicles* and over 100 conference and journal papers on guidance, control and navigation for single and multiple autonomous vehicles.

Nikos Tsourveloudis

Nikos Tsourveloudis is a Professor of Manufacturing Technology at the Technical University of Crete (TUC), Chania, Greece, where he leads the Intelligent Systems and Robotics Laboratory and the Machine Tools Laboratory. His research interests are mainly in the area of autonomous navigation of field robots. His teaching focuses on manufacturing and robotic technologies and he has published more than 100 scientific papers on these topics. Tsourveloudis serves on the editorial board of numerous scientific journals and conferences. He is a member of professional and scientific organizations around the globe, and several public organizations and private companies have funded his research.

Tsourveloudis' research group has been honored with several prizes and awards, among which the most recent are: the 3rd EURON/EUROR Robotic Technology Transfer Award (2009); the 1st ADAC Car Safety Award (2010 and 2011); and the Excellent Research Achievements Award by the TUC (2010). In 2010/2011 he held a Chair of Excellence in Robotics at the University Carlos III of Madrid (UC3M), Spain.

Kimon P. Valavani

Kimon P. Valavanis is currently Professor and Chair of the ECE Department at the School of Engineering and Computer Science, University of Denver (DU), and Director of the DU Unmanned Systems Laboratory. He is also Guest Professor in the Faculty of Electrical Engineering and Computing, Department of Telecommunications, University of Zagreb, Croatia.

Valavanis' research interests are focused in the areas of Unmanned Systems, Distributed Intelligence Systems, Robotics and Automation. He has published over 300 book chapters, technical journal/transaction and referred conference papers. He has authored, co-authored or edited 14 books, the two most recent ones being: *On Integrating Unmanned Aircraft Systems in to the National Airspace System: Issues, Challenges, Operational Restrictions, Certification, and Recommendations* (K. Dalamagkidis, K. P. Valavanis, L. A. Piegl), 2nd Edition, Springer 2012; *Linear and Nonlinear Control of Small Scale Unmanned Rotorcraft* (I. A. Raptis, K. P. Valavanis), Springer, 2011. Since 2006, he is Editor-in-Chief of the Journal of Intelligent and Robotic Systems. Valavanis has been on the organizing committee of many conferences, he is a senior member of IEEE and a Fellow of the American Association for the Advancement of Science. He is also a Fulbright Scholar.

Přemysl Volf

Přemysl Volf holds a Master's degree in Software Systems from the Faculty of Mathematics and Physics at Charles University, Prague. He is currently a researcher and PhD student at the Agent Technology Center of the Gerstner Laboratory, Department of Cybernetics, Czech Technical University. His research is focused on distributed cooperative algorithms used for collision avoidance in air traffic control and verification of these algorithms using theory and prototypes.

Rod Walker

Rod has degrees in Electrical Engineering, Computer Science and a PhD in Satellite Navigation and Electro-magnetics, the latter involving a year-long sabbatical at the Rutherford Appleton Laboratory, Oxford, UK. From 1997 to 2005 he was the program leader for the GNSS payload on 'FedSat', working closely with NASA's Jet Propulsion Lab in Pasadena, CA. From 1999 to 2009 he taught in QUT's Bachelor of Aerospace Avionics. He rose to the position of Professor of Aerospace Avionics at QUT in 2008. During this time he was involved in training over 300 aerospace engineers. He is the foundation director for the Australian Research Centre for Aerospace Automation (ARCAA).

Brian A. White

Brian A. White is Professor Emeritus at Cranfield University. His areas of expertise are robust control, non-linear control, estimation, and observer applications, navigation and path planning, decision making, guidance design, soft computing, and sensor and data fusion. He has published widely over his career in all of the areas with well over 100 papers. He has been invited for many keynote lectures, both in Universities and at International conferences, topics being mainly on autonomy, decision making, path planning in recent years. He has served on many editorial boards and working groups, both within the UK and Internationally. He was also a key member of the Stellar Team that won the MOD Grand Challenge, where many of the techniques mentioned in this proposal were implemented within an autonomous system comprising several UAVs and a UGV.

Michael Wilson

Michael Wilson is a Senior Researcher at Boeing Research and Technology – Australia, specialising in unmanned aircraft systems. Michael has worked on the Smart Skies project since 2007. During this time he was also involved in the first commercially-oriented trials of the ScanEagle in non-segregated civilian airspace.

Michael joined Boeing in 2000 and worked on the modelling and analysis of wireless and networked systems, the design and testing of signal and waveform detection algorithms and the modelling of antenna systems. Michael has also spent some time as a consultant and a part-time lecturer.

Michael started his career working on Australia's over-the-horizon radar programme. His research focussed on the effects of the radio wave propagation environment on the design and the performance of radar systems.

Michael gained his PhD in 1995, from the University of Queensland, where he used a phased-array radar to study ionospheric disturbances.

Andrew Zeitlin

Andrew Zeitlin leads the Sense & Avoid product team within RTCA SC-203, bringing this activity his experience with avionics standards and implementation. He is considered an eminent expert in collision avoidance, having devoted more than 30 years to spearheading the development and standardization of TCAS aboard commercial aircraft, and is currently co-chairing the Requirements Working Group of SC-147. He received the John C. Ruth Digital Avionics Award from the AIAA in 2007. He received a BSEE from the University of Pennsylvania, an MSEE from New York University, and a DSc from George Washington University.

Part I

INTRODUCTION

1

Introduction

**George Limnaios,* Nikos Tsourveloudis* and
Kimon P. Valavanis†**
*Technical University of Crete, Chania, Greece
†University of Denver, USA

1.1   UAV versus UAS

An unmanned aerial vehicle (UAV), also known as a drone, refers to a pilotless aircraft, a flying machine without an onboard human pilot or passengers. As such, 'unmanned' implies the total absence of a human who directs and actively pilots the aircraft. Control functions for unmanned aircraft may be either onboard or off-board (remote control). That is why the terms remotely operated aircraft (ROA) and remotely piloted vehicle (RPV) are in common use as well [1]. The term UAV has been used for several years to describe unmanned aerial systems. Various definitions have been proposed for this term, like [2]:

> A reusable[1] aircraft designed to operate without an onboard pilot. It does not carry passengers and can be either remotely piloted or pre-programmed to fly autonomously.

Recently, the most reputable international organizations – like the International Civil Aviation Organization (ICAO), EUROCONTROL, the European Aviation Safety Agency (EASA), the Federal Aviation Administration (FAA) – as well as the US Department of Defense (DoD), adopted unmanned aircraft system (UAS) as the correct official term. The changes in acronym are caused by the following aspects:

---

[1] The characterization reusable is used to differentiate unmanned aircraft from guided weapons and other munitions delivery systems.

Sense and Avoid in UAS: Research and Applications, First Edition. Edited by Plamen Angelov.
© 2012 John Wiley & Sons, Ltd. Published 2012 by John Wiley & Sons, Ltd.

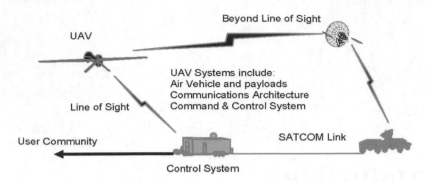

Figure 1.1   A typical UAS [62]

- The term 'unmanned' refers to the absence of an onboard pilot.

- The term 'aircraft' signifies that it is an aircraft and as such properties like airworthiness will have to be demonstrated.

- The term 'system' is introduced to signify that UAS is not just a vehicle but a (distributed) system consisting of a ground control station, communication links and launch and retrieval systems in addition to the aircraft itself.

A typical UAS comprises system elements in three major segments, as shown in Figure 1.1.

- **Air segment.** This includes one or more unmanned aircrafts (UAs) with their payloads. Each UA includes the airframe, the avionics and propulsion system components. The payload consists of sensor components that support mission requirements; sensors include cameras, radar, altimeter, inertial measurement unit (IMU), global positioning system (GPS), antennas, etc.

- **Ground segment.** This refers to the ground control station (GCS), which includes components like the payload control station/ground data terminal (GDT) and, if necessary, the launch and recovery system (LRS). The GCS includes all required equipment for the UA pilot, flight planning and mission monitoring. It also translates pilot inputs into the appropriate commands to be transmitted over the communication link to the aircraft.

- **Communications segment.** This is divided into the Command & Control data link, the Payload data link and External Communications. The term 'link' may be defined based on the distance the UAS is operating at: visual line of sight (VLOS), line of sight (LOS) or beyond line of sight (BLOS).

The FAA defines an *unmanned aircraft* or UA as [3]:

A device used or intended to be used for flight in the air that has no onboard pilot. This includes all classes of airplanes, helicopters, airships, and translational lift aircraft that have no onboard pilot. Unmanned aircraft are understood to include only those aircraft controllable in three axes and therefore, exclude traditional balloons.

As a comparison, the definition of *unmanned vehicle* given in the 2007–2012 Unmanned Systems Roadmap is also provided [4]:

> A powered vehicle that does not carry a human operator, can be operated autonomously or remotely, can be expendable or recoverable, and can carry a lethal or nonlethal payload. Ballistic or semi-ballistic vehicles, cruise missiles, artillery projectiles, torpedoes, mines, satellites, and unattended sensors (with no form of propulsion) are not considered unmanned vehicles. Unmanned vehicles are the primary component of unmanned systems.

Without loss of generality, the term UAV or UA is used to refer to an unmanned aircraft; the term UAS is used in instances where other parts of the system (like the control station) are relevant. The same terms will be used when referring to one or multiple systems.

1.2   Historical Perspective on Unmanned Aerial Vehicles

The best way to present the evolution of UAVs over the years is through a series of figures. The starting point is Ancient Greece and it continues to the beginning of the 21st century. An effort is made to arrange these figures chronologically; most have been taken from archives and other online sources. The layout and contents are similar to Chapter 1 of [5].

The first reported breakthrough work on autonomous mechanisms is attributed to Archytas from the city of Tarentum in South Italy, who was known as Archytas the Tarantine, also referred to as the Leonardo da Vinci of the Ancient World. He created the first UAV of all time in 425 B.C. by building a mechanical bird, a pigeon that could fly by moving its wings, getting energy from a mechanism in its stomach (see Figure 1.2). It is alleged that it flew about 200 meters before falling to the ground, once all its energy was used. The pigeon could not fly again [6], unless the mechanism was reset.

Figure 1.2 An artist's depiction of the flying pigeon, the first documented UAV in history. It is believed that it flew for 200 meters [5]

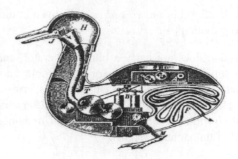

Figure 1.3 A similar 'flying bird' with a mechanism in its stomach, attributed to an engineer during the Renaissance [5]

During the same era, the Chinese were the first to develop the idea of vertical flight. The earliest version of a top consisted of feathers at the end of a stick. The stick was spun between the hands to generate enough lift, before being released into free flight.

More than seventeen centuries later, the initial idea attributed to Archytas surfaced again: a similar 'flying bird', credited to some unknown engineer of the Renaissance, was documented (see Figure 1.3). It is not known whether this new design was based on Archytas' idea; however, the concept was very similar.

Leonardo da Vinci, in 1483, designed an aircraft capable of hovering, called an aerial crew or air gyroscope, as shown in Figure 1.4. It had a 5-meter diameter and the idea was to make the shaft turn and, if enough force was applied, the machine could spin and fly. This machine is considered by some experts to be the ancestor of today's helicopter [7, 8]. Da Vinci also devised a mechanical bird in 1508 that could flap its wings by means of a double crank mechanism as it descended along a cable.

Many more flying machines were designed between 1860 and 1909, initially focusing on vertical take-off and landing aircraft because of the limitations of the steam-powered engines that were in use at the time. These machines led to the aircraft designs that are in use today. The first 'modern' UAV was manufactured in 1916 by the Americans Lawrence and Sperry [9]. It is shown in Figure 1.5. They developed a gyroscope to stabilize the body, in order to

Figure 1.4 Leonardo da Vinci's aerial crew (Hiller Aviation Museum [8])

Figure 1.5   The 'aviation torpedo' of Lawrence and Sperry [9]

manufacture an autopilot. This is known as the beginning of 'attitude control', used for the automatic steering of an aircraft. They called their device the 'aviation torpedo' and Lawrence and Sperry actually flew it a distance that exceeded 30 miles.

The main drive behind aircraft development has always been the fast and safe transportation of people and cargo. Nevertheless, the military soon realized the potential benefits of unmanned aircraft and efforts to adapt flying machines to operate without a pilot onboard started. Such systems were initially unmanned ordinance delivery systems, what would now be referred to as 'missiles' or 'smart bombs'. Another use for such systems was to operate as 'drones', to assist in the training of anti-aircraft gun operators.

Probably the first unmanned aircraft that can withstand today's definition of UAS was the Ryan Model 147 series aircraft shown in Figure 1.6. It was based on a drone design and was

Figure 1.6   Several variations of the Ryan Model 147 unmanned reconnaissance drone used in the 1960s and 1970s [5]

Figure 1.7   The MQ-1 Predator built by General Atomics Aeronautical Systems Inc. [10]

used for reconnaissance missions by the USA over China, Vietnam and other countries in the 1960s and 1970s.

After the Vietnam War, the USA and Israel began to develop smaller and cheaper UAVs. These were small aircraft that adopted small engines such as those used in motorcycles or snowmobiles. They carried video cameras and transmitted images to the operator's location. It seems that the prototype of the present UAV can be found in this period. The USA put UAVs into practical use in the Gulf War in 1991, and UAVs for military applications developed quickly after this. The most famous UAV for military use is the Predator, which is shown in Figure 1.7.

On the other hand, NASA is at the center of the research for civil use. The most typical example is the ERAST (Environmental Research Aircraft and Sensor Technology) project. It started in the 1990s, and was a synthetic research endeavor for a UAV that included the development of the technology needed to fly at high altitudes of up to 30,000 m, along with a prolonged flight technology, engine, sensor, etc. The aircraft developed in this project include Helios, Proteus, Altus, Pathfinder, etc., some of which are shown in Figures 1.8 and 1.9 [11]. These were designed to carry out environmental measurements.

Figure 1.8   The Helios UAV [11]

Figure 1.9   The Proteus by NASA [11]

1.3   UAV Classification

During recent decades, significant efforts have been devoted to increasing the flight endurance and payload of UAVs, resulting in various UAV configurations with different sizes, endurance levels and capabilities. This has led to attempts to explore new and somewhat unconventional configurations. Here, an attempt is made to classify UAVs according to their characteristics (aerodynamic configuration, size, etc.). Despite their diversity, UAV platforms typically fall into one of the following four categories:

- **Fixed-wing UAVs**, which refer to unmanned airplanes (with wings) that require a runway to take off and land, or catapult launching. These, generally, have long endurance and can fly at high cruising speeds (see Figure 1.10 for some examples).

- **Rotary-wing UAVs**, also called rotorcraft UAVs or vertical take-off and landing (VTOL) UAVs, which have the advantage of hovering capability and high maneuverability. These capabilities are useful for many robotic missions, especially in civilian applications. A rotorcraft UAV may have different configurations, with main and tail rotors (conventional helicopter), coaxial rotors, tandem rotors, multi-rotors, etc. (see Figure 1.11 for some examples).

- **Blimps** such as balloons and airships, which are lighter than air and have long endurance, fly at low speeds and generally are large in size (see Figure 1.12 for some examples).

- **Flapping-wing UAVs**, which have flexible and/or morphing small wings inspired by birds and flying insects (Figure 1.13).

- *Hybrid* **configurations** or *convertible* **configurations**, which can take off vertically and tilt their rotors or body and fly like airplanes, such as the Bell Eagle Eye UAV (Figure 1.14).

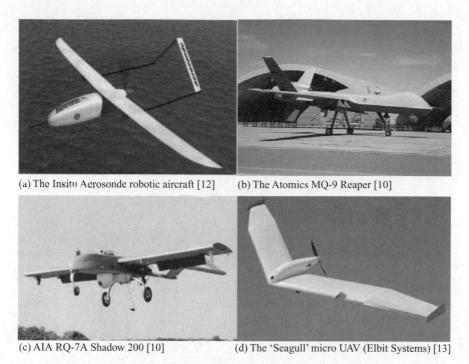

(a) The Insitu Aerosonde robotic aircraft [12]   (b) The Atomics MQ-9 Reaper [10]

(c) AIA RQ-7A Shadow 200 [10]   (d) The 'Seagull' micro UAV (Elbit Systems) [13]

Figure 1.10   Examples of fixed-wing UAVs

(a) Cypher II, Sikorsky Aircraft Corp. [31]   (b) RQ-8A/B FireScout, Northrop Grumman [10]

(c) Yamaha Rmax [63]   (d) Guardian CL-327, Bombardier Services [5]

Figure 1.11   Examples of rotary-wing UAVs

(a) High Altitude Airship (HAA)   (b) Marine Airborne Retransmission System
                                         (Lockheed Martin) (MARTS) (SAIC/TCOM LP) [10]

Figure 1.12   Examples of airship-design UAVs

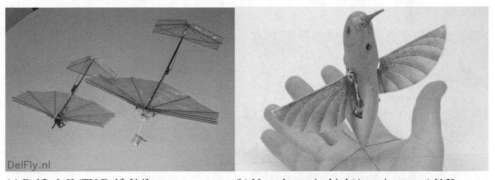

(a) Delfly I, II (TU Delf) [14]   (b) Nano-hummingbird (Aerovironment) [15]

Figure 1.13   Examples of micro flapping-wing UAVs

(a) T-wing (University of Sydney) [16]   (b) Bell Eagle Eye UAS (Bell Company) [17]

Figure 1.14   Examples of hybrid-configuration UAVs

Figure 1.15   NASA Global Hawk HALE UAV (Northrop Grumman) [18]

(a) MQ-1 Predator (US Air Force)   (b) IAI/Malat Heron UAV [19]

Figure 1.16   Examples of MALE UAVs

(a) RQ-7 Shadow (AAI Corporation) [20]   (b) RQ-2B Pioneer (AAI Corporation) [20]

Figure 1.17   Examples of tactical UAVs

(a) RQ-11 Raven (Aerovironment) [15]   (b) The Mikado UAV(EMT) [10]

Figure 1.18   Examples of man-portable UAVs

Another criterion used to differentiate between UAVs is size and endurance. The different categories used are [10]:

- High altitude long endurance (HALE) UAVs, as for example the *Northrop Grumman Ryan's Global Hawks* (65,000 ft altitude, 35 h flight time, 1900 lb payload) shown in Figure 1.15.

- Medium altitude long endurance (MALE) UAVs, as for example the *General Atomics' Predator* (27,000 ft altitude, 30/40 h flight time, 450 lb payload) shown in Figure 1.16.

- Tactical UAVs such as the *Hunter*, *Shadow 200* and *Pioneer* (15,000 ft altitude, 5/6 h flight time, 25 kg payload), see Figure 1.17.

- Small and mini man-portable UAVs such as the *Pointer/Raven* (*Aerovironment*), *Javelin* (*L-3 Communications/BAI*) or *Black Pack Mini* (*Mission Technologies*), see Figure 1.18.

- Micro aerial vehicles (MAV): these have dimensions smaller than 15 cm and in the last few years have gained a lot of attention. They include the *Black Widow* manufactured by *Aerovironment*, the *MicroStar* from *BAE* and many new designs and concepts presented by several universities, such as the *Entomopter* (*Georgia Institute of Technology*), *Micro Bat* (*California Institute of Technology*) and *MFI* (*Berkeley University*), along with other designs from European research centers like MuFly, Coax, etc. (see Figure 1.19).

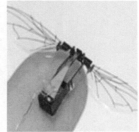

(a) The Wasp (Aerovironment) [10]   (b) Coax (Skybotix) [21]   (c) Harvard micro robotics fly

Figure 1.19   Examples of MAVs

1.4   UAV Applications

Currently, UAVs are being used primarily for military applications and the main investments are driven by future military scenarios. Most military unmanned aircraft systems are used for intelligence, surveillance, reconnaissance (ISR) and strikes. The main user is the US DoD, followed by the Israeli Military Forces. UAVs have been used in the recent conflicts over former Yugoslavia, Iraq, Afghanistan, Libya and elsewhere.

The next generation of UAVs will execute more complex missions, such as air combat, target detection, recognition and destruction, strike/suppression of an enemy's air defense, electronic attack, network node/communications relay, aerial delivery/resupply, anti-surface ship warfare, anti-submarine warfare, mine warfare, ship-to-objective maneuvers, offensive and defensive counter air and airlift. The trend is thus to replace manned missions, especially in 'dull, dirty and dangerous' tasks covering a significant part of warfare activity. The DoD goal is that by 2012–2013, one-third of the aircraft in the operational deep strike force should be unmanned [22]. The X-45 unmanned combat aerial vehicle (UCAV) (Figure 1.20), built by Boeing Corporation, incorporates the above-mentioned concept.

Nowadays, and after many years of development, UAS are reaching the critical point at which they could be applied in a civil/commercial scenario. Numerous UAS market forecasts portray a burgeoning future, including predictions of a $10.6B market by 2013 [23]. There are some corporations focusing on civil applications, for example two American organizations (Radio Technical Commission for Aeronautics (RTCA) and NASA) and one European organization (UAVNET), which have been applying research efforts in order to respond to the potential use of UAS for a variety of science and civil operational missions. Through a series of data-gathering workshops and studies, these organizations have developed some compendia [23–25] of potential UAS-based civil mission concepts and requirements. From these compendiums it is summarized that the potential civilian applications can be categorized into five groups [25]:

Figure 1.20   X-45 UCAV (Boeing Corporation) [31]

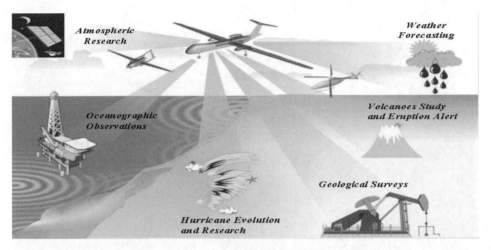

Figure 1.21   Environmental and earth science applications

- **Environmental (or earth science) applications (Figure 1.21).** These include remote environmental research (i.e., magnetic field measurement, ice thickness monitoring, etc.), atmospheric monitoring and pollution assessment (i.e., stratospheric pollution monitoring, CO_2 flux and volcanic dust measurements, etc.), weather forecast, geological surveys (i.e., mapping of subsidence and mineral distribution, oil search, etc.).

- **Emergency applications (Figure 1.22).** These include firefighting, search and rescue, tsunami/flood watch, nuclear radiation monitoring and catastrophe situation awareness, humanitarian aid delivery, etc.

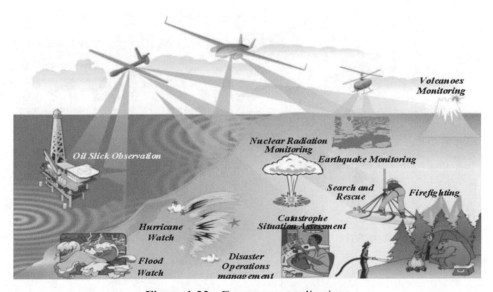

Figure 1.22   Emergency applications

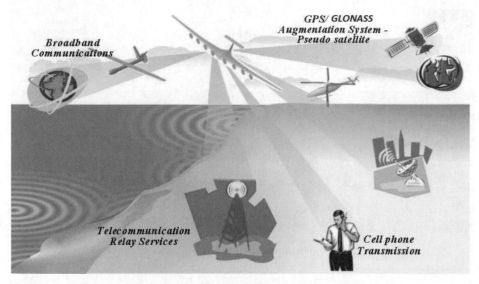

Figure 1.23   Communications applications

- **Communications applications (Figure 1.23).** Telecommunication relay services, cell phone transmissions or broadband communications are a few examples of communication applications.

- **Monitoring applications (Figure 1.24).** These include homeland security (marine and international border patrol, coastal monitoring, law enforcement, etc.), crop and harvest monitoring, fire detection, infrastructure monitoring (oil/gas lines, high-voltage power

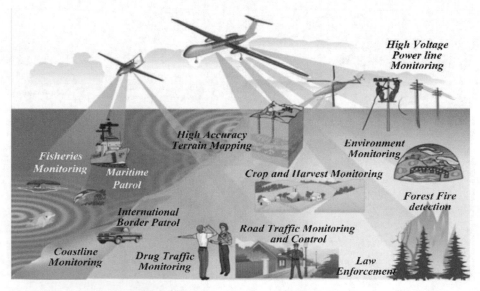

Figure 1.24   Monitoring applications

lines, pipelines, etc.) and terrain mapping (forest mapping, remote sensing of urban areas, etc.).

- **Commercial applications.** These include aerial photography, precision agriculture–chemical spraying, transportation of goods and post, etc.

1.5   UAS Market Overview

Summarizing, UAS can offer major advantages when used for aerial surveillance, reconnaissance and inspection in complex and dangerous environments. Today, there are several companies developing and producing hundreds of UAV designs. Indeed, major defense contractors are involved in developing and producing UAVs (like Boeing, BAE Systems, Lockheed-Martin and EADS). At the same time, newer or smaller companies have also emerged with innovative technologies that make the market even more vibrant. US companies currently hold about 63–64% of the market share, while European companies account for less than 7% [27]. In 2005, some 32 nations were developing or manufacturing more than 250 models of UAV, and about 41 countries were operating more than 80 types of UAV, primarily for reconnaissance in military applications [10].

Several market studies [26–29] have predicted that the worldwide UAV market will expand significantly in the next decade. As stated in [26], over the next 4–5 years (until 2015), the UAV market in the USA will reach $16B, followed by Europe, which is spending about $2B. In the USA, for example, development budgets increased rapidly after 2001, as shown in Figure 1.25, and UAV research and development was given a powerful push [10]. On the other hand, the R&D budgets in Europe have increased slowly, as seen in Figure 1.26.

Other studies are predicting the annual expenditure to reach 2.3 billion by 2017 [28], or 11.3 billion by 2020 [29] (Figure 1.27). According to [29], the most significant catalyst to this market has been the enormous growth of interest in UAVs by the US military, tied to the general trend toward information warfare and net-centric systems. Teal Group expects the military applications to dominate the market and the sales of UAVs to follow recent patterns of high-tech arms procurement worldwide, with Asia representing the second largest market (about 18% of the worldwide total), followed by Europe. A civil market for UAVs is

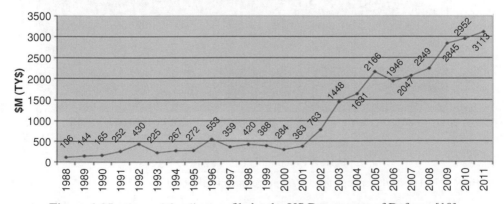

Figure 1.25   Annual funding profile by the US Department of Defense [10]

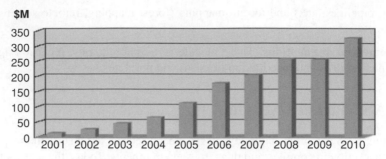

Figure 1.26   Annual funding profile in Europe [30]

expected to emerge slowly over the next decade, starting first with government organizations requiring surveillance systems similar to military UAVs, such as coast guards, border patrol organizations and similar national security organizations. A commercial, non-governmental UAV market is expected to emerge much more slowly. The significant differences between reports are caused by the fact that the UAV is an immature market and the lack of specific requirements, especially in the civil sector, makes forecasts problematic.

The true potential of a civil market for UAVs can be extracted by the example of Japan, where the incorporation of small unmanned helicopters used for agriculture–chemical spraying has increased tremendously the number of registered UAVs, as shown in Figure 1.28 [30].

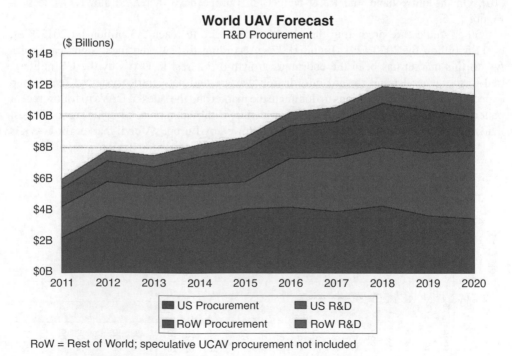

Figure 1.27   R&D and procurement costs forecast [29]

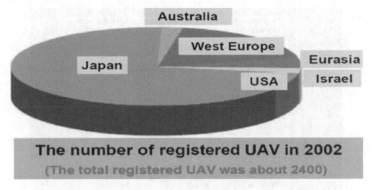

The number of registered UAV in 2002
(The total registered UAV was about 2400)

Figure 1.28   Number of registered UAVs per region [30]

As indicated in [25] and for a specific application (pipeline monitoring), the main drivers for UAV civil market expansion are:

- Increased capabilities (especially endurance, real-time deployment and full spectrum coverage) when compared with other technologies (Figure 1.29).

- Cost advantage (Figure 1.30).

- Technology maturation (due to military applications).

- New applications.

On the other hand, there are significant barriers to the emergence of a civil market for UAVs, both of technological and regulatory nature.

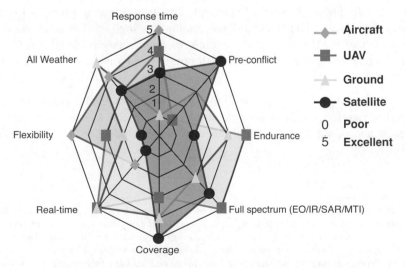

Figure 1.29   Strengths and weaknesses of UAVs [25]

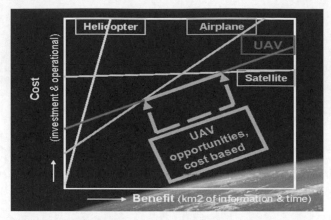

Figure 1.30   Cost/benefit of different technologies for pipeline monitoring application[2] [25]

1.6   UAS Future Challenges

The US DoD is the main contributor to the future evolution of UAS for military use, being the main R&D funder. The future requirements for military UAS, along with lessons learned and current issues, are included in the Unmanned Systems Integrated Roadmap reports published by the DoD [1, 4, 10, 22]. According to these reports, the performance envelope for unmanned systems must keep pace with the demands of the missions that will be expected of these types of system, thus performance attributes associated with unmanned systems must evolve significantly. Figures 1.31–1.33 provide a depiction of the projected evolution of key performance attributes unmanned systems must exhibit in order to enable the projected missions and tasks.

- First and foremost, the level of autonomy should continue to progress from today's fairly high level of human control/intervention to a high level of autonomous tactical behavior that enables more timely and informed human oversight (Figure 1.31). Thus, today's remotely controlled systems will turn to highly autonomous UASs (or groups of UASs).

- The focus of human interface with the machine should evolve from today's current physical interfaces such as joysticks, touch screens, etc. to interaction such as hand signals and, ultimately, to natural language understanding in order to be tasked for missions.

---

[2] The benefit can be expressed by the area coverage of the sensors as a function of time (usually the quantity km^2/h is used). Since this is a monitoring task, this area coverage is strongly connected with the information that can be gathered by the sensors. Thus a satellite has a big initial cost (especially due to investment costs but due to the high coverage capability (in km^2/h) of its sensors, the slope of the curve is very small. At the same time a UAV has higher investment cost compared to an aircraft but a lower operational cost as a function of area coverage per hour of operation (smaller slope). Sometimes the benefit is expressed as useful payload $\times$ endurance (hours of flight).

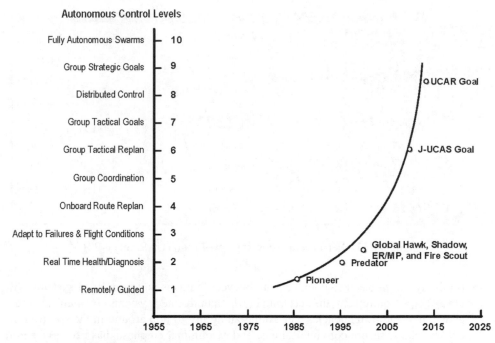

Figure 1.31   Trends in UAS autonomy [10]

	2009	Evolutionary Adaptation	2015	Revolutionary Adaptation	2034
Commands	Physical Human Machine Interfaces		Scripted Voice Command/Hand Signals		Natural Language Understanding
Collaboration	Individual System		Teaming w/in Domain Collaboration Across Domains		Teamed Collaboration
Frequency	Constrained RF		Frequency Hopping		Multi-Frequency Communications
Mission Complexity	Operator Controlled				Autonomous Adaptive Tactical Behaviors
Environmental Capability	Limited Environmental Difficulty		Expanded Environmental Difficulty		All-Weather Environmental Difficulty
Product Line	Mission Package Product Line Dependent				Product Line Independent
OPSEC	Signature High				Signature Low
Operational Control	1 Operator / Platform		1 Operator / Domain		1 Operator / Team
Bandwidth	Limited		Advanced Bandwidth Management		Autonomous Bandwidth
Mission Endurance	Hours		Days Months		Years
Maintenance	Operator				Automated
Awareness	Sensor Data		Situational Awareness		Actionable Information

Figure 1.32   Performance envelope evolution (common to all robotic systems) [22]

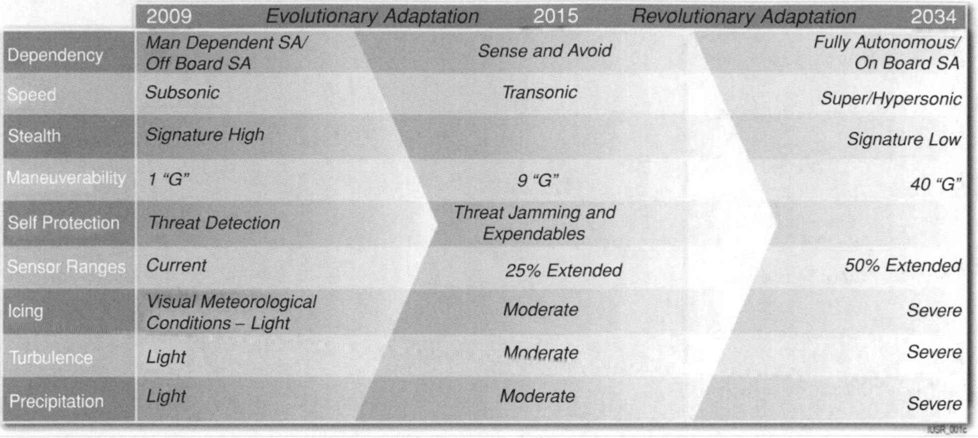

	2009 — Evolutionary Adaptation	2015 — Revolutionary Adaptation	2034
Dependency	Man Dependent SA/ Off Board SA	Sense and Avoid	Fully Autonomous/ On Board SA
Speed	Subsonic	Transonic	Super/Hypersonic
Stealth	Signature High		Signature Low
Maneuverability	1 "G"	9 "G"	40 "G"
Self Protection	Threat Detection	Threat Jamming and Expendables	
Sensor Ranges	Current	25% Extended	50% Extended
Icing	Visual Meteorological Conditions – Light	Moderate	Severe
Turbulence	Light	Moderate	Severe
Precipitation	Light	Moderate	Severe

Figure 1.33   Performance envelope evolution (UAS only) [22]

- Similarly, as the need to communicate between humans and unmanned systems will always be a requirement, the spectrum in which unmanned systems communicate must evolve past radio frequencies and exhibit an agility to hop around in the spectrum to ensure robust, secure communications. Today, minimal emphasis has been placed on operational security, thus most UAS exhibit fairly easily detectable acoustic, thermal, visual and communication signatures. In the future, unmanned systems will be required to carry out missions in a covert manner, thus low observable and signature management attributes will be desirable.

- Moreover, mission duration should increase. Today, mission endurance is measured in hours. In the future, it will be desirable for unmanned systems to conduct their missions in durations measured in days, weeks, months and possibly years. This is a key desirable attribute, as manned tasks are always constrained by the human body's need for food and sleep.

- Another key desirable feature will be mission equipment packages that can be interchanged between platforms and potentially even across domains. Today, most payloads are designed for integration with a single platform. By providing interchangeability across platforms and domains, commanders will be afforded a great flexibility in terms of available options for conducting specific missions in specific types of circumstance. Finally, performance should evolve from today's controller-to-platform ratio of many to one or, at best, one to one to a single controller being able to monitor multiple unmanned systems performing across domains as collaborating teams.

- Speed and maneuverability could also increase well beyond that of manned systems, where limitations are imposed by human physical limits. The human body can only sustain 9**g** of acceleration, whereas technology is the only limiting factor for unmanned systems being able to execute maneuvers that create forces reaching or exceeding 40**g** acceleration.

- The ability to operate in different weather profiles (all weather) and high turbulence will be required.

- Situational awareness is also a significant issue. In the air, UAS will need the ability to sense objects and avoid them, the biggest challenge being small objects moving at high speeds. The situation awareness capability is closely related to the availability of increased range sensors and highly intelligent processing algorithms.

- Finally, survivability, maintainability and reliability issues should be resolved if longer mission durations are to be accomplished. Minimally, unmanned systems must be reliable enough to keep up with mission endurance times.

Although the above guidelines apply for military UAS, a lot of them are common to civil applications as well (especially civil governmental applications). A unified roadmap has not yet been published, but there are general guidelines (for example, [23–25]). The major barrier to civil UAS expansion has been identified as their restricted operation in a segregated part of the airspace. This was the topic of investigation of a £62M UK-led project called ASTRAEA [61], the second phase of which is still under development. Virtually all of the civil applications discussed will require access to either a country's specific national airspace (NAS) and/or foreign airspace at some point in the flight pattern. Even missions intended for remote areas require access to get the aircraft to the area. This has not (for the time being) been the case for military use of UAVs that are deployed in conflict areas, where most civil aviation is ceased. However, civil use requires UAS to be operated in close proximity to human activity and to be fully integrated into the air traffic system (ATS). A similar interest has been expressed by military users. FAA (respectively CAA in the UK and EUROCAE and EUROCONTROL in Europe) standards are only now beginning to emerge for routine UAV operations in US airspace. This is also true worldwide with the exceptions of the UK (CAP722), Israel and South Africa, which have established UAV operating rules through their civil aviation administration. Up to now, in all the countries that have incorporated operational guidelines that allow limited operations in their respective NAS, UAS flight is segregated from the rest of the air traffic with the use of NOTAMs [31]. Even in these cases, there have been complaints by Israeli aviation over the interference experienced due to UAV operations. In 2006, there was an incident near Tel Aviv airport where a UAV came close to a passenger aircraft [29]. In Europe, the EASA issued a call in early 2006 for the formation of a new organization to coordinate the use of UAVs in Europe. The aim is to permit 'normalized' UAV flights in non-segregated airspace by the beginning of the next decade. In addition, efforts are underway to unify European standards with other standards such as those being developed in the USA by RTCA and ASTM (American Society of Testing and Materials), with EUROCAE (European Organization for Civilian Aviation Equipment) considering the development of mirror standards with FAA cooperation.

Closely related to the airspace access are a number of regulatory and technological issues similar to the ones considered for military UAS. These issues must be addressed before UAS are authorized for unrestricted and continuous access to the airspace. These issues are:

- The lack of prescriptive standards and regulations governing the routine operation of UAS in the civilian airspace system. The collaboration of air traffic controllers with UAS operators needs to be accurately defined. Also, automated separation assurance algorithms for seamless and safe operation of UAS in high-density operating environments are needed. As UAS are becoming more and more autonomous, allocation of roles and responsibilities between automation and humans in identifying conflicts and providing separation assurance is vital. Regulations are also needed for the certification of UAS operations and maintenance activities.

- Secure and reliable communications have to be established both between the UA and the control station and/or the ATS control station and the UAS. Minimum performance standards to be met by the communications and control system of the UAS need to be established.

- Reliability and airworthiness of UA is another important issue. Currently, there are strict regulations for the reliability of manned aircraft. The aircraft control system, for example, has been identified as a 'safety critical' system and as such it must be extremely reliable according to FAA regulation parts 23, 25 and 27. The probability of system failure should be less than 10^{-19} per flight hour. Moreover, other quality requirements are on top of the probabilistic assessment (i.e., a catastrophic consequence must not be due to a single failure – surface or pilot input jam). In order to meet requirements like the above, aerospace industry uses a combination of fault avoidance and removal, fault tolerance and fault detection and diagnosis. Every industry has developed its own methods (like the V-cycle implemented by Airbus [32]). Generally, increased reliability is accomplished by a combination of:

  ○ A stringent development process both for hardware and software, where reliability analysis, failure mode and effect analysis, risk classification, etc. are used to dynamically define equipment specifications.

  ○ Hardware (and software) redundancy (the use of triple or quadruple sensors and other equipment in safety critical systems is common to all aircraft [33]).

  ○ Dissimilarity and installation segregation of critical components.

According to the JAA/EUROCONTROL UAS Task Force, as well as the EASA, one of the guiding principles for UAS regulation should be the *equivalence*, or *ELOS* (equivalent level of safety) and based on that, they assert the following [31, 34]:

> Regulatory airworthiness standards should be set to be no less demanding than those currently applied to comparable manned aircraft nor should they penalize UAS systems by requiring compliance with higher standards simply because technology permits.

Since most UAS are based on military or general aviation aircraft, the increased risk stems from the separation of the pilot from the cockpit and the level of automation introduced, rather than the design and construction of the airframe of the UA itself. However, unlike manned aircraft, UAVs impose additional constraints in the above process due to limited payload and weight restrictions that prohibit the use of physical (hardware) redundancy in order to increase system reliability. Moreover, the cost involved in the use of high-reliability equipment could restrain the cost benefit of UAVs in civil applications. It is, thus, necessary to develop reliable algorithms for fault detection and isolation using the concept of analytical (or software) redundancy combined with algorithms that make it possible to control the vehicle in faulty situations (fault-tolerant control concept). In specific cases of faults (i.e., lost communication), the means to terminate the flight and/or recover the vehicle without causing an accident is extremely important, especially if the UAV is used in close proximity to humans.

The development of standardization procedures for airworthiness qualification is difficult for UAVs due to their immense diversity and roles. These standards are now beginning to emerge based on failure data obtained from military experience. A discussion of ELOS requirements for UAVs can be found in [31].

- Another major technological issue connected to airspace access is the need for 'sense and avoid' systems on UAVs operating in controlled airspace [35]. This system will allow UAVs to 'see' or detect other aircraft (piloted or uninhabited) and avoid them. The technology for this system is decomposed into two elements: 'see' and 'avoid'. The 'see' portion involves the detection of intruding aircraft through some type of sensor. The 'avoid' portion involves predicting if the intruding aircraft poses a danger and what course of action should be taken through a decision-making (software) algorithm. For sensors, the priority should be to detect aircraft at sufficient distance so that emergency maneuvering can be avoided. The first step in this development will be to implement a cooperative sensor for collision avoidance. Under the cooperative category, aircraft will have transponders or data links notifying other aircraft of their position. The second and more difficult portion is non-cooperative detection. In this case, the 'other' aircraft does not share its position (as would be the case for many general aviation aircraft) and must be detected with radar or optics. For avoidance, sensor information must be used to predict future positions of host (ownship) and intruder aircraft to determine collision potential. If a collision potential exists, a safe escape trajectory must be derived and automatically executed if the operator has insufficient time to react. The 'sense and avoid' concept is a complicated issue requiring both the design and development of lightweight long-range sensors and the algorithms required for information processing and reliable navigation. This issue is addressed in detail in the other chapters of this book.

Several other considerations for the future capabilities of UAVs have been introduced, focusing on civil applications [23]:

- A high level of autonomy in the mission management function is required to take advantage of using a UAV platform to support the missions. Less direct human interaction in flying the UAV allows less on-station personnel, less on-station support infrastructure, and one operator to monitor several vehicles at a given time. These goals must be balanced with the requirement for the operator and vehicle to respond to air traffic control in a timely manner. The mission management system should also allow redirection of the mission (including activating the contingency management system) from the ground. This would be useful, especially for dynamically changing operation environments which cannot be adequately foreseen at mission initiation. It is envisioned that the human interaction with the onboard mission manager system will occur at the mission objectives level.

- Just like military UAS, the use of swarms of UAVs is going to be necessary for the cost-effective application of UAS in many civil applications, especially those involved with monitoring.

- Longer durability and robustness to weather conditions and turbulence will also be a requirement depending on application.

1.7   Fault Tolerance for UAS

As described above, the increase of reliability and survivability is necessary for future UAS both for meeting airworthiness specifications and for cost-effective operation and longer missions. Both of these goals can be achieved in the context of fault tolerance. In this section the major concepts of fault tolerance for UAS are introduced with an overview of existing methods.

Incident analysis of UAS (Figure 1.34) has clearly shown that the major part of the accidents (nearly 60%) involve UA failures (propulsion and flight control), followed by communication failures and human factors [10]. With the removal of the pilot, the rationale for including the level of redundancy or for using man-rated components considered crucial for his/her safety, can go undefended in UA design reviews, and may be sacrificed for affordability. While this rationale may be acceptable for military missions, it cannot be adopted for civil UAS. On the other hand, aircraft reliability and cost are closely coupled, and unmanned aircraft are widely expected to cost less than their manned counterparts, creating a potential conflict in customer expectations. Less redundancy and lower-quality components, while making UA even cheaper to produce, mean that they become more prone to in-flight loss and more dependent on maintenance.

There are two widely accepted methods used in aircraft design to increase reliability: the use of higher-reliability components and the increase of hardware redundancy in the system. However, neither of them can be applied directly in the case of UAs because of additional cost and weight/payload constraints. Additional constraints also arise from faster dynamics and increased modeling uncertainty of UAS. Moreover, the reduced situation awareness of the operator located away from the cockpit makes the problem of failure handling particularly difficult for UAS. It seems that this technological gap can be covered by the use of analytical redundancy (software redundancy) techniques for fault detection and identification (FDI) and the incorporation of fault-tolerant control (FTC) techniques coupled with increased autonomy. FDI/FTC has been an active area of research for many years and many methods exist in the

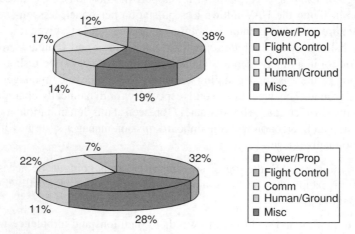

Figure 1.34   Failure sources for UAVs based on (upper) US military (194,000 flight hours) and (lower) IAI military (100,000 flight hours) [10]

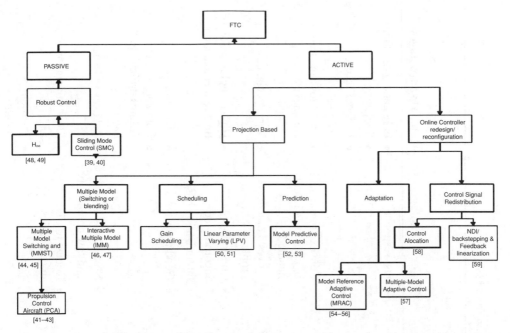

Figure 1.35 Classification of fault-tolerant flight control methods with recent examples

literature ([36, 37] provide thorough overviews). Some of these methods have been applied to aerospace systems, especially military [38] and large transport aircraft [33].

The fault-tolerant control can, generally, be classified as passive and active. In passive techniques the controller is designed to tolerate not only uncertainty but also a class of faults leading to conservative designs while in the active approach, the controller relies heavily on an FDI module and the information it provides. The classification of some recent and popular techniques to design fault-tolerant flight control systems along with example books/papers is shown in Figure 1.35.

Some of the methods described above were applied on a detailed simulation of the Boeing 747-100/200 aircraft [33] as part of the Group of Aeronautical Research and Technology in Europe (GARTEUR) program. Table 1.1 summarizes the major results of comparison of the several methods found in the literature.

Despite the large variety of methods, FDI/FTC techniques are not widely adopted in the aerospace industry and only some space systems have incorporated these techniques in the final design. The reason for this is the immaturity of the methods, especially for non-linear systems, as well as the complexity of the designs and the possibility of high levels of false alarms in case of large modeling uncertainties and/or disturbances. What's more, the high risk of human lives in manned aircraft, along with the mature and tested alternative of hardware redundancy, makes the incorporation of the above methods less attractive.

This is not the case for UAs, where the reduced payload prohibits (or restricts) the use of existing hardware redundancy schemes. The faults that an FDI/FTC system must detect and compensate are similar to those of a manned aircraft (Table 1.2), however, the specifications that an FDI/FTC system must meet for a UA are much more strict. The design of FDI/FTC algorithms for UAs must focus on robustness to modeling uncertainties, simple design and

Table 1.1 Comparison of different methods for FTC (Information partially taken from [33])

FTC Methods	System		Complexity	Robust[1] Tech	Adaptive[2] Tech	Major Advantages	Major Disadvantages
	Linear	Nonlinear					
Multiple Model Switching and Tuning (MMST)	✓		low		✓	• fast responce	• only anticipated faults • Switching • Failure space segregation
Interactive Multiple Model (IMM)	✓		moderate to high		✓	• Convex combination of anticipated faults handled	• Speed loss and higher complexity
PCA	✓	✓	low	○[3]		• Solves a practical problem	• Very special case
Model Predictive Control (MPC)	✓	✓	high to extremely high	○	○	• Constraints easily incorporated • Optimization based	• High complexity for real time implementation
Sliding Mode Control (SMC)	✓	✓	low	✓		• Good robustness	• Chattering • Complete actuator failures difficult to handle
Feedback Linearization (FL)		✓	low		✓	• Nonlinear systems and low complexity	• Robustness Issues
Model Reference Adaptive Control (MRAC)	✓		moderate	○	✓	• Robust to FDI uncertainties	• Slow Adaptation to faults • Transient period stability issue

Method		Complexity	Advantages	Disadvantages
Backstepping Based Adaptive Fuzzy Control	✓	moderate	• Robust to FDI uncertainties	• Slow Adaptation to abrupt • Transient period stability issue
Eigentstructure Assignment (EA)	✓	low	• Low complexity	• Robustness Issues
Control Allocation (CA)	✓	moderate to high	• Controller structure unchanged after fault	• Optimization required adds complexity
Pseudo-Inverse Method (PIM)		low	• Simple and fast	• The system may not be stabilized
Modified PIM	✓	high	• Stabilization quaranted	• Constrainted Optimazation (high complexity)

[1] Robust techniques are part of the method
[2] Adaptive techniques are part of the method
[3] The feature can be incorporated in the method

Table 1.2   Aircraft/UA failure modes [60]

Sensor	Actuator	Structural	Failure	Effect
√			Sensor loss	Minor if it is the only failure
	√		Partial hydraulics loss	Maximum rate decrease on several control surfaces
	√		Full hydraulics loss	One or more control surfaces become stuck at last position for hydraulic driven aircraft, or float on light aircraft
	√		Control loss on one or more actuators due to internal fault (not external damage)	One or more control surfaces become stuck at last position
	√	√	Loss of part/all of control surface	Effectiveness of control surface is reduced, but rate is not; minor change in the aerodynamics
	√	√	Loss of engine	Large change in possible operating region; significant change in the aerodynamics
		√	Damage to aircraft surface	Possible change in operating region; significant change in aerodynamics

low complexity. For small UAS with very limited computational power, passive techniques seem especially attractive due to the avoidance of the FDI module. In any case, the comparison of many different techniques to highlight the pros and cons for every category of platform is necessary.

In order to reach these goals, there is a need to develop realistic benchmark models to assist the research. These models should include actuator and sensor dynamics as well as modeling uncertainties and disturbances. Also, issues like fault-tolerant navigation and decision making under health state uncertainty must be addressed as well.

Finally, it should be pointed out that the FDI/FTC methods found in the literature cannot provide a complete solution to the reliability improvement problem for UAs. These methods focus on increasing fault tolerance for a given degree of redundancy and, thus, they are limited to the degree of redundancy selected. On the other hand, reliability improvement is a multi-objective optimization problem that involves reliability specifications, redundancy, fault-tolerance evaluation and cost. A schematic representation of a possible design cycle is shown in Figure 1.36.

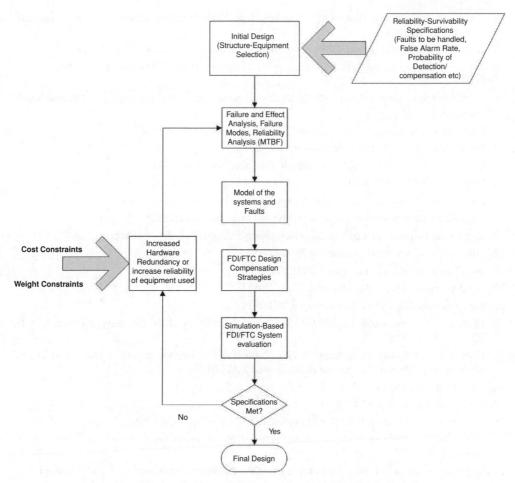

Figure 1.36   Design cycle of UAS for reliability improvement

References

1. Unmanned Systems Roadmap 2002–2027, Report of the Office of the Secretary of Defense, DoD, USA, 2002.
2. STANAG 4671 – Unmanned Aerial Vehicle Systems Airworthiness Requirements (USAR), Joint Capability Group on Unmanned Aerial Vehicles (Draft), NATO Naval Armaments Group, 2007.
3. Unmanned aircraft systems operations in the U.S. national airspace system, Federal Aviation Administration, Interim Operational Approval Guidance 08-01, 2008.
4. Unmanned Systems Roadmap 2007–2032, Report of the Office of the Secretary of Defense, DoD, USA, 2007.
5. Castillo, P., Lorenzo, R., and Dzul, A.E. *Modeling and Control of Mini-Flying Machines*, Springer, 2005.
6. Guedj, D. *Le Théorème du Perroquet*, Editions du Seuil, 1998.

7. Helicopter History Site, History of helicopters, June 2004. Available at http://www.hiller.org (last accessed 27/4/2011).

8. Hiller Aviation Museum. Available at http://www.hiller.org (last accessed 27/4/2011).

9. Stoff, J. *Historic Aircraft and Spacecraft in the Cradle of Aviation Museum*, Dover Publications, 2001.

10. Unmanned Systems Roadmap 2005–2030, Report of the Office of the Secretary of Defense, DoD, USA, 2005.

11. http://www.dfrc.nasa.gov (last accessed 27/4/2011).

12. http://www.aerosonde.com (last accessed 17/4/2011).

13. http://www.unmanned.co.uk/keyword/elbit/ (last accessed 28/4/2011).

14. http://www.dclfly.nl/ (2/4/2011).

15. http://www.avinc.com/media_gallery/images/uas/ (last accessed 26/4/2011).

16. http://sydney.edu.au/engineering/aeromech/uav/twing/ (last accessed 28/4/2011).

17. http://www.bellhelicopter.com/en/aircraft/military/bellEagleEye.cfm (last accessed 28/4/2011).

18. http://www.NASA.org (last accessed 29/4/2011).

19. http://www.iai.co.il/ (last accessed 18/4/2011).

20. http://www.aaicorp.com/ (last accessed 20/4/2011).

21. http://www.skybotix.com/ (last accessed 24/4/2011).

22. Unmanned Systems Roadmap 2009–2034, Report of the Office of the Secretary of Defense, DoD, USA, 2009.

23. Civil UAV capability assessment, NASA, 2006. Interim Report, http://www.nasa.gov/centers/dryden/research/civuav/index.html (last accessed 21/3/2011).

24. http://www.rtca.org/ (19/3/2011).

25. http://www.uavnet.org/ (29/4/2011).

26. The unmanned aerial vehicles (UAV) market 2009–2019, Visiongain, 2009.

27. Dickerson, L. 'UAV on the rise', *Aviation Week Space Technology*, Aerospace Source Book, 166(3), 2007.

28. http://www.researchandmarkets.com/research/afdcf8/homeland_security_and_commercial_unmanned (last accessed 28/4/2011).

29. Zaloga, S.J., Rockwell, D., and Finnegan, P. 'World Unmanned Aerial Vehicle Systems: Market Profile and Forecasts', 2011 edition, Teal Group Corporation, 2011.

30. Nonami, K., Kendoul, F., Suzuki, S., Wang, W., and Nakazawa, D. *Autonomous Flying Robots*, Springer Science and Business Media, 2010.

31. Dalamagidis, K., Valavanis, K., and Piegl, L.A. *On Integrating Unmanned Aircraft Systems into the National Airspace*, Springer Science and Business Media, 2009.

32. Goupil, P. 'Airbus state of the art and practices on FDI and FTC', 7th IFAC Symposium on Fault Detection, Supervision and Safety of Technical Processes, Barcelona, Spain, 30 June–3 July, pp. 564–572, 2009.

33. Edwards, C., Lombaerts, T., and Smaili, H. *Fault Tolerant Flight Control – A Benchmark Challenge*, Spinger-Verlag, 2010.

34. European Aviation Safety Agency (EASA), A-NPA, No. 16/2005, Policy for unmanned aerial vehicle (UAV) certification, 2005.

35. Davis, B. 'UAS in the national airspace: the NTSB takes a look', *Unmanned Systems*, 26(6):40–41, 2008.

36. Zhang, Y. and Jiang, J. 'Bibliographical review on reconfigurable fault-tolerant control systems', *Annual Reviews in Control*, 32:229–252, 2008.

37. Venkatasubramanian, K. *et al.* 'A review of process fault detection and diagnosis Part I (Quantitative model-based methods), Part II (Qualitative models and search strategies, Part III (Process history based methods)', *Computers and Chemical Engineering*, 27:293–346, 2003.

38. Urnes, J., Yeager, R., and Steward, J. 'Flight demonstration of the self-repairing flight control system in a NASA F-15 aircraft', National Aerospace Electronics Conference, Dayton, OH, USA, May 1990. Report 90CH2881-1.

39. Hess, R.A. and Wells, S.R. 'Sliding mode control applied to reconfigurable flight control design', *AIAA Journal of Guidance, Control and Dynamics*, 26:452–462, 2003.

40. Alwi, H. and Edwards, C. 'Fault detection and fault tolerant control of a civil aircraft using a sliding-mode-based scheme', *IEEE Transactions on Control Systems Technology*, 16(3):499–510, 2008.

41. Harefors, M. and Bates, D.G. 'Integrated propulsion-based flight control system design for a civil transport aircraft', Proceedings of the 2002 IEEE International Conference on Control Applications, pp. 132–137, 2002.

42. Burcham, F.W., Fullerton, C.G., and Maine, T.A. 'Manual manipulation of engine throttles for emergency flight control', Technical Report NASA/TM-2004-212045, NASA, 2004.

43. Tucker, T. 'Touchdown: the development of propulsion controlled aircraft at NASA Dryden. Monographs in Aerospace History, 1999.

44. Boskovic, J.D. and Mehra, R.K. 'A multiple-model-based reconfigurable flight control system design', Proceedings of the 37th IEEE Conference on Decision and Control, Tampa, FL, December, pp. 4503–4508, 1998.

45. Aravena, J., Zhou, K., Li, X.R., and Chowdhury, F. 'Fault tolerant safe flight controller bank', Proceedings of the IFAC Symposium SAFEPROCESS '06, Beijing, pp. 8908–8912, 2006.

46. Rago, C., Prasanth, R., Mehra, R.K., and Fortenbaugh, R. 'Failure detection and identification and fault tolerant control using the IMM-KF with applications to the Eagle-Eye UAV', Proceedings of the 37th IEEE Conference on Decision and Control, Tampa, FL, December, pp. 4503–4508, 1998.

47. Zhang, Y. and Jiang, J. 'Integrated active fault-tolerant control using IMM approach', *IEEE Transactions on Aerospace and Electronic Systems*, 37:1221–1235, 2001.

48. Zhou, K. and Ren, Z. 'A new controller architecture for high performance, robust and fault tolerant control', *IEEE Transactions on Automatic Control*, 46:1613–1618, 2008.

49. Ye, S., Zhang, Y., Li, Y., Wang, X., and Rabbath, C.-A. 'Robust fault-tolerant tracking control with application to flight control systems with uncertainties', Proceedings of the 10th IASTED International Conference on Control and Applications, 2008.

50. Shin, J.-Y. and Gregory, I. 'Robust gain-scheduled fault tolerant control for a transport aircraft', Proceedings of the 16th IEEE Conference on Control Applications (CCA 2007), 1–3 October 2007.

51. Ganguili, S., Marcos, A., and Balas, G.J. 'Reconfigurable LPV control design for Boeing 747-100/200 longitudinal axis', Proceedings of the American Control Conference, pp. 3612–3617, 2002.

52. Maciejowski, J.M. and Jones, C.N. 'MPC fault-tolerant control case study: flight 1862', Proceedings of the IFAC Symposium SAFEPROCESS '03, Washington, DC, pp. 119–124, 2003.

53. Campell, M.E., Lee, J.W., Scholte, E. and Rathbun, D. 'Simulation and flight test of autonomous aircraft estimation, planning and control algorithms', *AIAA Journal of Guidance, Control and Dynamics*, 30(6):1597–1609, 2007.

54. Shin, Y., Calise, A.J., and Johnson, M.D. 'Adaptive control of advanced fighter aircraft in nonlinear flight regimes', *AIAA Journal of Guidance, Control and Dynamics*, 31(5):1464–1477, 2008.

55. Tao, G., Chen, S., Tang, X., and Joshi, S.M. *Adaptive Control of Systems with Actuator Failures*, Springer-Verlag, 2004.

56. Shore, D. and Bodson, M. 'Flight testing of a reconfigurable control system on an unmanned aircraft', *AIAA Journal of Guidance, Control and Dynamics*, 28(4):698–707, 2005.

57. Fekri, S., Athans, M., and Pascoal, A. 'Issues, progress and new results in robust adaptive control', *International Journal of Adaptive Control and Signal Processing*, 20(10):519–579, 2006.

58. Ducard, G.J.J. *Fault-tolerant Flight Control and Guidance Systems – Practical Methods for Small UAVs*, Springer-Verlag, 2009.

59. Lombaerts, T.J.J., Huisman, H.O., Chu, Q.P., Mulder, J.A., and Joosten, D.A. 'Flight control reconfiguration based on online physical model identification and nonlinear dynamic inversion', Proceedings of the AIAA Guidance, Navigation and Control Conference,18–21 August, Honolulu, HI, 2008. AIAA 2008-7435, pp. 1–24.

60. Jones, C.N. Reconfigurable Flight Control – First Year Report, Control Group, Department of Engineering, University of Cambridge, 2005.

61. http://www.projectastraea.co.uk/ (last accessed 27/4/2011).

62. 'Civil UAV capabilities assessment', Interim Status Report, NASA, 2006.

63. http://www.barnardmicrosystems.com/L4E_rmax.htm (27/4/2011).

2

Performance Tradeoffs and the Development of Standards[1]

Andrew Zeitlin
MITRE Corporation, McLean, VA, USA

2.1 Scope of Sense and Avoid

The purpose of a sense and avoid (S&A) function is to act in the place of a human pilot to detect and resolve certain hazards to safe flight. These hazards consist of other traffic or objects presenting a risk of collision. Air traffic encompasses aircraft, gliders, balloons and even other unmanned aircraft system (UAS). Other hazards include terrain and obstacles (e.g., buildings, towers, power lines).

As there is no human pilot aboard a UAS, the motivation of S&A is not necessarily to preserve the aircraft but it is certainly needed to prevent collisions with other traffic, with persons on the ground, or collateral damage to property. S&A must operate for emergency and diversionary events as well as throughout normal operations.

On a manned aircraft, the human pilot is required to see and avoid hazards. The pilot's duties include regular visual scans across the forward field of view in order to detect other

---

[1] This work was produced for the US Government under Contract DTFAWA-10-C-00080 and is subject to Federal Aviation Administration Acquisition Management System Clause 3.5-13, Rights In Data-General, Alt. III and Alt. IV (October 1996). The contents of this material reflect the views of the author and/or the Director of the Center for Advanced Aviation System Development. Neither the Federal Aviation Administration nor the Department of Transportation makes any warranty or guarantee, or promise, expressed or implied, concerning the content or accuracy of the views expressed herein. Approved for Public Release: 11-3338. Distribution Unlimited.

Sense and Avoid in UAS: Research and Applications, First Edition. Edited by Plamen Angelov.
© 2012 John Wiley & Sons, Ltd. Published 2012 by John Wiley & Sons, Ltd.

aircraft. The scanning may be more focused toward areas where operations are expected, or as informed by hearing radio traffic or by an electronic display. Whenever traffic is seen, the pilot must make a judgment about its trajectory relative to own aircraft motion, and determine whether any risk of collision might necessitate a maneuver. The 'see and avoid' process can be difficult in some conditions of poor visibility, confusing backgrounds or high workload. The premise that UAS S&A need only be as good as human see and avoid is looked upon unfavorably by airspace regulators.

Some UAS are too small to feasibly carry S&A equipment onboard. One solution for these aircraft is to operate them only within direct radio communication of the pilot and to maintain a visual line of sight between pilot and aircraft. This form of operation may prove sufficiently safe. Further restrictions may be imposed to limit risk, such as precluding operation over densely populated areas and limiting the mass of the aircraft that need not be capable of S&A.

This chapter discusses many considerations concerning the design of S&A, and presents important tradeoffs that need to be made. In contrast to the specific designs chosen by each implementation for their own purposes, the chapter also addresses the method of developing standards that present requirements that all implementations must meet, regardless of their design specifics.

2.2   System Configurations

There are numerous variations of S&A configurations. The main components are the aircraft and systems onboard; the off-board control station; and communication link(s) between these. The key distinctions involve two factors:

1. Whether the S&A surveillance system consists of sensors located onboard the aircraft, off-board, or distributed among both of these, and

2. Whether the S&A decisions are made at the off-board control station or onboard the aircraft by its automation.

Several example configurations are illustrated in Figures 2.1–2.3.

Figure 2.1   Sensor and decisions on the ground

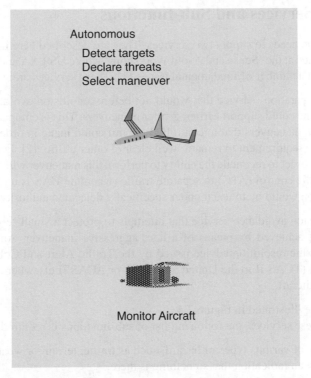

Figure 2.2   Sensors and decisions located aboard the aircraft

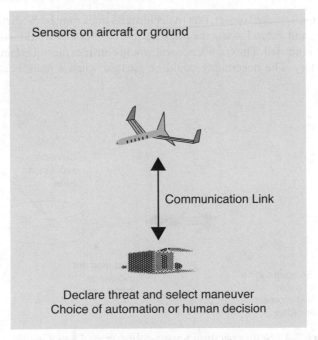

Figure 2.3   Sensors aboard the aircraft, decisions made on the ground

2.3   S&A Services and Sub-functions

The S&A function needs to supply two services. They are described here in accordance with agreements reached at the 'Sense and Avoid Workshops' where US FAA and Defense Agency experts discussed a number of fundamental issues [1]. These services are:

(a) A 'self-separation' service that would act before a collision avoidance maneuver is needed, and could support earlier, gentler maneuvers. This is comparable to the visual separation maneuvers that uncontrolled aircraft could make in order to carry out the regulatory requirement to remain 'well clear of other traffic' [2]. Further definition is needed in order to reconcile the ability to perform this maneuver with the responsibility of air traffic control (ATC) to separate traffic (when the UAS is under ATC control). One option would be to use it under specifically delegated authority from ATC.

(b) The collision avoidance service that attempts to protect a small 'collision zone' and usually is achieved by means of a late, aggressive maneuver. An example of this service is the resolution advice issued by the Traffic Alert and Collision Avoidance System II (TCAS II in the United States [4], or ACAS II elsewhere [5]) used aboard manned aircraft.

These services are illustrated in Figure 2.4.

To achieve these services, the following list of sub-functions is required [1]:

1. **Detect** any of various types of hazard, such as traffic, terrain or weather. At this step, it is merely an indication that something is there.

2. **Track** the motion of the detected object. This requires gaining sufficient confidence that the detection is valid, and making a determination of its position and trajectory.

3. **Evaluate** each tracked object, first to decide if its track may be predicted with sufficient confidence and second to test the track against criteria that would indicate that a S&A maneuver is needed. The confidence test would consider the uncertainty of the position and trajectory. The uncertainty could be greatest when a track is started, and again

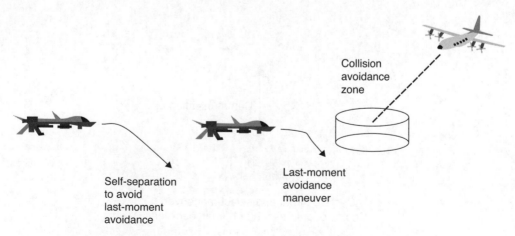

Figure 2.4   Self-separation versus collision avoidance maneuvering

whenever a new maneuver is first detected. A series of measurements may be required to narrow the uncertainty about the new or changed trajectory. Also, when a turn is perceived, there is uncertainty about how great a heading change will result.

4. **Prioritize** the tracked objects based on their track parameters and the tests performed during the evaluation step. In some implementations, this may help to deal with limited S&A system capacity, while in others prioritization might be combined with the evaluation or declaration steps. Prioritization can consider criteria for the declaration decision that may vary with type of hazard or the context of the encounter (e.g., within a controlled traffic pattern).

5. **Declare** that the paths of own aircraft and the tracked object and the available avoidance time have reached a decision point that does indeed require maneuvering to begin. Separate declarations would be needed for self-separation and collision avoidance maneuvers.

6. **Determine** the specific maneuver, based on the particular geometry of the encounter, the maneuver capabilities and preferences for own aircraft, and all relevant constraints (e.g., airspace rules or the other aircraft's maneuver).

7. **Command** own aircraft to perform the chosen maneuver. Depending upon the implementation of the S&A, this might require communicating the commanded maneuver to the aircraft, or if the maneuver determination was performed onboard, merely internal communication among the aircraft's sub-systems.

8. **Execute** the commanded maneuver.

If any aspects of the **Evaluate** or **Determine** sub-functions are to be performed at the control station, air–ground communication becomes critical. It is a matter of design as to whether all tracked targets are sent to the ground for evaluation, or if the aircraft sub-system performs some further sub-functions beyond **Detect** (e.g., **Evaluate**, **Prioritize**) in order to reduce the amount of air–ground communications. The latency and bandwidth of the data link are important considerations, as well as the feasibility of placing substantial processing capability aboard the UAS.

2.4 Sensor Capabilities

The surveillance system that detects hazards can be implemented in various forms. Some technologies could be carried onboard the UAS, while another approach is to exploit sensing on the ground, such as a radar. These choices have extremely different capabilities, ranging from their coverage volume to the types of measurements made and the respective accuracies, update rates and probabilities of false detection.

2.4.1 Airborne Sensing

Technologies available for airborne sensing of other aircraft are best divided into two groups, termed cooperative and non-cooperative.

Table 2.1   Typical sensor coordinate systems

Sensor Technology	Coordinate System
Active interrogation of Mode A/C transponder	Relative range, absolute altitude
ADS-B	Latitude, longitude, altitude, velocity
Electro-optic	Bearing (azimuth and elevation)
Laser/LIDAR	Relative range
Onboard radar	Relative range, bearing (azimuth and elevation)
Ground-based radar	Range and bearing from ground reference
Acoustic	Bearing

Cooperative technologies are those that receive radio signals from another aircraft's on-board equipment. The leading cooperative technologies are:

(a) ATC transponder. A large number of aircraft carry a transponder, which has long been used to respond to ground-based secondary radar interrogations for air traffic control usage. The same technology has been exploited for the manned aircraft Traffic Alert and Collision Avoidance System (TCAS).[2] Aircraft are required to be equipped with transponders to operate in some classes of airspace (ICAO classes A, B and C; with altitude encoding in A and B).

(b) Automatic Dependent Surveillance – Broadcast (ADS-B). This technology utilizes the Global Positioning System (GPS) or an alternate navigation source, and broadcasts own aircraft position, velocity and other data without needing to be interrogated. Standards for ADS-B are in place, and although its equipage is limited as of this writing, its widespread use is contemplated in NextGen and SESAR in the USA and Europe respectively, as well as certain other international locations. In the USA, those aircraft currently required to carry a transponder must equip with ADS-B for broadcast by 2020 [3].

Since the detection of a cooperative signal should be reliable within its intended range, these technologies should be superior for target detection and track association. However, these can only detect suitably equipped aircraft that choose to 'cooperate' with ATC or other aircraft by equipping and operating that equipment. Some, but not all, classes of airspace mandate carriage of this equipment. In airspace where non-cooperative traffic is allowed, other technologies would be needed to detect traffic. Candidate technologies include electro-optic or infrared cameras, primary radar, laser range finding and acoustic processing. Each of these non-cooperative technologies has different limitations, particularly when miniaturized for mounting onboard an aircraft. No single approach appears to provide all the necessary measurement coordinates (angles and range to target) with good accuracy; optical and acoustic measurements are best for angular measurement, while radar and lasers are best for ranging (see Table 2.1). Equipping a UAS with a combination of S&A technologies might serve to combine the strengths of each. The combination could include both cooperative and

---

[2] The system is known as Airborne Collision Avoidance System (ACAS) outside the United States.

non-cooperative elements. Measurements from the separate sensors would need to be associated. A design might utilize the existing knowledge base of data fusion techniques, and may use Kalman filtering to account for differences in accuracies between the sources. Further complication would arise if sensors differed in their update rates, as they would measure targets at different times, and thus in different locations.

2.4.2   Ground-Based Sensing

For smaller UAS, the size, weight and power required to equip with multiple sensors may be prohibitive. Ground-based sensing may be attractive for these UAS, although these sensors also have limitations, such as their accuracy and update rate. These limitations may be reflected in aircraft operations, using more conservative separation measures. These sensors' field of view could also preclude long-range, low-altitude coverage. At present, radar is used for S&A in a small number of locations. The cost of a radar is likely suitable only for governmental use; private UAS operators would need to arrange access to the surveillance data.

Since the sensor technologies vary in their range and surveillance coverage, a system trade is needed to determine these requirements to meet required safety levels, considering also the timeline for declaration and avoidance maneuvering (Section 2.7) and evaluating the likelihood that a target initially outside the sensor's range or field of view would not be detected in time to avoid it. An operational trade may limit flights to some lesser volume contained within the surveillance volume, ensuring that entering targets can be acquired with high probability and sufficient warning time.

2.4.3   Sensor Parameters

Sensor technologies can be evaluated using standard parameters which can provide a basis for comparison, as well as characterizing the performance of the entire surveillance system.

- **Field of view.** This describes the angular sector within which the sensor makes measurements. When a target is outside the field of view, this sensor cannot detect or update it.

- **Range.** A distance measured from the sensor, within which some good probability of detection of targets may be expected.

- **Update rate.** This is the interval at which the sensor provides its measurements. If it does not detect the target at every interval, its effective update rate will be lower.

- **Accuracy.** This parameter describes the uncertainty of the sensor position measurement. It often addresses a single dimension, so that evaluation of the surveillance system must combine accuracy values for different dimensions to create a volume of uncertainty.

- **Integrity.** This represents the probability that a measurement falls beyond some limit characterizing its normal operation.

For cooperative sensors and targets, an additional parameter is relevant:

- **Data elements.** Specific data provided by the cooperative target to enhance the measurement or knowledge of surveillance. Examples include position, trajectory, identity, intent.

2.5   Tracking and Trajectory Prediction

The surveillance system needs to associate successive measurements with specific targets. Over time, a track is formed and updated for each target that is believed to be real. Various technologies are susceptible to varying degrees of noise, resulting in false detections. Another effect of measurement noise is stochastic variation of the position measurement, potentially making it difficult to associate one measurement of the target with the next. Both of these effects may require a design associating several consistent measurements before a valid track can be declared. The track then would be updated each time another consistent measurement is received. A measurement would be associated with an established track if its position agreed with the expected position (equal to the previous position plus estimated velocity times the update interval) within some predetermined margin of error. This margin must account for measurement and estimation uncertainties, as well as feasible maneuvers by the target. The update interval would depend on the technology, and typically would lie within 1 to 5 seconds.

The tracking function should be capable of maintaining the track for a certain time even in the absence of an update, as the detection function is likely to be designed for a good trade between valid and false detection; thus some updates would be missed. In this case, the track can be projected ahead to an expected position, but its uncertainty would grow. After too many updates are missed, the track would need to be dropped. In particular, any maneuver that began after the last detection would be unseen (Figure 2.5).

If the surveillance system is combining measurements from different technologies, their respective data should be aligned in time, properly compensated for uncertainties that differ by technology or dimension, and given appropriate weight. The function should strive to avoid creating duplicate tracks for the same target, but if this occurs, the receipt of additional data should enable the separate tracks to be associated and merged. Cooperative technologies should be more successful at uniquely identifying targets, as Mode S and ADS-B equipment provide a unique aircraft address within their reply formats.

The track should develop a velocity vector as the basis for predicting the trajectory of each target (except fixed objects or terrain). Additional data such as turn rate may enhance the projection. Features that attempt to detect the start or end of maneuvers are useful, especially in modulating the amount of uncertainty associated with the prediction. Even a track with large uncertainty can be of some use. The subsequent resolution decision may need to avoid a large volume to account for this uncertainty.

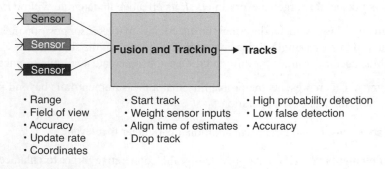

Figure 2.5   Surveillance system requirements

2.6   Threat Declaration and Resolution Decisions

Aircraft operations are conducted to perform some mission that could be disrupted by making unplanned maneuvers. Therefore it is important that S&A distinguish threatening from non-threatening traffic or other hazards, and call for a maneuver only when required for safety. The threat declaration function then must balance two primary requirements: to determine that a hazard poses a threat such that some maneuver is required, and to minimize these declarations for targets that are actually non-threatening. The timeline discussed below places further requirements upon the timing of the declaration.

A measurement of the range between UAS and a hazard is the basis for predicting the time to collision. For a true, linear collision trajectory, the time to collision is given by τ in:

$$\tau = -\frac{r}{\dot{r}}$$

Note that $\dot{r}$ is negative when the UAS and the hazard converge. When the trajectory is not leading to a collision (i.e., a 'miss distance' would result), the task of threat determination is to decide whether some protected volume of airspace would be penetrated. Since other aircraft may make unforeseen maneuvers in the future, some additional margin may be provided to protect against adverse maneuvers (discussed further below).

2.6.1   Collision Avoidance

The need for resolution must begin with a defined volume to be avoided. The usual choice for the collision avoidance function would equate to the 'critical near-midair collision' definition [4] of a truncated cylinder ±100 ft in height and 500 ft in radius (see Figure 2.6). This fixed volume is a surrogate for the actual dimensions of a threatening aircraft, as those dimensions are difficult to measure in the dynamic conditions of flight. The choice of resolution maneuver also must consider the latencies involved in deciding, communicating and executing the maneuver, the capabilities of the airframe in accelerating laterally, vertically or changing speed, the ultimate climb or descent rates or bank angle to be achieved, and other constraints deriving from airspace rules and other detected proximate traffic or hazards such as terrain.

The resolution choice also needs to consider compatibility with avoidance maneuvers to be made by the other aircraft in the encounter. Threat aircraft maneuvering visually would be expected to follow the customary right-of-way rules [2], while those threats making use of their own collision avoidance system would use its prescribed set of rules. TCAS,

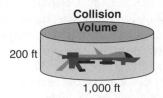

Figure 2.6   Collision avoidance zone

for example, uses vertical maneuvers. As of this writing, TCAS is the only system aboard manned aircraft that generates resolution advisories, but future airspace could see a variety of systems implemented, perhaps starting with systems placed onboard various UAS. These onboard technologies might migrate to some segments of manned aircraft that do not yet have collision avoidance capability (e.g., general aviation), if the operational and cost attributes become attractive. Therefore, since no single behavior can be expected from other aircraft, it will remain essential to coordinate maneuver choices between two equipped aircraft in an encounter, much as TCAS does at present. The TCAS process sends addressed messages from each aircraft to the other to explicitly exchange intent data. This process uses a priority protocol to resolve ties (e.g., when each aircraft has selected 'climb'). Alternate means have been proposed to avoid transmitting these messages:

- *Implicit coordination by observing the other aircraft maneuver*. This would need to overcome difficulties in detecting the start of a maneuver, and still would need some mutual protocol to break ties or overcome differences of observation.

- *Implicit coordination by design*. This method might attempt to pre-select maneuvers based upon closing angle or closest approach point; or may restrict maneuvers to the lateral dimension against TCAS aircraft that presumably would only make vertical maneuvers. Difficulties would involve differences between each aircraft's observation of one another, as well as limitations from using pre-selected choices.

Using a different method of coordination also would introduce difficulties in remaining inter-operable with existing TCAS-equipped aircraft. The first difficulty is proving that compatible advisories will be selected with extremely high reliability. Second is the limitation in current TCAS design where coordination uses the Mode S data link and only is performed against another TCAS-equipped aircraft.

The magnitude of the avoidance maneuver must be sufficient to overcome errors. One such error is the position measurement. Another is the uncertainty in predicting a target's trajectory. A margin accounting for each of these would effectively expand the volume to be avoided (Figure 2.7) so that the maneuver would seek the sum of the desired separation plus the error margin. The ADS-B technology assists this provision, as the data provided by an aircraft is accompanied by an indication of the accuracy and integrity associated with it. For any other technology, the implementation needs to be coordinated with the algorithm design so that the error margin used for a target corresponds with the characteristics of the sensor measuring that target.

A long-standing challenge for collision avoidance algorithms is the ability to deal with maneuvers made by the threat aircraft. These can range from maneuvers begun shortly before a collision avoidance decision to maneuvers in the future that cannot be foreseen. Any algorithm must be tested against a credible range of maneuver possibilities and found to be robust, within the physical limits of aircraft maneuverability. Desired features include the ability to monitor the encounter as its resolution evolves and reconsider the original maneuver decision. Changed maneuvers may include stronger or weaker maneuvers in the same direction, a reversal within the same (horizontal or vertical) plane, a change to maneuver in the other plane, or resolution combining both horizontal and vertical avoidance. For some aircraft, speed changes also could be feasible for resolution. Depending on the aircraft maneuverability and the encounter specifics, some of these choices may not be available.

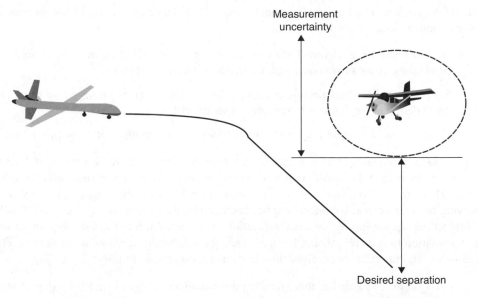

Figure 2.7 Accounting for uncertainty in determining maneuver

2.6.2   Self-separation

The self-separation maneuver described in Section 2.3 must evaluate targets and begin maneuvers earlier than collision avoidance, as its purpose is to avoid the more dangerous state
requiring prompt and vigorous action. Its threat declaration function must anticipate future
collision avoidance initiation, and determine when a maneuver is required to remain outside that state. In most respects, the principles of self-separation parallel those for collision
avoidance. Coordination of maneuvers may not be required.

A potential constraint on the use of self-separation involves conditions that may be placed
on the assignment of authority to make the maneuver. These likely will vary according to the
UAS flight regime (under ATC control or not), and whether specific delegation authority to
maneuver is received.

A system tradeoff between S&A surveillance capability and the self-separation maneuver
involves the sensor system accuracy described above, whether the S&A surveillance range
is sufficient to initiate track on the worst-case threat geometry, the magnitude of the self-
separation maneuver itself, and both the communications latency and human latency (when
decisions are not made entirely onboard the UAS).

2.6.3   Human Decision versus Algorithm

For the functions described in this section, it is a matter of design whether a human decision
is involved or the process is partially or completely automated. A totally non-automated
decision would show traffic or other hazard information to a human decision-maker, who
would consider the UAS position and trajectory and would decide whether and when to make

a threat declaration. The human also would select an avoidance maneuver. The arguments in favor of human decision could be:

- A human can have a current view of the 'big picture' including mission or airspace constraints, other surrounding traffic and own aircraft capabilities.

- Some may believe that alerting criteria and/or resolution priority rules are so situation-dependent that they cannot be captured in an algorithm.

- Concern that UAS pilot may not understand why an algorithm took a specific action.

Beyond merely providing traffic symbols on a graphic display, there is a range of further automation assistance that could be provided to a human decision-maker (normally the UAS pilot). These range from distance- or time-measuring tools on the display to projections showing whether a threat volume would be penetrated by the estimated trajectory, and finally to projections displaying hypothetical resolution outcomes. An intermediate step might use the automation to declare a threat, but then leave the resolution decision to the human. The arguments favoring reliance on algorithms and automation are as follows:

(a) It is yet to be established that a remote pilot can effectively perform S&A using a traffic display for the full range of feasible encounter situations. Particularly for collision avoidance where time is short, the timeline may allow little margin for pondering whether action is necessary, and for interpreting graphical or numerical information to safely resolve the more difficult conflict geometries. One of the challenges for acceptance of a S&A system will be demonstrating the reliability of pilot decision-making. This will require sufficiently clear and complete information to be provided on the traffic display, optionally augmented by automation aids, and may also involve setting requirements for pilot qualification and training.

(b) Another consideration is connected to the architecture decision. If a judgment is made that the communication link between aircraft and control station cannot be made sufficiently capable for the timely and reliable resolution of collisions, the aircraft may need to be equipped for automation of both detection and resolution functions.

Some people envision some similarities between the human roles in S&A and in air traffic control. With this line of thinking, crucial issues are overlooked. In fact, the human pilot role in the collision avoidance portion of S&A is only remotely like that of an air traffic controller. The latter role involves training with specific equipment to enforce authorized separation standards with relatively large distances compared to collision avoidance. The S&A collision avoidance task includes time-critical actions that directly control the flight of an aircraft. There is no relevant data on a controller's ability to perform collision avoidance, in part due to the limitations of the ATC traffic display accuracy and update rate, as well as the latency in communicating an instruction to an onboard pilot who then must execute it.

2.7   Sense and Avoid Timeline

Figure 2.8 depicts a notional timeline for S&A. The surveillance system needs to provide sufficient detection range so that a threat or hazard can be detected and the subsequent steps performed in time to resolve a collision. The surveillance range and the timeline of subsequent

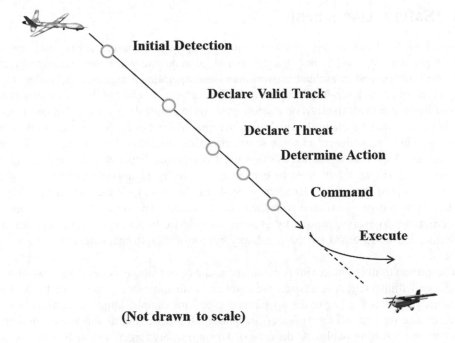

Initial Detection

Declare Valid Track

Declare Threat

Determine Action

Command

Execute

(Not drawn to scale)

Figure 2.8   S&A sub-function timeline

sub-functions need to act upon own and threat's relative trajectories with closing rates up to some designated maximum value. Several of the following steps potentially increase the time required to resolve a collision encounter:

- Sensor technologies that require multiple measurements to determine a valid detection.

- Declaration of a threat, which could be delayed by factors such as measurement uncertainties, or the need to confirm that a candidate threat aircraft is maneuvering.

- Human decision time in determining the action to take.

- Communication delays in transmitting the action to the aircraft.

- Aerodynamic response of the aircraft in performing the intended maneuver.

In the author's opinion, the design of this timeline's components represents the greatest tradeoff challenge within S&A. The stochastic nature of several components, the complexity of a threat declaration algorithm, the potentially complex interaction of diverse sensor elements and the character of human delays make this a fascinating, complex problem.

The worst-case surveillance detection range then is the maximum combined closing speed multiplied by the sum of the processing times and delays. It might be argued that this worst case is unlikely, as all the maximum times should not occur in the same encounter. It then would be appropriate to calculate how likely lesser delays and lesser speeds would occur and design the timeline components accordingly, so long as the specified safety targets were met.

2.8   Safety Assessment

Approval to fly UAS in any airspace requires approval from the appropriate authorities. At present, approval is not straightforward, as regulations are very scarce regarding UAS. Authorities tend to evaluate airworthiness and operational approvals according to existing regulations for aircraft, but some shortfalls emerge as a result of the unmanned nature. Each of these approvals depends on a safety case, which must demonstrate that any foreseeable hazards would be controlled within an acceptable frequency. Some stakeholders tend toward proving 'equivalence' in some sense, compared to manned aviation. One difficulty with this approach is that manned aviation itself comprises disparate users and operations, so there is no single safety level to benchmark. A different approach follows the basic principles adopted for Safety Management Systems [6], which dictate that hazard frequencies be controlled in accordance with their consequences. The regulatory environment for UAS continues to evolve, pressed by growing user desire to access airspace with minimal restriction, but constrained by the regulatory responsibility to maintain safe airspace for all users.

The current thinking places S&A within the realm of operations rather than airworthiness. Operations within a region of airspace are regulated with some safety target as a basis. These targets may differ according to the operational use. For example, larger air carriers bear the burden of meeting a higher level of safety than do small private aircraft, due to their obligation to protect the traveling public. At the time of writing, safety targets for UAS have not been determined, and it is not clear that a single safety level would apply to all types of UAS or all operations.

Various methods of demonstration will be required to prove that a safety target is satisfied. Flight testing cannot suffice, as time and cost would preclude a large number of flights, and thus limit the amount of data collected. Instead, fast-time simulation should be used to prove the breadth of performance of S&A. This method has been used with great success for the development and standardization of TCAS [7]. Flight testing retains a role in validating the simulation results and ensuring effective integration of system components.

A critical step toward developing performance requirements will be performing safety assessment of S&A. This will involve the determination of operational hazards – an example would be a midair collision – that could arise from the failure or incorrect performance of each function or data flow constituting S&A.

Examples of failure events arising from the surveillance system could include:

- Aircraft not detected by surveillance sub-function.

- Aircraft detected late by surveillance sub-function.

- Aircraft detected with incorrect position or velocity.

The latter two of these involve complex analysis, since a 'late' or 'incorrect' detection may not cause a hazardous outcome with certainty, but would increase its likelihood. The relationship between the cause and the effect might depend on the design or technology involved.

The analysis should consider not only failures in resolving collisions, but also incorrect maneuvers that 'induce' a collision where none would have otherwise occurred. This is a very real hazard, which could arise from various causes, including measurement error, human decision, limitations in the algorithm, or even the communications link to the aircraft.

2.9   Modeling and Simulation

Several of the elements within S&A are stochastic in nature. Their interactions need to be fully explored so that adverse combinations of events are uncovered and considered in the proper context. The practical means of obtaining sufficient data involves modeling and simulation [8].

Monte Carlo simulation is an established technique that repeatedly simulates a situation with its components containing independently selected values from their respective models or probability distributions. For example, in a S&A simulation, an encounter geometry would be run repeatedly, each time using a different value of measurement error, pilot delay, communication link delay and other component values as appropriate. This would be repeated for a comprehensive set of encounter geometries and other factors (e.g., targets with or without cooperative equipage). The selection of encounter features and component values would correspond to their likelihood, and the results of the simulations might compare the distribution of separations achieved by the S&A process in repeated trials versus the separation if no avoidance action were taken. Models need to be obtained for each element in the architecture. Figure 2.9 shows an example using several onboard sensors, a communication link to a ground-based control station, and a pilot who either determines or evaluates a resolution and sends commands to the aircraft.

For this example, the onboard surveillance system consists of several sensor technologies whose measurements are to be combined within a track fusion function. This surveillance system needs to be modeled, with each sensor represented by its appropriate field of view and range, its measurement parameters such as accuracy, update rate and probability distribution of measurement error, and the track fusion process. This model would be able to simulate the attempted detection and tracking of threat aircraft whose positions change as they and the UAS follow their respective flight trajectories.

Another model is needed to simulate encounters. Statistics of encounters are needed so that a comprehensive set may be generated to represent encounters expected by the UAS in the intended airspace. The encounters must provide for horizontal and vertical profiles, both constant rate and maneuvering. The range of realistic speeds and approach angles need to be evaluated. It is not sufficient to record and replay observed encounters. First, there is very little UAS operation taking place in non-segregated airspace. Second, a data collection effort

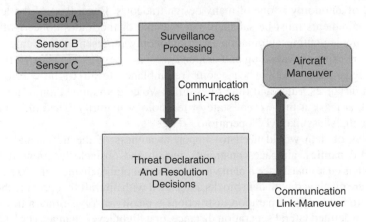

Figure 2.9   Example S&A implementation for simulation

is unlikely to observe enough of the more challenging geometries (e.g., maneuvering) that must be simulated in large numbers in order to assure robust performance.

Whereas models for manned aircraft encounters have been created from observed flight data, UAS operations should primarily differ in many respects due to their aerodynamics as well as the types of mission flown. An agreed method of modeling the UAS flight locations and profiles would need to be combined with the profiles of other traffic, to create synthetic encounter statistics.

In this example, tracks of detected aircraft are sent to the ground for evaluation by the pilot. The threat detection and resolution decisions could be performed either onboard the aircraft or on the ground; this example assumes that the pilot would evaluate the traffic picture before approving a resolution choice and sending it to the aircraft for execution. The communication downlink of traffic and uplink of maneuver instructions both need to be modeled. Communication links exhibit latency, availability and error rates. Each of these may be stochastic and may vary according to conditions (e.g., air–ground range, or blockage by terrain or buildings).

If a candidate algorithm is under consideration, it should be used within the simulation, processing the simulated inputs in order to represent the event and timing of declaring a threat and selecting the maneuver. Pilot actions, whether involved in threat declaration and maneuver choice, or merely in approving an algorithm's choice, need to be modeled, as human decision time and performance (e.g., accuracy of maneuver response) will exhibit variation.

Finally, the aircraft maneuverability must be modeled so that the maneuver performance truly represents the UAS. The aerodynamic performance may vary according to factors such as altitude, weight, speed and any restrictions placed upon maneuvering.

2.10   Human Factors

The performance of a human pilot has many aspects across UAS operations. There are differences of opinion as to whether the pilot needs to be trained or even experienced in piloting manned aircraft, or whether the specific operations of the UAS constitute sufficient qualification. Regarding S&A, specific functions would dictate the skills required of the human and the data and formats of presentation supporting those functions.

The level of autonomy is one of many design tradeoffs. Regardless of the level chosen, the safety requirements must be satisfied. An example of the trades concerns the cost and reliability of implementing automation (with autonomy representing extreme automation) versus the complexity of evaluating and demonstrating successful human performance.

The self-separation function bears some resemblance to air traffic control separation tasks, though distances, rules and time lines differ. Voice communication to the UAS is not contemplated, at least at the present state of technology maturity. The collision avoidance function, though, is less like ATC operation.

The display of data would need to support awareness of the traffic and fixed hazard positions and dynamics, airspace constraints, terrain and obstacle locations if known. Air traffic controllers customarily see a plan view (horizontal plane) display of a fixed geographic area. Targets are displayed with data blocks associated with aircraft that provide their altitude, identity and airspeed. ATC separation instructions typically involve either a heading change that preserves a defined lateral separation distance, or a flight level change. The TCAS cockpit displays likewise use plan view, but unlike the ATC display show the (moving) host aircraft depicted at a fixed reference point and other traffic displayed so as to indicate their relative lateral position and motion. In the case of TCAS, its automation determines the vertical avoidance maneuver and no lateral maneuvering based on its display is authorized.

Unlike ATC, which uses standard separation minima, the UAS self-separation criteria would need to be adjusted according to appropriate margins calculated to account for measurement and prediction errors, plus sufficient margin to account for delays in delivering maneuver commands to the aircraft. These margins depend on the specific architecture and system design, and could vary dynamically depending on such factors as the sensors or communication links in use.

If the display is augmented with automation aids, these should be chosen so that they assist in reliable decision-making and ensure conspicuity when timely decisions are required. Likewise, the means of communicating maneuver instructions to the aircraft must be easily accessible and supportive of correct and accurate input. Some interest has been expressed in equipping a pilot to control multiple UAS. In this situation, it is essential to avoid confusion in associating pilot actions with the desired aircraft, as well as directing the pilot's attention to any situation requiring prompt evaluation.

It is possible to envision a wide variety of display and automation possibilities supporting S&A. A challenge will be to characterize the human use of each distinct implementation, to assure that specified performance targets are achieved, and then demonstrate that results obtained in a simulation environment will translate to equivalent safety for actual operations [9].

2.11   Standards Process

Standards are expected to play an important role supporting the certification of UAS to enable operations in civil airspace without the extreme restrictions imposed today. The S&A function is a notable area where the use of an approved standard can provide fundamental support to an applicant's safety case, since the function would replace one or more functions traditionally performed by the human pilot aboard manned aircraft. While some aspects of human performance are extremely difficult to quantify, the UAS version will need to be demonstrated to meet safety targets yet to be prescribed.

Standards typically provide direction to system developers toward achieving system certification and operational approvals. By demonstrating conformance to the standard's requirements, much of the work to prove the safety case would be accomplished. The standard must provide a comprehensive set of requirements that would assure safety and operational compatibility, as well as a prescribed suite of tests for demonstrating its requirements are met by a candidate implementation.

The standards process has traditionally begun with evidence of a mature technology. In the case of UAS, many aspects of aircraft and system design and operation have been demonstrated, but S&A remains in the research and experimental phases. The following series of steps describe the process planned for standards within the RTCA, Inc. Special Committee 203.

2.11.1   Description

Since UAS standards will encompass many aspects of the system and operations, the process begins with an operational system and environment definition. This aims at describing the system functional elements and concepts of operation, the range of performance attributes for unmanned aircraft, and the airspace structure and its management system. These descriptions will support the subsequent analyses of safety, performance and interoperability.

2.11.2   Operational and Functional Requirements

The next step of the process is to enumerate operational and functional requirements. These describe 'what' needs to be done, although at this stage they do not state 'how well'.

2.11.3   Architecture

System architecture defines functional elements and data flows that are envisioned to perform the main functions. There can be more than one variation. A good example involves pilot-in-the-loop versus autonomous S&A. The latter would not require communication links to and from the control station for the purpose of the S&A operations.

The architecture work within the standards process involves the development of a number of documentation artifacts, sometimes called 'views' of the UAS system. These range from operational concepts and use cases to system connection diagrams at the functional level, allowing enumeration of all sub-functions and data flow requirements from one sub-function to another. While this work involves a great amount of analysis and documentation, it ensures that all cases are thoroughly considered and that each analysis covers a documented definition of the system.

2.11.4   Safety, Performance, and Interoperability Assessments

The safety assessment analyzes the potential failures of each function and data flow from the architecture step. It identifies the consequence of the failure and the corresponding likelihood for acceptance of that risk. This step provides crucial information for the allocation of risks (to follow) and identifies areas where additional mitigation of risks can be considered. The measures for specifying event likelihoods need to relate to an overall safety target. In aviation today, safety targets differ between air carriers and certain other user types. It must be determined where UAS fall within this spectrum and if some UAS may be treated differently than others. It remains a matter of discussion as to whether this could depend on aircraft size and weight, airspace, mission type or other factors.

Some requirements do not have a direct bearing on safety, but are needed for other reasons such as ensuring compatible operations with other airspace users. A performance assessment determines the quantitative requirements for functions in this case. An example might be to measure the successful performance of the self-separation function. Its failure may have some safety impact, though less so than a collision avoidance failure; meanwhile, the same failure of self-separation may be required to be infrequent for operational reasons. Thus, the performance assessment also can influence quantitative requirements.

An interoperability assessment may take two forms for S&A. The first may address technical interoperability. A good example would be the specific equipment, signals and formats required to interoperate with cooperative aircraft or ATC, or to coordinate resolutions with other collision avoidance systems. The second form involves interoperability with ATM. Its scope is still to be determined, but might impose constraints to ensure compatibility with airspace structure or rules of the air.

The steps to this point are illustrated in Figure 2.10(a).

2.11.5   Performance Requirements

At the conclusion of these several assessments, performance requirements for functions and data flows will have been identified. It is possible that certain requirements may appear

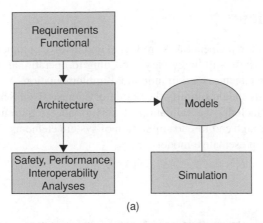

(a)

Figure 2.10(a) Initial steps toward standards development

infeasible or at least undesirable for economic reasons. A process of allocation of requirements between functions would allow certain tradeoffs to be made so as to make certain requirements less onerous without compromising overall safety or operational performance. For example, achieving a long sensor range might be difficult, so maneuvering could be planned to begin later. Another example concerns data link performance. Instead of providing an extremely robust link, threat declaration might be adjusted to begin maneuvers earlier and operate safely despite a brief link outage.

2.11.6   Validation

A standard is expected to present validated requirements. The modeling and simulation work should be used to evaluate S&A metrics over the expected encounter space. Some flight testing should be performed to reinforce the simulation results.

These important steps are shown in Figure 2.10(b). With these completed, the way is clear to write the standard.

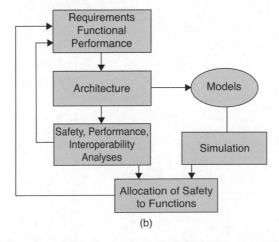

(b)

Figure 2.10(b) Completing standards development

2.12   Conclusion

The choices available to implement S&A are varied and involve complex trades between subsystem requirements. These will likely vary according to operational needs and the choices of technologies and architecture used in individual implementations.

The use of standards should simplify the regulatory approvals of S&A systems, inasmuch as their development includes the validation of performance requirements. This validation needs to include a thorough end-to-end simulation of system elements. The simulation in turn depends upon use of representative models.

References

1. Federal Aviation Administration, 'Sense and Avoid (SAA) for Unmanned Aircraft Systems (UAS)', October 2009.
2. US Code of Federal Regulations – Title 14 Aeronautics and Space; Part 91 General operating and flight rules; Section 111, 'Operating Near other Aircraft' and Section 113, 'Right-of-way rules: Except water operations'.
3. US Code of Federal Regulations – Title 14 Aeronautics and Space; Part 91 General operating and flight rules; Section 225, 'Automatic Dependent Surveillance – Broadcast (ADS-B) Out equipment and use'.
4. Minimum operational performance standards for traffic alert and collision avoidance system II (TCAS II) version 7.1, DO-185B, RTCA, Inc., June 2008.
5. *Airborne Collision Avoidance System (ACAS) Manual*, Doc 9863, 1st edn, International Civil Aviation Organization, 2006.
6. Federal Aviation Administration Safety Management System manual, Version 1.1, May 2004.
7. McLaughlin, M., Safety Study of the Traffic Alert and Collision Avoidance System (TCAS II) – Final Version, MTR97W32, The MITRE Corporation, June 1997.
8. Zeitlin, A.D., Lacher, A.R., Kuchar, J., and Drumm, A., *Collision Avoidance for Unmanned Aircraft: Proving the Safety Case*, MP060219, The MITRE Corporation, 42-1017, MIT Lincoln Laboratory, McLean, VA and Lexington, MA, October 2006.
9. Tadema, J., *Unmanned Aircraft Systems HMI and Automation*, Shaker Publishing, Maastricht, 2011.

3

Integration of SAA Capabilities into a UAS Distributed Architecture for Civil Applications

Pablo Royo, Eduard Santamaria, Juan Manuel Lema, Enric Pastor and Cristina Barrado
Technical University of Catalonia, Spain

3.1 Introduction

In this chapter the integration of 'sense and avoid' (SAA) capabilities into a distributed architecture for unmanned aircraft systems (UAS) is discussed. The presented UAS architecture provides a framework that enables rapid development and integration of hardware and software components required for a wide range of civil missions. This framework includes a number of common services,[1] which are envisioned as necessary for most missions – such as flight plan management, mission management and contingency management, among others.

One of the most important issues that needs to be addressed when developing a UAS architecture is SAA. This chapter tackles SAA from two perspectives. Firstly, it discusses integration of SAA capabilities in mission-oriented architecture. In contrast with commercial aviation, where missions mostly consist of flying from an initial location to a given destination, unmanned systems may be used in a wide variety of situations. The UAS response in case of

---

[1] Over the different distributed nodes of the UAS, one can deploy software components, called services that implement the required functionalities.

Sense and Avoid in UAS: Research and Applications, First Edition. Edited by Plamen Angelov.
© 2012 John Wiley & Sons, Ltd. Published 2012 by John Wiley & Sons, Ltd.

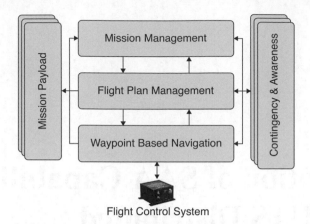

Figure 3.1   Organization of main UAS services

conflict has to be defined in a way that minimizes its impact on the mission being carried out. Secondly, system design and operations considerations aimed at making the UAS behavior more predictable and facilitating conflict prevention are also provided.

The proposed architecture provides a basis for building flexible UAS that can operate in multiple scenarios. It includes a wide set of services (see Section 3.3) to select from according to the needs of each specific mission. These services can be reconfigured so that there is no need to create a new implementation for each scenario. In this way, by combining components that can easily be added, removed and reconfigured, the user can benefit from faster preparation and deployment of the system. Operating on top of commercial off-the-shelf (COTS) components, cost-effectiveness is further improved.

Figure 3.1 shows how the main UAS services are organized. At the bottom of the figure, a COTS flight control system (also known as autopilot) is found. The autopilot provides waypoint-based navigation. Interaction with the autopilot takes place through an intermediary layer that provides a hardware-independent interface. A flight management layer is responsible for performing the flight plan. In the proposed system, an XML-based language is used to specify the flight plan [1]. This language improves on the waypoint-based navigation method used in most commercial autopilots by providing higher-level constructs, with richer semantics, that enable adaptation to mission progress. Some ideas are based on current practices in commercial aviation industry for the specification of area navigation (RNAV) procedures [2]. Alternative flight paths for conflict and emergency situations can also be included in the flight plan. The mission management layer is in charge of orchestrating operation of the different components. The mission management layer interacts with payload-related services and can also adapt the flight plan to the mission needs. All these flight and mission components are complemented by contingency and awareness services. Contingency services continuously evaluate the health of the system and trigger reactions when something goes wrong. Awareness services, which implement the SAA capabilities, provide conflict detection and management related to the environment where the UAS is operating.

Nowadays, there is a strong focus on providing technical solutions to the SAA problem. But for these solutions to be fully effective, they need to be integrated into an architecture that takes into account the UAS flight and mission as a whole.

Table 3.1 Tactical versus strategic conflict handling

	Tactical	Strategic
Source of conflict	Aircraft collision, terrain avoidance	Weather conditions, ATC constraints, predictable mid-air conflicts
Response	Direct commands to UAS control surfaces and throttle	Change flight plan or execute an alternative flight path

There are two types of conflict: (a) tactical conflicts, which require immediate reaction and (b) strategic conflicts, where selection of a predefined alternative path is possible. Table 3.1 provides some examples of potential conflicts and the UAS behavior during the reaction. A rapid response to a tactical conflict will be achieved by pre-empting flight and mission layers and directly operating on the UAS throttle and control surfaces. In a strategic conflict, with less stringent constraints on the reaction time, the flight plan can be modified or a predefined alternative flight path can be selected.

An additional way to prevent conflict situations consists of making the UAS behavior as predictable as possible and making this information known to other airspace actors. The UAS follows predetermined depart and approach patterns and restricts its flight trajectory to the one described in its flight plan, which also contains possible alternatives. Doing so will also make the system more likely to obtain an airworthiness Certificate of Authorization (COA), since regulation authorities will not allow unmanned systems with unpredictable behavior to operate in non-segregated airspace.

The next sections describe the proposed architecture and discuss how flight and mission layers interact with awareness components. Sections 3.2 and 3.3 outline the distributed architecture of the system and describe the UAS service abstraction layer (USAL), which defines the interfaces and responsibilities of the system's components. A more detailed view of the main embarked and on-ground components follows in Section 3.4. After that, in Section 3.5, we describe how the system deals with conflict situations. Some strategies for facilitating conflict prevention are discussed in that section. The chapter concludes with the final remarks provided in the conclusions section.

3.2   System Overview

The main goal of our UAS architecture is to provide a platform that enables rapid and flexible development of UAS systems for a wide range of applications [3–6]. Issues that need to be considered when developing such architecture include air-to-ground and air-to-air communications, payload abstraction for a wide range of sensors and actuators, UAS services interconnection, flight and mission management and, of course, contingency and awareness services to, respectively, monitor the system's health and provide information about the environment the UAS operates in. As will be seen, awareness services include the sensing, processing and reaction components required for implementing the SAA capabilities. Definition and development of the UAS services need to be approached taking into account that legal implications and regulatory aspects for the operation of UAS aren't clear yet. Therefore, the architecture itself needs to be flexible enough to be able to evolve and adapt to regulatory frameworks that may also vary from country to country.

3.2.1   Distributed System Architecture

Access to UAS technology is getting easier and cheaper, but important efforts are still required to leverage all its potential in complex surveillance and remote sensing missions. More so, if the platform has to be able to perform a wide range of missions in a diverse set of environments. Current research mainly focuses on UAS flight control and autopilot optimization. Besides that, specific proposals on UAS platforms selection and payload interaction are found for specific missions [7, 8]. In contrast, our architecture does not focus on a specific mission. This diversity of scenarios demands SAA capabilities be integrated into the architecture. In this section, we describe our general-purpose architecture for executing UAS civil missions that will accommodate such SAA services.

In recent years, there has been a clear trend in various fields to move from centralized and monolithic systems to networked and distributed ones. As complexity increases, it seems a good idea to divide functionality into simpler components that cooperate to achieve the overall task. These components are usually interchangeable and can sometimes be redundant, to improve the fault tolerance of the whole system. In addition, in some fields in which specialized and expensive networks are commonly used, for example manufacturing or avionics, there has been a gradual increase in acceptance of common local area networks, specifically Ethernets. Ethernets have been used extensively since the mid-1980s, and are an affordable and proven solution.

In this context, our vision is of an architecture in which low-cost devices are distributed throughout the system and form networks of smart peripherals or intelligent sensors. That is, the proposed architecture is built as a set of embedded microprocessors, connected by a local area network (LAN), in a distributed and scalable architecture. Over the different distributed elements of the system we deploy software components, which we refer to as 'services' that will implement the required functionalities (see Figure 3.2). These services cooperate for the accomplishment of the UAS mission. They rely on a communications middleware [9] for exchanging data and commands. The communication primitives provided by the middleware promote a publish/subscribe model for sending and receiving data, announcing events and executing commands among services.

As seen in Figure 3.2, the different services of the UAS are interconnected by a COTS Ethernet network that makes data communication between them very flexible and efficient. Services like the Image Acquisition, Storage Module, Autopilot Management, Real-Time Data Processing and Mission Management are independent and can be executed in different nodes located on the aircraft. The idea behind this is to increment the interoperability, flexibility and extensibility of the system and its individual components. We want to be able to reuse components of the existing system in the implementation of any new system.

In the proposed architecture, several services can be executed on the same node. While this may be seen as a source of runtime and development overhead and excessive in terms of functional decomposition, it is intended to encourage careful design of service interactions and proper separation of concerns. It also provides advantages, such as: independence of the service from deployment and hardware configuration, service interoperability (e.g., different sensors or algorithms for the same functionality), fault tolerance (a service could be replicated

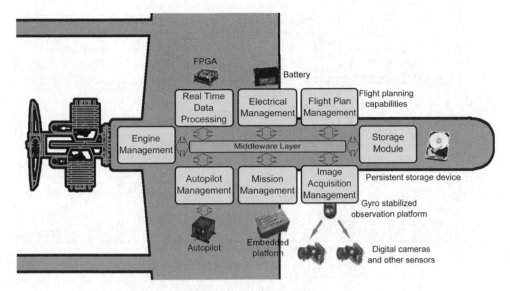

Figure 3.2   Overview of the UAS distributed system architecture

in different hardware nodes for redundancy) and service migration (between nodes in case of changes in the system's needs), among others.

To sum up, the proposed architectural approach offers important benefits in our application domain.

- **Development simplicity:** inspired by Internet applications and protocols, the computational requirements can be organized as services that are offered to all possible clients connected to the network.

- **Extreme flexibility:** we are free to select the actual type of processor to be used in each node. Different processors can be used according to requirements, and they can be scaled according to computational needs of the application.

- **Simple node interconnection:** an Ethernet-based architecture provides a much simpler alternative to the complex interconnection schemes needed by end-to-end parallel buses.

- **Easier adaptation to changes:** the distributed nature of the architecture facilitates integration of new types of sensors and techniques, which is of special importance to adapt to the technical and legal requirements of SAA.

3.3   USAL Concept and Structure

The previous section outlined our distributed architecture and its main benefits. On top of that, a collection of reusable services is provided that will facilitate mission development. We refer to these as the UAS service abstraction layer (USAL).

The USAL is the set of services needed to give support to most types of UAS civil mission. The USAL can be compared to an operating system: computers have hardware devices used for input/output operations. Every device has its own particularities and the operating system

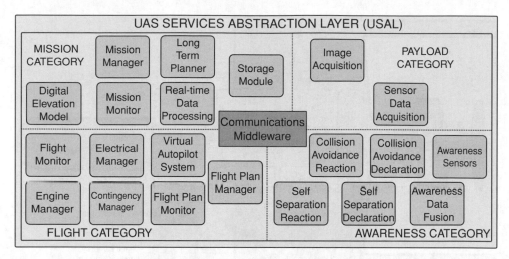

Figure 3.3   USAL services by category

offers an abstraction layer to access such devices in a uniform way. Basically, it publishes an Application Program Interface (API) which provides end-users with simplified access to hardware elements. In the same way the USAL considers sensors and, in general, all payload as hardware devices of a computer. It is a software layer that gives facilities to end-users' programs to access the UAS flight, mission and payload components. The idea is to provide an abstraction layer that allows the mission developer to reuse components and that provides guiding directives on how the services should interchange avionics information with each other. The USAL services cover an important part of the generic functionalities present in many missions. Therefore, to adapt our aircraft for a new mission it will be enough to reconfigure the services deployed in the UAS boards. Even though the USAL is composed of a large set of services, not all of them have to be present in every UAS mission. Only those services required for a given configuration/mission should be deployed in the UAS. As shown in Figure 3.3, USAL services have been classified into four categories: Flight, Mission, Payload and Awareness.

The first element that should be noted in the flight category is the autopilot. Although not exactly a *service*, it provides the core functionality that enables autonomous flight. The autopilot operation is supervised by the Virtual Autopilot System (VAS), which provides a standardized autopilot-independent interface for other *services* to interact with. On top of the VAS, the Flight Plan Manager (FPM) offers flight plan specification and management capabilities that go beyond classical waypoint navigation. Contingency-related services help improve the security and reliability of operations. Some of these services are complemented by on-ground counterparts that enable supervision and control by ground personnel. SAA services from the awareness category will interact with services in the flight category to adapt the flight to deal with conflict situations.

The main service in the Mission category is the Mission Manager (MMa). Its role is to orchestrate operation of the overall mission. This service executes a mission specified using a state-machine-like notation. The MMa will be accompanied by other mission-related components, like Storage, Real-Time Data Processing and Communication Management services, among others. If a conflict resolution results in a deviation from the initial flight plan, the MMa needs to be notified so that operation of mission payload can be configured according to the new situation. As an example, some sensors could be switched off while away from the mission area.

The Payload category includes those services that facilitate interaction with embarked devices, especially sensors and actuators. Input devices consist of aircraft state sensors, such as GPS, IMU or anemometers, earth/atmosphere observation sensors, such as visual, infrared and radiometric cameras, not excluding other sensors that may be required for the mission (chemical and temperature sensors, radars, etc.). Output devices are less common, but flares, parachutes or loom shuttles are possible examples of UAS actuators.

The remaining category used to conceptually organize services is awareness. This category deals with those services required to manage interaction of the UAS with its surroundings. Successful integration of UAS into non-segregated airspace requires a number of features to be included in the system architecture, such as interaction with cooperative aircraft and detection and avoidance of non-cooperative aircraft. In these situations the pilot in/on command should stay continuously informed, automatic reactions should also be considered. The awareness category is covered in depth in Section 3.5.

3.4   Flight and Mission Services

In the previous section a general view of the UAS architecture and the USAL concept was provided. In this section we focus on the key services that enable automated mission execution and its supervision. Flight and mission automation services will usually be onboard and, therefore, belong to the air segment. Supervision services UAS operators interact with belong to the ground segment.

3.4.1   Air Segment

The three services that will be covered herein are the Virtual Autopilot System (VAS), the Flight Plan Manager (FPMa) and the Mission Manager (MMa). All three of them have a major role in governing the UAS flight and mission behavior. The VAS and the FPMa, which belong to the flight services category (see Figure 3.4), are the ones with the most relevant role in supporting SAA.

The VAS operates as an interface between the autopilot and USAL services. At one end, the VAS interacts directly with the onboard autopilot and needs to be adapted to the autopilot peculiarities. At the other end, the VAS faces other USAL services providing an interface that hides autopilot implementation details from its actual users. The VAS also offers a number of information flows to be exploited by other USAL services. All the information provided by the VAS is standardized independently of the actual autopilot being used.

The VAS works in conjunction with the FPMa. The FPMa is the USAL service responsible for processing and executing flight plans. In order to execute a flight plan, the FPMa sends navigation commands to the VAS. These commands mainly consist of waypoints which the aircraft has to fly through. Since the flight plan is specified in terms of legs, a certain translation process is needed to convert them into the waypoint sequences expected by the

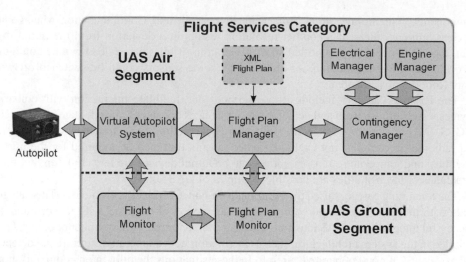

Figure 3.4   Architecture of the USAL flight category

VAS. This flow of waypoint commands is the main form of interaction between the FPMa and the VAS.

Other services belonging to the flight category that also appear in Figure 3.4 are the contingency services, which monitor the health of the electrical and engine sub-systems (more details in [10]). The flight monitor and the flight plan monitor, shown at the bottom of the figure, are part of the ground services and will be covered later.

The functionality implemented by the VAS is divided into four main areas: flight state management, flight telemetry, navigation and status/alarm information.

Flight state management relates to the different operational modes implemented by the VAS. Figure 3.5 displays all possible VAS states. As can be seen, they are organized in different groups according to the flight phase they belong to. The initial state inside each group is shown with an arrow on the top right box state corner. The other arrows show the transitions between different states. The diagram descends from the beginning of the mission to the end, although, in some cases, there are horizontal transitions. When equivalent states are provided by the underlying autopilot, the VAS will act as a proxy that enables access to them. The VAS implements those states not found in the autopilot.

The main VAS states with regard to the SAA capabilities are auto take-off, approach and safe reaction. The first two implement standardized terminal operations that provide predictable behavior and reduce the risk of conflicts. The safe reaction state implements the actual collision avoidance maneuvers.

The other three areas that complete the implementation of the VAS, namely, flight telemetry, navigation and status/alarm information fulfill the following roles. Telemetry refers to the VAS capability to extract telemetry data from the autopilot and provide it to USAL services for exploitation. The navigation information area focuses on VAS input/outputs available during the navigation states. It implements the main navigation commands and provides information regarding the current waypoint and some other flight-related data. Finally, the status/alarm area gives information about current autopilot and VAS status or alarms. Fight telemetry and status/alarm information are outgoing flows, while navigation and state management have an input/output direction.

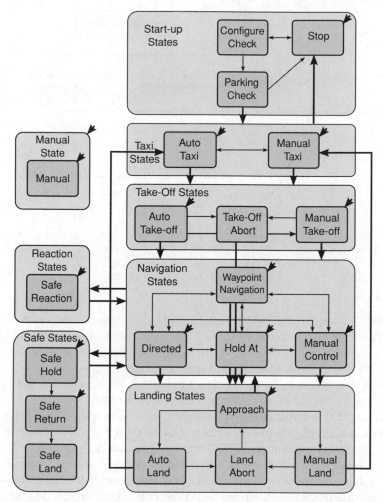

Figure 3.5 VAS state diagram with arrows indicating allowed transitions

The inclusion of the VAS greatly improves the flexibility of the USAL framework because:

- The autopilot unit can be replaced by a new version or a different product, and this change will have no impact on the rest of the UAS system. In other words, the VAS implements an abstraction layer that isolates the system from the particular autopilot solution in use.

- An increased level of functionality is provided. Besides providing a set of states that the UAS can take advantage of, the VAS also helps overcome typical limitations of UAS autopilots, such as limitations on the number of waypoints the autopilot can handle.

- In addition, commercial autopilots mainly focus on waypoint navigation, however, the UAS operation may require considering a global perspective, from take-off to the mission and back to landing. The VAS promotes standardized mission-oriented states in order to cope with more elaborate operational requirements.

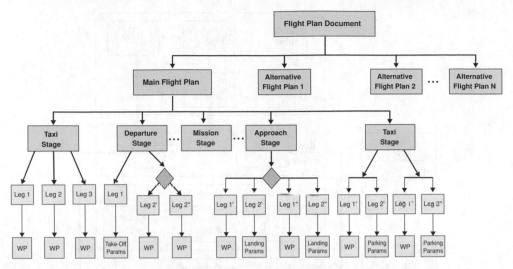

Figure 3.6   Flight plan structure

The FPMa is a service designed to provide flight management functions that go beyond following a predefined sequence of waypoints. The FPMa offers structured flight-plan phases with built-in alternatives, leg-based navigation and constructs to enable forking, repetition and generation of complex trajectories. All these elements are combined in a description of the flight plan that is sent to the FPMa. The FPMa takes this description and dynamically translates it into a sequence of waypoints that are sent to the VAS.

In our system, UAS navigation instructions are represented by means of an XML document that contains a main flight plan plus a number of alternatives. Each one of them is composed of stages, legs and waypoints hierarchically organized as seen in Figure 3.6.

Stages group together legs that seek a common purpose and organize them into different phases that will be performed in a sequence. Legs specify the path that the plane must follow in order to reach a destination waypoint. A waypoint is a geographical position defined in terms of latitude/longitude coordinates that may also be accompanied by target altitude and speed indications.

There are four different kinds of leg: basic, iterative, intersection and parametric. Basic legs specify primitives such as fly directly to a fix, follow a track to a fix, holding patterns, etc. They are based on existing ones in RNAV. Iterative legs are used to indicate that parts of the flight plan should be flown several times. Intersection legs are used in situations where there is more than one possible path to follow and a decision needs to be made. Finally, with parametric legs complex trajectories can be automatically generated from a reduced number of input parameters.

The FPMa is responsible for processing and executing flight plans. It can be seen as a translator of legs to waypoints. This translation process enables leg-based navigation of systems that only support waypoint navigation. The FPMa does not directly interact with the autopilot but with the VAS. From the VAS or autopilot perspective, the FPMa can be seen as a provider of waypoints to fly to. From a mission-related services perspective, the FPMa is the service to talk to in order to control the flight progress and make it adapt to the mission needs. There are multiple possibilities of interaction with the FPMa, the primary ones being setting condition values, sending updates that modify the flight plan and triggering execution of alternative plans.

In the context of SAA, the FPMa enables self-separation in two ways. First, it provides update commands that can be used to add/remove legs or modify existing ones (e.g., changing its altitude). The update commands to perform these changes must be issued by the on-ground pilot on command using the flight plan monitor. If this is not sufficient, during preflight a number of alternative flight plans can be defined, so that the UAS can switch from the main plan to an alternative one in order to avoid a potential conflict. While the system will provide a recommended choice depending on the design of the flight plan during preflight, the final decision will always rely on the on-ground command (pilot).

Alternative flight plans may lead the aircraft to a safe area where a holding pattern can be executed without disturbing other traffic. After that, the mission may be resumed once the mission area is deemed clear. Another possibility consists of having a complete flight plan for executing the rest of the mission to its end. If one wants to resume the main flight plan, the pilot in command will need to specify which leg the mission should continue from. Finally, if we are facing a completely unexpected conflict and feel the mission should be cancelled altogether, these alternative plans can be used to fly to a close airfield.

The last service covered in this section is the MMa, whose role is to extend the UAS automation capabilities by being able to execute a specification of the UAS behavior. The specification determines how operation of embarked services is orchestrated during a mission. The language chosen for describing the UAS behavior is an XML-based representation of state charts [11]. State charts extend traditional state machine diagrams with support for hierarchy and orthogonality, which respectively enable modular descriptions of complex systems and provide constructs for expressing concurrency.

The MMa listens to messages coming from the FPMa, the VAS and other services. These messages translate into events that feed the MMa execution engine. As a result of these incoming events, the MMa may change its current state, generating some kind of response in the process. A response will, usually, consist of a message modifying the behavior of some UAS service; for instance, a message switching on or off some payload according to the current mission phase. In this way, UAS services operate according to the current mission state in a coordinated manner.

The MMa does not directly involve in the response to a conflict situation. However, the choice of an alternative flight plan to perform the mission may require the embarked payload to be configured and operated in a different way. It is the responsibility of the MMa to perform these changes.

3.4.2   Ground Segment

Our vision is to provide a highly capable system and tools to support development of complex UAS-based missions. Automation capabilities are built into the UAS to reduce operators' workload and cope with situations that require immediate response and cannot wait for external input. Despite its automation capabilities, the system will rely on on-ground personnel to supervise and control the evolution of the mission.

Figure 3.8 displays the main positions at the ground control station. The Flight Monitor (FM) is the system used by the pilot in command to control the UAS flight. To do so, it

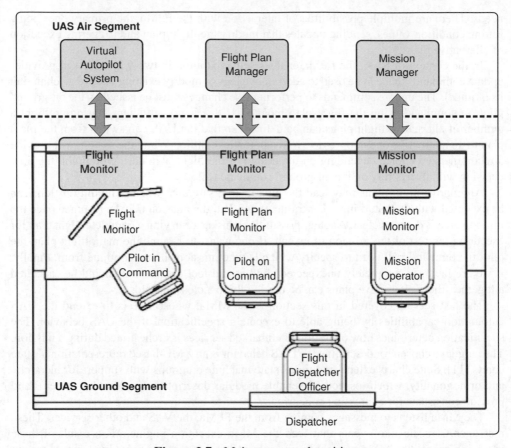

Figure 3.7   Main on-ground positions

communicates with the embarked VAS. The Flight Plan Monitor (FPMo) is used to monitor and modify the flight plan. It complements the functionality provided by the onboard FPMa. Finally, the Mission Monitor (MMo) is used to supervise and manage the mission and payload operation.

Generally speaking, current UAS autopilots offer manual and/or assisted piloting plus basic waypoint navigation capabilities. Although manual piloting is not the main form of expected interaction, the HMI interfaces are designed to maintain this capability (this is discussed further in the next chapter of the book). The FM is the on-ground service that interacts with the VAS and enables real-time supervision and control of the UAS flight. Using this front-end service, the pilot in command can change the VAS state directly on a displayed diagram. It also displays telemetry emitted by the VAS. The FM offers two separate screens; one of them, focusing on manual piloting, which is always available, and a multifunction screen to exploit higher-level VAS capabilities. With the multifunction screen, the pilot can switch between different views to display information such as VAS states, configuration parameters, telemetry, etc. Figure 3.8 shows the FM primary screen; in this example the FM has been configured to illustrate the VAS states, but in that location of the screen, the pilot can show the VAS states, electrical information or engine information.

Figure 3.8 Flight Monitor primary screen

The FPMo provides flight plan visualization and tracking capabilities and also permits modification of the flight plan and submission of changes to the FPMa. The FPMo is the ground service responsible for the onboard FPMa. The capabilities required by the FPMo are related to inherent dynamic behaviors offered by the FPMa.

Similarly to the FM human–machine interface, the FPMo interface is divided into two separate screens that work in coordination. The primary screen displays the flight plan and additional annotations in a graphical way. The secondary screen provides different views according to the different operational modes. The different views available in the primary screen are 'flight plan tracking', 'departure tracking', 'approach tracking' and 'flight plan validation'. These views are complemented by sub-views in the secondary screen that provide additional information, enabling modification of leg properties and configuration of FPMo parameters. More details about FPMo can be found in [12].

Figure 3.9 shows an overview of the FPMo; the left of the picture illustrates the main screen where the flight plan is monitored. The gray box shows the mission area where the UAS is going to develop a scan pattern (remote sensing mission). The right part of the figure exposes the secondary screen where the different legs of the flight plan can be updated, skipped or replaced.

The MMo is used to supervise the progress of the mission as defined in the MMa and display mission-related information in a friendly way. For example, during a wildfire monitoring mission, it may present the current state of the fire front on a map. The information produced by the UAS payload, such as sensors or cameras, will be shown in this workstation. The MMo should be highly configurable to fit the specific requirements of each mission.

3.5   Awareness Category at USAL Architecture

UAS are highly instrumented and, thus, the most suitable flight rules for them are IFR.[2] However, for civil applications, one of the advantages of UAS is precisely its capability to fly at low altitude, where VFR[3] aircraft are found. Therefore, the UAS has to be integrated into non-segregated airspace. A critical enabler for UAS integration into non-segregated airspace are the SAA systems, which have to ensure an equivalent capability to 'see & avoid' in manned aviation (performed by flight crew, usually with the support of onboard tools and systems like TCAS). A possible definition of the SAA system is the onboard, self-contained ability to:

- detect traffic, terrain, weather and obstructions that may lead to a conflict;

- determine the right of way of traffic; and

- maneuver according to the rules.

The USAL awareness category is responsible for being 'aware' of what is happening outside the UAS in order to manage suitable responses to any awareness conflict. Following

---

[2] An aircraft flying under IFR rules uses several navigation instruments which provide the pilot with the information needed to follow its trajectory or navigation route with no need for external visual references.

[3] VFR navigation is based on visual references which the pilot picks from the outside, such as rivers, mountains, roads, etc. This kind of navigation is strictly constrained to the existing meteorology with some minimum conditions measured in terms of visibility and minimum separation from clouds.

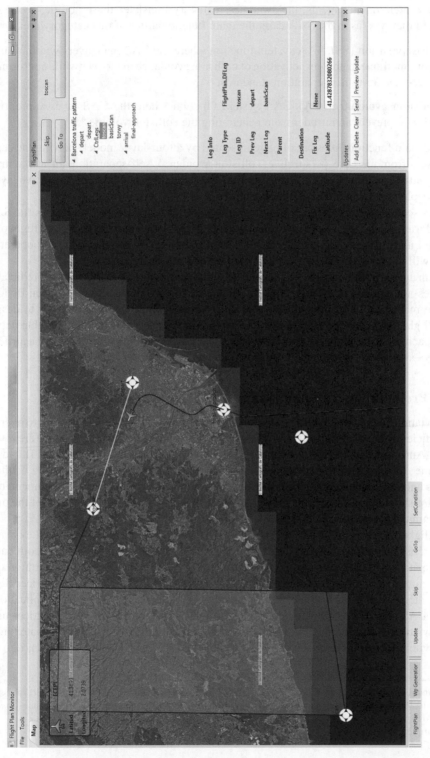

Figure 3.9   Flight Plan Monitor

these SAA intention statements, the awareness category translates these abilities to USAL services. In that way, as in the SAA case, the main functionalities of this category are:

- **Self-separation (SS).** SAA system function where the UAS maneuvers within a sufficient timeframe to prevent activation of collision avoidance maneuver while conforming to accepted air traffic separation standards.

- **Collision avoidance (CA).** SAA system function where the UAS takes appropriate action to prevent an intruder from penetrating the collision volume.[4]

These main functionalities are often represented by a four-layer model showing the separation layers from the UAS perspective: collision avoidance, self-separation, air traffic management and procedural [13, 14]. That is, air traffic management and procedural are two layers which complement the main SAA functionalities.

In this section, we are going to describe how the USAL architecture addresses the SAA concept during the different UAS mission phases. The SAA concept includes a good flight plan design and its communications under a procedural point of view. In that way, a cooperative aircraft[5] will know our flight path and can predict the UAS behavior.

To sum up, the SAA concept goes beyond SS and CA, making standardized and predictable procedures a complementary safety layer. For these reasons, this section has been divided into three main parts. The first sub-section outlines preflight procedures in order to design suitable flight plans and corresponding alternative flight plans. After that, predictable depart and approach procedures are described in order to facilitate conflict prevention on the airfield. Finally, SS and CA are discussed during en-route and mission operations.

3.5.1 Preflight Operational Procedures: Flight Dispatcher

In civil aviation, a set of procedures and standardized practices are followed in order to operate safely, efficiently and regularly all kind of aircraft. Criteria of safe operating practice are found in ICAO Annex 6, Part I [15] for commercial air transport operators, while Parts II and III of the same Annex deal with general aviation and helicopter operations respectively. In these standards and recommended practices one can find, for instance, what kind of documentation an operator should provide to flight crews, what are the responsibilities and duties of the pilot in/on command before, during and after a flight, etc.

The flight operations officer, also known as the flight dispatcher, is one of the key actors during aircraft operations, sharing duties and responsibilities with the pilots in/on command. Flight dispatchers assist pilots in/on command with all tasks related to flight preparation (for example, aircraft load planning, meteorological briefing, operational, air traffic services flight planning, etc.). We propose a new dispatching methodology focused on UAS civil applications; assisting UAS operations following the same philosophy of flight dispatching practices used in civil aviation. However, due to the singularities of the UAS systems, flight dispatching is merged with pilot in/on command duties as well as mission analysis and operation, i.e., the

---

[4] A cylindrical volume of airspace centered on the UA with a horizontal radius of 500 feet and vertical height of 200 feet (±100 feet) within which avoidance of a collision can only be considered a matter of chance [*Source:* RTCA DO-185A].

[5] Aircraft that have an electronic means of identification (i.e., a transponder) aboard and operating.

mission and flight plan dispatching. The overall process is mission-centric, focusing on all the requirements needed to properly implement the assigned tasks, but also integrating the traditional dispatching requirements. The full process is described in [16], however, we are going to summarize the whole process in order to understand why the dispatcher is important to SAA concepts.

The proposed dispatching process is built upon the USAL architecture and is introduced to characterize:

- The UAS mission and flight plan: its objectives, payload requirements, operation, flight plan, flight plan alternatives, etc.

- The UAS airframe: its various characteristics, performance, USAL services required for managing the flight and the mission, available payload bays, fuel and electrical architecture.

- Required sensors and other payload, etc.

All these elements are combined together in an iterative dispatching flow as shown in Figure 3.10. Starting from the mission objectives, a UAS has to be selected that initially fits the mission requirements. After that, the overall set of services required to implement the mission, and all types of payload (communication, computation, sensors, batteries, etc.), have to be selected. Then, the payload has to be organized in the airframe, services assigned to computation payload modules, and the initial flight plan template instantiated into an operative one.

The result of the process is the actual UAS configuration in terms of fuel, electrical system, payload configuration, flight plan, etc.; but also the operational flight plan, alternative routes and landing sites in case of deroutings and/or emergencies. In addition the detailed USAL service architecture is configured, how services are assigned to payload modules, and even the set of rules and system reactions that define the contingency planning.

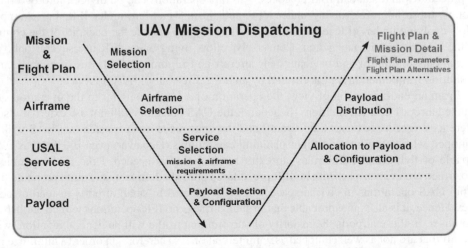

Figure 3.10 Organization of the UAS mission dispatching process

In order to integrate the UAS into non-segregated airspace, it's critical to know the UAS flight plan intentions. Not only in terms of the mission flight plan but also in terms of flight reactions during the mission. In a realistic environment, the UAS cannot decide on its own the proper reaction in a given SAA conflict. Therefore, it is really important at the dispatcher phase to define the different flight plans, alternative routes and landing sites in case of deroutings and/or emergencies. To sum up, we believe that the dispatch phase has to be included as part of a UAS mission. That is, instead of starting the UAS mission at taxi or depart operations, it will start at the dispatcher process.

Going back to Figure 3.10, only the Flight Plan & Mission Detail phase of the dispatch process is related to SAA, specifically with the term 'Avoid'. With the help of a Geographic Information System (GIS), the pilot in/on command and the flight dispatcher officer will design the flight plan and the flight plan alternatives taking into account flight rules, weather changes, cities, terrain, etc.

The result of this part of the dispatch process is the Flight Plan Document structured in different levels as described in Section 3.4.1, where the mission flight path and all its flight path alternatives can be found. In that way, this file can be distributed or transmitted by the data link in order to disclose our flight intentions. Thus, cooperative aircraft can know our flight path and predict SAA reactions.

3.5.2   USAL SAA on Airfield Operations

The depart operations are extensible to approach and landing operations; both cases involve operating UAS in airports. Regarding SAA, the authors believe that the use of standardized and predictable depart/approach procedures for UAS would be a complementary safety layer, potentially decreasing the complexity of SAA systems. Inspired by some existing procedures for (manned) general aviation, some automatic and predefined procedures for UAS are proposed. In this sub-section, we are going to summarize the main aspects of the airfield operations; a full version of this work is detailed in [17, 18].

The safe, efficient and regular operation of all kinds of aircraft relies mainly on a set of procedures and standardized practices. Aircraft operations can be divided into two main groups: those aircraft evolving under VFR and those which are under IFR. For example, pilots evolving under VFR rely entirely on what they see outside the cockpit for the control of the aircraft and its navigation. Conversely, pilots flying under IFR use several onboard instruments to control and navigate their aircraft and in almost all classes of airspace, the separation from other aircraft is ensured by an air traffic control (ATC) service.

From an end-user point of view, the operation of a UAS is similar to the operation of a manned aircraft in IFR conditions. In general, the UAS operators will not use external visual references in order to control the aircraft and navigate since it is expected that UAS will be equipped with autopilots and flight planning capabilities. However, even if a UAS is fully capable of flying under IFR rules, an extra functionality is needed if the UAS operations performed are in an airport with no IFR procedures published. In fact, it is quite probable that initial UAS operations in civil airspace will be conducted in small airports instead of busy ones. Hence, it is also quite probable that in such airports no IFR operations will be published. Moreover, in these airports the majority of surrounding traffic will be general aviation, with aircraft that are not as well equipped as commercial ones. Therefore, in order to minimize the risk of mid-air collisions, it is necessary to add an extra safety layer by introducing procedures that are predictable and well known by all users [19]. We propose the integration of UAS for the departure, arrival and approach phases and in particular, for VFR environments under the USAL architecture. Thus, some specific UAS procedures are proposed for these environments in order to operate safely while at the same time minimizing the interference with other traffic.

As commented before, VFR operations are based on visual cues that the pilot takes from the cockpit. For unmanned flight, one may think that a possible solution for VFR operations would be to install a set of cameras in the aircraft and transmit all the video signals to the ground control station, where the UAS pilot would remotely fly the aircraft in visual conditions. Even more elaborate 'sense' techniques have already been proposed by other researchers (see, for instance, [20–24]), but some of these approaches would not be a feasible solution for many UAS implementations (especially for those using medium to small platforms). Moreover, besides the potential 'sense' system(s) used for a specific UAS, the use of standardized and predictable procedures for the UAS would be a complementary safety layer, which would potentially decrease the complexity of these SAA systems.

These complementary procedures are conceived, aiming to minimize the interference with surrounding traffic and also the pilot in/on command work load, which will be connected with the UAS by using some kind of data-link communications. As mentioned earlier, the mid-air collision risk due to the use of UAS is reduced if procedures are clearly defined. This means that a UAS equipped with one (or more) SAA systems, which is also executing some kind of standardized procedures, will have more chance of being certified. The use of these standardized procedures becomes even more important when the aircraft evolves close to an airport with VFR traffic, because this is where the major risk for mid-air collision exists [25]. In addition, the procedures may facilitate the coordination with eventual ATC or, in the non-controlled case, with the rest of the pilots operating in the same area.

In this sub-section, we have considered that the UA has similar performance to other aircraft flying at the same airport. It is clear that for UAS with different performances (such as mini/micro UAS) than other aircraft, separate airfield traffic patterns may be considered (such as is done nowadays in airfields with small ULM or gliders). On the other hand, the operations described here are oriented more toward tactical and large UAS.

Depart Operations

It is clear that a manual take-off is always possible, especially if the pilot in/on command is present in the departing airfield and has visual contact with the aircraft. In this case, the UAS would fly up to a point and/or height where the navigation phase can be initiated and the autopilot system engaged. Yet, the authors propose an automatic depart phase to execute this process more easily, more predictably and therefore, in a safer way. Thus, the take-off phase will automatically fly the aircraft from the departing runway to an end of departure waypoint (EDWP). These waypoints are located close enough to the airport in order to avoid complex navigation paths, but far enough to reduce as much as possible conflicts with surrounding traffic. Once at the EDWP, the UAS will start the navigation phase (see Figure 3.11).

These maneuvers are implemented at the USAL architecture by services such as the VAS and the FPMa. The VAS must load at the dispatcher phase all the information related to the departure (such as runways, EDWP, altitude to navigation state) before the flight. When the UAS is calibrated and ready to fly, the pilot in command will ask to change to 'taxi' state

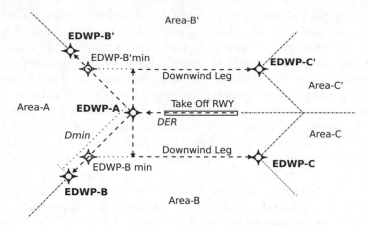

Figure 3.11   EDWP and associated departure areas

(as described in Section 3.4.1). In this state the UAS will have to address the runway, by means of the taxi procedure suitable for each airport in order to not disturb other aircraft.

When the pilot in command switches to 'auto take-off', the VAS will have to perform this technique with the control of the autopilot supervised by the pilot in/on command. When the UAS reaches the safe navigation altitude at the EDWP, it will automatically change its state. If there is a flight plan charged on the VAS queues, the UAS will start the navigation of those waypoints. If not, it will perform a waiting maneuver at a 'save hold' state until new waypoints are uploaded.

In order to test the departure maneuver proposed, and the good working of the USAL services that implement the action, a simulation platform has been implemented [26]. Figure 3.12 shows the simulation of a UAS in departure mode with its EDWP powered by the Google Earth® API. As can be seen, the departure procedure does not obstruct air traffic that flies at a lower altitude level.

Figure 3.12   Departure procedure tested at the simulation platform

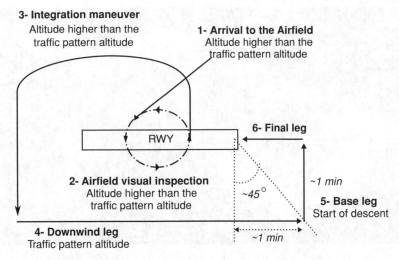

Figure 3.13 Standardized procedure for the arrival and approach operations in non-controlled VFR airfields

Approach Operations

Following the same philosophy as with the departures, we propose some standardized procedures that may be performed by the UAS in the approach phases to a given airport. These procedures are inspired by what is flown currently by manned aircraft operating in VFR and in non-controlled airfields (see Figure 3.13). We think that these procedures will allow us to improve the predictability of UAS trajectories, so they might also be used in case of flying to an airport even with ATC services, but with no IFR procedures published.

As with the depart operations, for the development of the arrival procedures the VAS and the FPMa implement the maneuvers. There is also a preflight configuration needed at the dispatching state. The VAS must know the land pattern used at each runway selected.

The FPMa or the pilot in command will command the VAS to start approaching the runway in order to begin the landing pattern. When the UAS arrives at the airfield, it starts a hold pattern over the airfield higher than the traffic pattern. This technique is used as a waiting procedure while there is a visual inspection of the field. When the UAS is able to begin the landing it starts the integration maneuver by incorporating into the traffic at the down wing leg. These legs are defined by waypoints charged by the FPMa and commanded by the VAS. The base leg is for reducing altitude and entering the final leg with the UAS ready to land.

This proposal has been tested in the simulation scenario. Figure 3.14 shows the procedure described in this sub-section, where the UAS holds over the airfield. Traffic 1, 2 and 3 simulates the landing maneuvers effected by other aircraft. The UAS has to wait until aircraft 3 starts its landing procedure for the UAS to integrate into the arrival traffic.

3.5.3   Awareness Category during UAS Mission

Flight services are in charge of aircraft management and UAS airworthiness under nominal conditions. However, the awareness services are able to administer UAS control in a critical

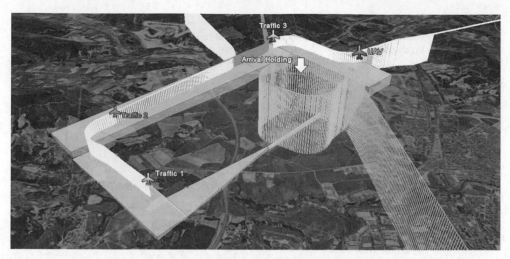

Figure 3.14   Arrival procedure tested in the simulation scenario

awareness situation since air traffic, or civilian lives, may be in danger. In this case, mission and payload services take second place until the flight conditions return to normal.

The awareness services category is a system capable of detecting problems, finding solutions and reacting to them. The complexity of this system lies in translating to programs and sensors the ability of a pilot to sense and react to fast avoiding obstacles. There are some actions that a human brain carries out in milliseconds that must be translated to a UAS. These SAA actions have been categorized by the FAA [13] into eight sub-functions: detect, track, evaluate, prioritize, declare, determine action, command and execute.

To implement the eight sub-functions, the awareness services are divided into different services that implement several responsibilities and interact with each other. Figure 3.15 introduces these services and their interactions. The category is defined by:

- awareness sensors

- awareness data fusion

- self-separation declaration/collision avoidance declaration

- self-separation reaction/collision avoidance reaction.

The arrows in the picture indicate the system flow. Arrows that have two directions indicate that the following sub-system can ask on demand the data available. The awareness sensors gather all the data from the environment through the sensors onboard, and/or the air collision avoidance systems. The sensor information is collected and preprocessed to be sent to the awareness data fusion service. This service links the awareness information with specific data, from the rest of the USAL services, such as flying state, current mission tasks, flight telemetry, etc. After evaluating the situation, the awareness data fusion transfers responsibility to the collision avoidance declaration or self-separation declaration. These modules declare the current situation of the risk, and after determining which action to follow, they call their reaction service to command and execute its decision. Next, we are going to explain in detail each part of the category.

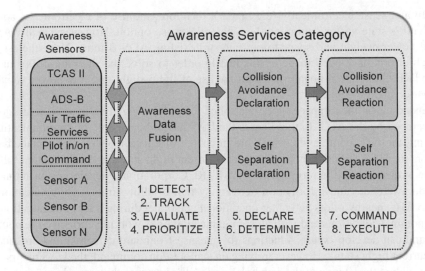

Figure 3.15 Composition of awareness service category

Awareness Sensors

This module is the sensors/system input of the awareness category. The goal here is to feed the UAS with at least the same amount of information a pilot on board has. There are several sensor types and technologies which can be used in the 'sense' sub-system of a SAA, for example: radar sensors, electro-optical sensors, laser sensors, infrared sensors, cooperative systems or a combination of these.

The sense scenario can be sub-divided into passive or active techniques applicable in cooperative or non-cooperative traffic environments. On the one hand, the active cooperative scenario involves an interrogator monitoring a sector ahead of the UAS to detect oncoming traffic by interrogating the transponder on the other aircraft. The active non-cooperative scenario relies on a radar sensor scanning a sector ahead of the UAS to detect all traffic, whether transponder-equipped or not. On the other hand, the passive cooperative scenario relies on everyone having a transponder, but with everyone's transponder broadcasting its position, altitude and velocity data. The passive non-cooperative scenario is the most demanding one. It is also the most analogous to the human eye. A system in this scenario relies on a sensor to detect and provide azimuth and elevation to the oncoming traffic.

However, all these sensors have not yet shown enough maturity to be considered for every UAS size. For instance, a radar sensor or TCAS II or ADS-B systems would not be suitable for mini UAS. Therefore, as each UA may present a different sensor configuration, we are abstracted of a particular sensor solution for a specific UA.

The first two awareness sensors illustrated in Figure 3.15 are the Traffic Alert and Collision Avoidance System (TCAS II) and Automatic Dependent Surveillance – Broadcast (ADS-B) systems. The TCAS II and ADS-B systems monitor the airspace around an aircraft, independent of air traffic control, and warn the UAS of the presence of other aircraft which may present a threat of mid-air collision. For example, TCAS II can also generate collision warnings in the form of a 'traffic advisory' (TA) and also offers the pilot direct instructions to avoid danger, known as a 'resolution advisory' (RA). Small UAS are difficult to see visually and sense electronically (e.g., radar), therefore, the use of electronic broadcast of the aircraft's state vector data might be a potential sense option. However, small UAS have limited power, weight and space to put onboard existing ADS-B units. Nevertheless, some research efforts have been made in this line in order to solve that limitation. For instance, the MITRE Corporation began (in 2006) the use of lightweight and low-power versions of ADS-B for small UAS [27].

Depending on the airspace type (controlled airspace), SAA functionalities are going to be developed by the ATC. In these cases the UAS has to be capable of communicating and following the ATC commands, which are in charge of ensuring 'well clear' maneuvers. In addition, the UAS should be able to broadcast current flight status and future intent following the standardized operation procedures. The MITRE Corporation is developing an autonomous situation awareness and information delivery system to allow timely communication from UAS to ATC, nearby pilots and other operational personnel during a lost-link situation [28].

In nominal conditions and with all communication links working correctly, the pilot in/on command can develop 'sense' functions (by means of onboard cameras) and broadcast the UAS current flight status. However, relying on a pilot in/on command, the UAS incurs human latency, adding the latency of the data link bringing the image to the ground for a decision and the avoidance command back to the UAS. This added latency can range from less than a second for line of sight (LOS) links to more time for satellite links. Latency and the possibility of a lost-link situation are sufficiently important inconveniences to develop an awareness situation with the pilot in/on command.

Sensor A, Sensor B and Sensor N illustrate the possibility of using sensors such as radars or cameras for the 'sense' part. For example, radar uses electromagnetic waves to identify the range, altitude, direction or speed of both moving and fixed objects. It is a very mature technology, 'all-weather' and provides accurate range, direction and closing speed. Unfortunately, the size and weight of microwave radar sensors are substantial; thus, use in aviation (especially in smaller UAS) is limited. Some research efforts have been made in this field in order to integrate radars in a UAS. In 2003, NASA equipped a UA with a radar system able to detect non-cooperative targets and a traffic advisory system to detect cooperative ones. With this UA configuration and traffic surrounding the aircraft, NASA made some successful flight tests [29, 30]. Kemkemian [31, 32] presents low-cost multiple-input multiple-output (MIMO) radar for UA airframes which allows the location of the radar without any moving parts.

The optical sensors (visual sensors) use the natural waves coming from the intruding aircraft to detect it. Software programs analyze incoming streams of pixilated digital horizon for aberrations in the flow of the pixels, which generally mark an intruding object. Important research efforts have been addressed to the development of detection algorithms, e.g. [24, 33]. This technology is also mature and relatively low cost, however, atmospheric disturbances can interfere and hinder its ability to detect oncoming traffic. In addition, to achieve the required field of view, sensors have to be located in different positions of the aircraft. Karhoff [34] developed an interesting assessment of SAA technology for the army's 'Warrior' UAV, declaring that visual technology offers the best chance for FAA approval, waiting on further maturity and improvements in other technologies. In addition, the study discusses other potential sense technologies such as laser radar, infrared sensors and bistatic radar.

As has been mentioned in the previous paragraphs, there are various sensor technologies which can be used in the 'sense' sub-system of SAA. Each technology presents positive and negative attributes to be integrated in UAS. The suitable technology to be onboarded will depend on the UAS size and the airspace class where the UA is going to fly.

Going back to Figure 3.15 and from an architecture point of view, we are going to be abstracted of the awareness sensor implementation details. One service is provided for each sensor which operates as a sensor driver. Each of these drivers is in charge of interacting with a specific awareness sensor. This service operates similarly to the way that drivers work in operating systems, removing the implementation details from actual sensor users. These awareness sensor drivers are where the first preprocess data algorithms (such as detection algorithms) will be located. In that way, just relevant data will feed the awareness data fusion service.

Awareness Data Fusion

Once the sense of the awareness category is satisfied by means of the awareness sensor services, the UAS must collect all this information and use it efficiently. The detection information provided by several awareness sensors should be evaluated and prioritized taking into account the rest of the UAS information such as telemetry, flight plan and mission information.

The goal of the awareness data fusion (ADF) is to implement the first four of the eight sub-functions: detect, track, evaluate and prioritize. With the detect function, we have to determine the presence of aircraft or other potential hazards, while track is to keep this information during time and estimate the future position of the intruder.

The ADF collects current air traffic, meteorological and terrain state (digital elevation model, DEM) of the aircraft as seen by the different sensors of the aircraft. The ADF fuses and centralizes all data received to offer an awareness overview. All of this data is put side by side with the information given by other USAL services: UAS flight telemetry provided by the VAS (current position and bearing), flight plan mission information provided by the FPMa (whole mission flight plan and its alternatives), AGL altitude from DEM, etc.

Figure 3.16 illustrates the main ADF interaction inside the USAL architecture. The VAS provides flight telemetry for the rest of the services. This data is important to compare the UAS current position, bearing and UAS speed with other airspace users. The FPMa offers a flight path and alternatives to compute future conflicts and ensure self-separation during the mission. In the future, this service has to incorporate 4D navigation in order to know at what

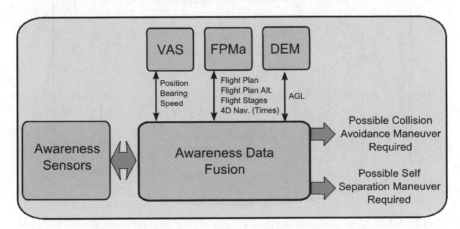

Figure 3.16   Awareness data fusion service interaction

moment the UAS is going to arrive at the different flight plan waypoints. The FPMa stages the UAS goes through to perform a mission are also provided (such as on-ground, departure, en-route, mission or arrival). This data is going to be important in selecting a suitable conflict reaction. The reaction can be different depending on the mission stage.

After grouping the data and validating the state of the mission, the position in the air and all the factors that a pilot onboard would compare, it must evaluate the collision risk based on these predictions and the current data of the UAS, and prioritize the reaction – e.g., TCAS-II RA messages or ATC indications. As a result of this prioritization, the ADF service may choose between two possible outputs: collision avoidance maneuver required or self-separation maneuver required. In other words, the system has detected a hazardous situation which requires treatment.

Self-separation and Collision Avoidance Declaration

SAA is the capability of a UAS to remain 'well clear' and avoid collision with other air-borne traffic. The two functions of SAA are self-separation and collision avoidance. In the USAL architecture we have divided these functions into two different steps: the collision/self-separation declaration and collision/self-separation reaction. In order to explain the system, we have to understand the different volumes of risk.

Figure 3.17 explains the different dangerous zones and the condition of awareness. As can be seen, the ATC separation service volume is where there is no threat, but an aircraft detected far away can be turned into an obstacle. In controlled airspace, ATC ensures a safe distance is maintained from other aircraft, terrain, obstacles and certain airspace not designated for routine air travel. The ATC is in charge of giving the order to reconfigure the path and avoid crossing trajectories. The second volume is called the self-separation threshold. In this situation, the ATC may have failed, or another unresolved alarm may have occurred, and the

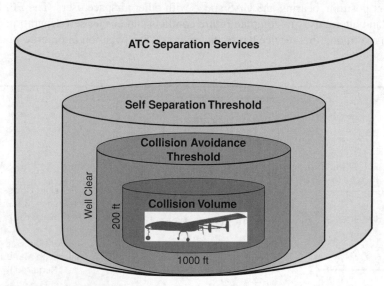

Figure 3.17 UAS ATC separation, self-separation and collision avoidance [13]

aircraft detected turns out to be a threat. The SS could be the only function provided given that the safety analysis demonstrates that the target level of safety (TLS) can be met with SS alone [13]. However, when all forms of SS have failed, we are now at the third volume and the CA takes appropriate action to prevent a threat aircraft from penetrating the collision volume. On rare occasions the UAS SS functions may conflict with ATC separation services, and as a general rule the UAS will follow ATC separation services.

Back to Figure 3.15, where the awareness USAL services are defined. It can be seen that the SS and CA are divided into four different services:

- Services to manage the declaration (collision avoidance declaration, CAD and self-separation declaration, SSD).

- Services to command and execute the reaction (collision avoidance reaction, CAR and self-separation reaction, SSR).

Once the ADF resolves between CA and SS, the declaration services must declare if any action is required to keep the intruder from penetrating the collision volume. If the declaration services announce that an action is required, they have to determine and propose what sort of response is required to maintain the 'well clear'. This determination says whether a SS or CA reaction is needed. In the SS case, where the UAS has enough time to react, the action required to maintain the 'well clear' has to be supervised by the pilot on command and previously defined in most cases. An aspect that should be considered during this process is establishing who has the right of way. At the moment, the UAS regulation material is not clear on which category to place the UAS in and therefore know its right of way.

Self-separation and Collision Avoidance Reaction

After the declaration services have determined the risk, the reaction services will take the lead and implement the command and execute functions. The reaction services must command the action to the UAS, and the UAS has to execute the commanded action. In order to do so, we propose using FPMa alternatives. During the aircraft dispatcher phase, the dispatcher officer and pilots in/on command have designed the flight plan mission and flight plan alternatives in order to give response for any UAS contingency.

As has been explained in controlled airspace, the first cylinder is managed by the ATC separation services. In this case, the ATC separation service commands have to be followed, updating the flight plan through the FPMa updates. These changes in the flight path have to be supervised by the pilot on command who has the ability to immediately affect the trajectory of a UA if necessary.

When the UAS is flying in non-controlled airspace, once the UAS self-separation threshold has been breached, the UAS has to react to ensure 'well clear' of the threat. In this case, the SSR service receives the required action to solve the conflict. If any CAR is required, the reaction is executed by the VAS by means of safe reaction state, for instance making an immediate change of bearing. A change in this state means stopping the current VAS operation to react as soon as possible. In safe reaction state, the VAS accepts several parameters to react to conflicts. It needs the heading to avoid the obstacle and the operation priority. Through the priority parameter the VAS knows how severe the maneuvers have to be. These changes of the heading have to develop following ICAO Annex 2 'Rule of the air'.

3.6   Conclusions

In this chapter, integration of SAA capabilities into a distributed architecture for UAS civil applications has been presented. The USAL architecture takes into account the UAS flight, mission, payload and awareness as a whole. Special attention has been paid to the composition of the awareness category where 'pure' SAA functions, such as self-separation and collision avoidance, are developed. Particular techniques for SS or CA are important for the UAS civil integration; however, these techniques have to be integrated inside the UAS architecture. SS and CA systems have to cooperate with the UAS autopilot and flight plan manager to ensure safe flight. This chapter describes the awareness services definition, responsibilities and interactions with other services of the USAL architecture.

Another important issue tackled is prevention operations in order to anticipate and avoid SAA conflicts. Suitable flight plan designs during the dispatch process should prevent future conflicts. On the other hand, the use of standardized and predictable maneuvers for the UAS, such as depart and approach procedures, should be a complementary safety layer for avoiding hazardous traffic situations. Preflight process and predictable airfield operations are examples of that statement.

Acknowledgments

This work has been partially funded by the Ministry of Science and Education of Spain under contract CICYT TIN 2010-18989. This work has been also co-financed by the European Organization for the Safety of Air Navigation (EUROCONTROL) under its CARE INO III programme. The content of the work does not necessarily reflect the official position of EUROCONTROL on the matter.

References

1. Santamaria, E., Royo, P., Barrado, C., Pastor, E., and Lopez, J., 'Mission aware flight planning for unmanned aerial systems', in Proceedings AIAA Guidance, Navigation and Control Conference (GN&C), Honolulu, HI, August 18–21, 2008, pp 1–21.

2. EUROCONTROL, 'Guidance material for the design of terminal procedures for area navigation', 2003.

3. SC-203 RTCA, 'Guidance material and considerations for unmanned aircraft systems', Radio Technical Commission for Aeronautics, Document Do-304, Washington, DC, March 2007.

4. RTCA, 'Operational services and environmental definition (OSED) for unmanned aircraft systems (UAS)', Radio Technical Commission for Aeronautics, Document Do-320, Washington, DC, June 2010.

5. Cox, T., Somers, I., and Fratello, D., 'Earth observations and the role of UAVs: a capabilities assessment', Technical Report, NASA 20070022505, August 2006.

6. UAVNET, 'European civil unmanned air vehicle roadmap, action plan and overview', Technical Report, 2005.

7. Iscold, P., Pereira, S., and Torres, A., 'Development of a hand-launched small UAV for ground reconnaissance', *IEEE Transactions on Aerospace and Electronic Systems*, pp 335–348, January 2010.

8. NASA Ames Research Center, SIERRA project 2009, Earth Science Division (WRAP): http://www.espo.nasa.gov/casie/.

9. Lopez, J., Royo, P., Pastor, E., Barrado, C., and Santamaria, E., 'A middleware architecture for unmanned aircraft avionics', in ACM/IFIP/USENIX International Conference on Middleware Companion, New Port Beach, CA, November 2007, pp 1–6.

10. Pastor, E., Royo, P., Santamaria, E., Prats, X., and Barrad, C., 'In-flight contingency management for unmanned aerial vehicles', in AIAA Unmanned . . . Unlimited Conference, Seattle, WA, April 6–9, 2009, pp 1–15.

11. Harel, D. and Politi, M., *Modeling Reactive Systems with Statecharts: The STATEMATE Approach*, McGraw-Hill, New York, 1998.

12. Pastor, E., Santamaria, E., Royo, P., López, J. and Barrado, C., 'On the design of a UAV flight plan monitoring and edition system', in Proceedings of the IEEE Aerospace Conference, AIAA/IEEE, Big Sky, MT, March 2010, pp 1–20.

13. FAA, 'Sense and avoid (SAA) for unmanned aircraft systems (UAS)', Sense and Avoid Workshop, Federal Aviation Administration, October 2009.

14. Chen, D.W.-Z., 'Sense and avoid (SAA) technologies for unmanned aircraft (UA)', National Cheng Kung University (NCKU), December 2008, http://ord.ncku.edu.tw/ezfiles/3/1003/img/467/20081204_ppt.pdf.

15. ICAO, I.C. International Standards and Recommended Practices, Operation of Aircraft, Annex 6 to the Convention on International Civil Aviation, 1998.

16. Prats, X., Pastor, E., Royo, P., and Lopez, J., 'Flight dispatching for unmanned aerial vehicles', in Proceedings of AIAA Guidance, Navigation and Control Conference and Exhibit (GN&C), Honolulu, HI, August 2008, pp 1–21.

17. Delgado, L., Prats, X., Ferraz, C., Royo, P., and Pastor, E., 'An assessment for UAS depart and approach operations', 9th AIAA Aviation Technology, Integration, and Operations Conference (ATIO), Hilton Head, SC, September 21–23, 2009, pp 1–16.

18. Prats, X., Delgado, L., Royo, P., Pérez-Batlle, M., and Pastor, E., 'Depart and approach procedures for UAS in a VFR environment', *AIAA Journal of Aircraft*, in press, 2011.

19. Weibel, R.E. and Hansman, J., 'Safety considerations for operation of different classes of UASs in the NAS', 4th AIAA Aviation Technology, Integration, and Operations Conference (ATIO), Chicago, 2004, pp 1–11.

20. Fasano, G., Accardo, D., and Moccia, A., 'Multi-sensor-based fully autonomous non-cooperative collision avoidance system for unmanned air vehicles', *Journal of Aerospace Computing, Information, and Communication*, 5(10), 338–360, 2008.

21. Korn, B. and Edinger, C., 'UAS in civil airspace: demonstrating "sense and avoid" capabilities in flight trials', 27th Digital Avionics Systems Conference, Orlando, FL, October 2008, pp 4.D.1-1–4.D.1-7.

22. Kephart, R.J. and Braasch, M.S., 'See-and-avoid comparison of performance in manned and remotely piloted aircraft', *IEEE Aerospace and Electronic Systems Magazine*, 25(5), 36–42, 2010.

23. Tadema, J. and Theunissen, E., 'A concept for UAS operator involvement in airborne conflict detection and resolution', IEEE/AIAA 27th Digital Avionics Systems Conference, St. Paul, MN, October 2008, pp 4.C.1-1–4.C.1-12.

24. Carnie, R., Walker, R., and Corke, P., 'Image processing algorithms for UAV sense and avoid', in Proceedings of IEEE International Conference on Robotics and Automation (ICRA 2006), Orlando, FL, June 2006, pp 2848–2853.

25. Simon, J.N. and Braasch, M.S., 'Deriving sensible requirements for UAS sense-and-avoid systems', 28th Digital Avionics Systems Conference (DASC), Orlando, January 2008, pp 6.C.4-1–6.C.4-12.

26. Royo, P., Lopez, J., Tristancho, J., Lema, J.M., Lopez, B., and Pastor, E., 'Service oriented fast prototyping environment for UAS missions', 47th AIAA Aerospace Sciences Meeting and Exhibit, Orlando, FL, January 2009, pp 1–20.

27. Strain, R.C., DeGarmo, M.T., Moody, J.C., 'A lightweight, low-cost ADS-B system for UAS applications', MITRE Technical Papers and Presentations, Case Number: 07-0634, January 2008.

28. Hu, Q. and Jella, C., 'Intelligent UAS situation awareness and information delivery', 29th IEEE/AIAA Digital Avionics Systems Conference (DASC), Salt Lake City, December 2010, pp 5.C.3-1–5.C.3-6.

29. Wolfe, R., 'NASA ERAST non-cooperative DSA flight test', in Proceedings of the AUVSI Unmanned Systems Conference, Baltimore, MD, July 2003.

30. Schaeffer, R.J., 'A standards-based approach to sense-and-avoid technology, 3rd AIAA 'Unmanned Unlimited' Technical Conference, Workshop and Exhibit, Paper AIAA 2004-6420, Chicago, IL, September 2004.

31. Kemkemian, S., Nouvel-Fiani, M., Cornic, P., and Garrec, P., 'A MIMO radar for Sense and Avoid function: a fully static solution for UAV', 11th IEEE International Radar Symposium (IRS), Vilnius, Lithuania, August 2010, pp 1–4.

32. Kemkemian, S., Nouvel-Fiani, M., Cornic, P., and Garrec, P., 'MIMO radar for sense and avoid for UAV', IEEE International Symposium on Phased Array Systems and Technology (ARRAY), Waltham, MA, October 2010, pp 573–580.

33. Mejias, L., Ford, J.J., and Lai, J.S., 'Towards the implementation of vision-based UAS sense-and-avoid', in Proceedings of the 27th International Congress of the Aeronautical Sciences, Acropolis Conference Centre, Nice, September 2010, pp 1–10.

34. Karhoff, B.C., Limb, J.I., Oravsky, S.W., and Shephard, A.D., 'Eyes in the domestic sky: an assessment of sense and avoid technology for the army's "Warrior" unmanned aerial vehicle', IEEE Systems and Information Engineering Design Symposium, Charlottesville, VA, January 2007, pp 36–42.

Part II

REGULATORY ISSUES AND HUMAN FACTORS

4

Regulations and Requirements

Xavier Prats, Jorge Ramírez, Luis Delgado and Pablo Royo
Technical University of Catalonia, Spain

In civil aviation, several mechanisms are present to minimize the probability of collision with other aircraft, objects or terrain. Generally speaking, they are categorized as separation assurance and collision avoidance.

The first category aims to keep aircraft separated according to minimum separation distances both in the lateral and vertical planes. These minimum values depend on several factors, such as the airspace class, the flight rules, the flight phase, the air traffic control (ATC) surveillance means (if any), the performance of the onboard navigation systems, etc. Roughly speaking, lateral minimum separation between aircraft can range from 3 nautical miles in terminal areas with ATC radar separation service to up to 60 nautical miles for two aircraft at the same altitude in a North Atlantic track. Yet, in non-controlled airspaces minimum separation does not involve precise separation minima and aircraft must remain *well clear* from each other. *Well clear* is a qualitative rather than a quantitative term used in current regulations when referring to the minimum miss distance between two aircraft that are avoiding a collision.

On the other hand, collision avoidance is considered as a last resort manoeuvre to prevent a collision in case of a loss of separation. In some cases, collision avoidance between aircraft is performed cooperatively, meaning that two conflicting aircraft use common systems and procedures which have been designed to jointly detect an imminent collision with enough time to react and avoid it. However, not all aircraft are equipped with these systems and obviously neither are other flying obstacles, such as birds, or terrain. Thus, whenever visibility conditions permit, every pilot in manned aviation is expected to *see and avoid* these hazards. This means that in these conditions, flight crew is ultimately responsible for ensuring aircraft safety by preventing and avoiding collisions.

Sense and Avoid in UAS: Research and Applications, First Edition. Edited by Plamen Angelov.
© 2012 John Wiley & Sons, Ltd. Published 2012 by John Wiley & Sons, Ltd.

Unmanned aircraft systems (UAS) do not have the flight crew onboard and therefore, the see and avoid capability is essentially lost. Yet, UAS can be equipped with several sensors and mechanisms that can replace this unavoidable functionality. Thus, the more appropriate term *sense and avoid* (SAA) is used for UAS and in [1], it is simply defined as: '*the process of determining the presence of potential collision threats, and manoeuvring clear of them; the automated equivalent to the phrase "see and avoid" for the pilot of a manned aircraft*'. Briefly, the bare minimum features for an SAA system can be summarized as follows:

- detect and avoid mid-air collisions with other flying traffic according to the right-of-way rules;
- detect and avoid other flying objects (such as birds);
- detect and avoid ground vehicles (when manoeuvring on ground);
- detect and avoid terrain and other obstacles (such as buildings or power-lines);
- avoid hazardous weather; and
- perform functions such as maintaining separation, spacing and sequencing, as done visually in manned aviation.

Several issues arise when trying to apply current regulations (developed for manned aviation) to UAS, and SAA is obviously one of the most challenging. Significant operational differences exist between UAS and manned aircraft which have to be addressed before UAS can be safely integrated into civil and non-segregated airspace. In this context, an excellent review on existing manned and unmanned regulations world-wide, along with valuable thoughts and recommendations on this UAS integration, is given in [2].

UAS operations in civil airspace are asked to provide at least the same level of safety as that of manned aviation. This chapter focuses on regulations, requirements and open issues specific to SAA systems for UAS. First, some background information is given on separation and collision avoidance mechanisms, flight rules and airspace classes, UAS categorization and how safety levels could be defined prior to establishing minimum system requirements for SAA. Then, in Section 4.2, the existing regulations and standards on SAA are presented while Section 4.3 highlights the possible requirements that could be demanded of a SAA system. Finally, Section 4.4 is devoted to some discussion on human factors and situational awareness, while Section 4.5 concludes this chapter.

4.1   Background Information

Separation assurance and collision avoidance include several layers of protection against collisions that use different systems, procedures, airspace structure and human actions. Figure 4.1 depicts these different mechanisms, which are summarized as follows:

- Non-cooperative collision avoidance is the lowest-level mechanism to prevent an imminent collision with any type of aircraft, obstacles or terrain. In manned aviation, this relies entirely on the ability of the crew members to see and avoid. Conversely, for UAS this functionality must be assumed by an SAA system.

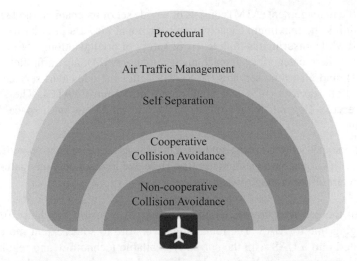

Figure 4.1 Separation and collision avoidance mechanisms

- Cooperative collision avoidance includes all the systems and procedures between cooperative aircraft that can avoid imminent collisions. The standard for an airborne collision avoidance system (ACAS) is specified by the International Civil Aviation Organisation (ICAO) in [3], defining it as an aircraft system based on secondary surveillance radar (SSR) transponder signals, which operates independently of ground-based equipment to provide advice to the pilot on potential conflicting aircraft that are equipped with SSR transponders. The traffic collision avoidance system (TCAS) is a particular ACAS implementation widely used in commercial aviation. ACAS/TCAS-I systems just provide traffic alerts (TA) when a collision threat is detected. In addition to TAs, ACAS/TCAS-II systems provide the pilot with resolution advisories (RA), proposing an avoidance manoeuvre in the vertical plane. Future TCAS versions also foresee horizontal manoeuvres in the resolution advisories.

- Self-separation mechanisms are the lowest layer that can guarantee a minimum safe separation distance. In manned aviation, see and avoid mechanisms are again widely used for this purpose, especially in non-controlled airspace under visual meteorological conditions. Besides this, self-separation can be significantly improved with different kinds of airborne separation assistance systems (ASAS), which consist of highly automated systems that present the pilot with information to enhance their situational awareness. Moreover, ASAS can even provide a set of explicit solutions to guarantee separation with other aircraft while reducing the workload of the crew. The majority of ASAS applications are based upon the automatic dependent surveillance (ADS) concept (where each aircraft transmits its position and likewise receives the positions transmitted by other aircraft or vehicles using the same system); and some sort of cockpit display traffic information (CDTI). Thus, these types of application are expected to dramatically enhance the situational awareness of the pilot and consequently, the safety levels in non-controlled airspace; although they are also aimed at delegating separation tasks from controllers to pilots in some controlled airspaces.

- Air traffic management (ATM) consists of a wide set of mechanisms and services aimed at providing the maximum capacity to airspace and airports in order to accommodate demand while ensuring the high levels of safety of civil aviation. ATM can be divided into three main categories: airspace management (ASM); air traffic flow management (ATFM); and air traffic services (ATS). The latter includes alert services (AS), flight information services (FIS) and finally air traffic control (ATC). The availability of these services depends mainly on the flight rules and class of airspace the aircraft is evolving.

- Operational procedures are the outermost layer in assuring separation with other aircraft (along with known obstacles and terrain). Here, we find not only navigation procedures but also aircraft operating procedures.

Among all the previous layers, the non-cooperative collision avoidance function is the most challenging one for UAS. The remaining layers are to some extent more likely to be easily integrated into a UAS with the currently available technology and regulations. Thus, SAA is one of the main issues that must be addressed before integrating them into civil and non-segregated airspace.

As already commented, the particular mechanisms available for each of the previous layers depend on several factors, such as the type of aircraft, airspace, meteorological conditions, flight rules, etc. For example, in non-controlled airspace the ATM layer is hardly present; in instrument meteorological conditions (IMC) the ability to see and avoid will be drastically reduced for manned aircraft; self-separation mechanisms will be undoubtedly different if ADS is available or not for all the aircraft, etc. Furthermore, other considerations specific to UAS operations exist: like the automation level of the UAS (autonomous, automated or remotely controlled); the type of communications relay with the control station; or even the presence of UAS operators in the airfield of operations. Moreover, flights over populated areas also raise increased safety issues as minimum safety figures are usually derived from the number of fatalities that an accident may cause [2].

4.1.1   Flight Rules

The ICAO specifies in its 2nd Annex to the Convention on International Civil Aviation [4] the rules of the air and the right-of-way rules. Each state is responsible for accepting and eventually adapting these rules for their national regulations. For example, in the United States, flight rules are defined in the Federal Aviation Regulations (FAR), Part 91. Two types of flight rule are established for manned aviation: visual flight rules (VFR) and instrument flight rules (IFR). VFR operations are based on visual cues that the pilot takes from outside the cockpit, not only for aviating the aircraft but also for navigating and avoiding collisions with other aircraft, obstacles and terrain. Yet, in certain classes of airspace, separation instructions may be provided by the ATC. Nevertheless, instructions will remain as simple headings, altitude changes or position reports asking to identify visual reference landmarks or relative visual positions inside the airfield traffic pattern. Visibility and cloud ceiling are the most important factors for safe VFR operations. These minimum weather conditions vary depending on the type of airspace in which the aircraft is operating and whether the flight is conducted during day or night time. Meteorological conditions that allow VFR flight are referred to as visual meteorological conditions (VMC) [4].

Conversely, IMC require pilots to fly solely by reference to flight instruments. Thus, pilots flying under IFR use several onboard instruments to aviate and navigate their aircraft and in almost all classes of airspace, the separation with other aircraft is ensured by an ATC service. ATC instructions may also be in the form of heading and/or altitude changes but, since IFR aircraft are always following navigation procedures, position reports may be based on radio-navigation aids and fixes. In some states, there exist a third category of flight rules; the special VFR (SVFR), which allows aircraft to fly in visual conditions with meteorological conditions below VMC minimums and up to a certain level of weather degradation. The actual values of SVFR minimum weather conditions vary on the type of airspace and state regulations and, in general, they are conducted only in controlled airspace and prior to an ATC clearance.

On the other hand, basic right-of-way rules state that the least manoeuvrable aircraft has always the right of way. For example, balloons have priority over gliders, which have priority over airships, which have priority over motorized heavier-than-air aircraft. In case of conflict between two aircraft of the same category, the aircraft on the right has the right of way. In case of a potential frontal collision, both aircraft must turn to their right and in case of an aircraft which is being overtaken, the last has the right of way and the overtaking aircraft must remain clear. Finally, aircraft in distress always have right of way over others.

4.1.2   Airspace Classes

Blocks of airspace are classified alphabetically into seven different categories (classes A, B, C, D, E, F and G) and are found in the ICAO in its 11th Annex to the Convention on International Civil Aviation [5]. This classification is defined in terms of flight rules (as seen above), interactions between the aircraft and the ATS and as a function of the airspace class, different operating rules apply.

In airspaces of class A to E, some level of ATC is provided and therefore these kinds of airspace are referred to as controlled airspaces. Classes A to E are ordered from most restrictive to least restrictive and consequently different levels of ATS are given, ranging from full separation assurance to all flights in classes A (where VFR flights are not allowed) and B; to separation only between IFR/IFR and IFR/SVFR flights in class E (where a clearance to enter is not even required for VFR flights). Conversely, airspaces F and G are non-controlled and only flight information services and eventual IFR/IFR separation are provided whenever possible in class F, while in class G there is no traffic information at all if not explicitly requested and possible. Table 4.1 summarizes these services for all types of airspace class.

At national level, each state determines how the above-mentioned ICAO classifications are used and implemented in the design of their national airspace, according to their needs. Thus, not all ICAO classes are adopted by all the countries and some national aviation authorities even slightly modify their definition in order to fit former airspace rules and ATS that existed before the ICAO standardization came into effect. For example, in the USA class F airspace is not used (FAA Part 71), whereas it is defined in Canada; or in France neither classes B nor F are implemented. Furthermore, as a function of the state, other requirements may apply to different airspace classes such as the need to be equipped with appropriate communications, navigation equipment, transponder or collision-avoidance system; the minimum separation from clouds and visibility conditions for VFR flights; the maximum airspeed; or even minimum requirements on pilot certificates.

Table 4.1 Summary of airspace classes and their basic characteristics

Airspace	Controlled					Non-controlled	
Class	A	B	C	D	E	F	G
IFR allowed	Yes	Yes	Yes	Yes	Yes	Yes	Yes
SVFR allowed	Yes	Yes	Yes	Yes	Yes	No	No
VFR allowed	No	Yes	Yes	Yes	Yes	Yes	Yes
Separation	For all aircraft	For all aircraft	IFR/IFR IFR/VFR	IFR/IFR	IFR/IFR	IFR/IFR if possible	None
Traffic information	–	–	VFR/VFR	IFR/VFR VFR/VFR	For all aircraft if possible	For all aircraft if possible and requested	For all aircraft if possible and requested
Clearance required	Yes	Yes	Yes	Yes	Only for IFR	No	No

In Europe, Eurocontrol is proposing to simplify this classification and reduce the number of airspace classes to only three, which would roughly correspond to current classes C, E and G. According to [6], it is proposed to create three traffic environment 'airspace categories' as follows:

- Category N: airspace within which all of the traffic and all the intentions of the traffic are known to ATC.

- Category K: airspace within which all of the traffic is known, but not all of the intentions of the traffic are known to ATC.

- Category U: airspace where not all of the traffic is known to ATC.

4.1.3   Types of UAS and their Missions

There are several ways to categorize UAS based on different aspects of the system. For instance, UAS may be grouped as a function of the weight of the unmanned aircraft (UA); its performance; level of autonomy; altitude of operation; communications data link; or the type of operations or missions carried out. The most relevant examples of categorization of UAS are presented as follows.

Weight Categorizations

The UK Civil Aviation Authority divides the UAS according to vehicle weight [7]. The first category is called 'Small Aircraft', which includes aircraft weights lower than 20 kg. The next category is 'Light UAV' and comprises aircraft between 20 and 150 kg. The last category is just called 'UAV' and includes aircraft of 150 kg or more. Another typical categorization relates directly aircraft size (or weight) and type of expected mission. For instance, in [8] four different categories are given, the last one being split as a function of the operating altitude. Table 4.2 shows these types of UAS, along with some high-level parameters commonly found in each category.

Flight Performances

On the other hand, the Radio Technical Commission for Aeronautics (RTCA) proposes a categorization based on flight performance characteristics, remarking that ATC already uses flight performance for managing flows and maintaining separation. Thus, in their document Do-320 [9] the following categories are proposed:

- turbojet fixed-wing (e.g., Global Hawk, N-UCAS);

- turboprop fixed-wing (e.g., Pedator B);

- reciprocating/electric engine fixed-wing (e.g., Predator A, Shadow 200);

- VTOL (vertical take-off and landing) (e.g., Firescout, RMAX Type II);

- airship (e.g., SA 60 LAA).

Table 4.2 UAS classification as a function of UA weight [8]

Type	Weight	Operating scenario	Typical operating altitude	Typical operating cruise speed	Endurance
Micro	Less than 2 lb	Local	Near surface to 500 ft	–	Minutes
Mini	2–30 lb	Local	100–10,000 ft	30–90 kts	Several hours
Tactical	30–1000 lb	Regional	1400 ft–FL180	80–110 kts	5–10 hours
Medium altitude	1000–30,000 lb	Regional, national	FL180–FL600	100–200 kts	10 hours–days
High altitude	1000–30,000 lb	Regional, national, international	Above FL600	20–400 kts	10 hours–days

However, within each of the five UAS categories, there is still a huge variability of UAS types and four additional sub-categories are also defined in [9] to further differentiate UAS:

- Standard category, representing those UA resembling manned aircraft.

- Non-standard small category, representing those UA with physical sizes and weight considerably smaller than the smallest manned aircraft.

- Non-standard high-altitude long-endurance (HALE) category, with altitudes and endurances that go beyond those of manned aircraft.

- Conversion category, which represents UAS that are converted from manned aircraft to operate as UA.

UAS Missions

Regarding the UAS missions, there is no doubt today that a huge market is emerging from the potential applications and services that will be offered by unmanned aircraft. UAS perform a wide variety of functions. Many UAS missions are described in the literature (see for instance [9–12]) and according to these references, civilian applications can be summarized in four groups: communications applications, environmental applications, emergency applications and monitoring applications. Within these categories, a wide range of application scenarios exist. For instance:

- **Communications applications:** telecommunication relay services, cell phone transmissions and broadband communications are just a few communications applications.

- **Environmental applications:** with the UAS capability for remote sensing, applications like atmospheric research, oceanographic observations and weather forecasting can be more effective.

- **Emergency applications:** this group includes fire fighting, search and rescue missions, oil slick observation, flood watch, hurricane watch and volcano monitoring.

- **Monitoring applications:** forest fire detection, international border patrol, fisheries monitoring or high-voltage power line monitoring are among the most important missions in this category.

Operational Behaviour

In [9], the RTCA describes another interesting categorization related to operational behaviours of UA once airborne. Three different flight profiles are presented, which represent generic operational UA behaviours. These are:

- Point-to-point UAS operations: direct flight and do not include aerial work or delays that may occur during the en route portion (transport of passengers or cargo).

- Planned aerial work: refer to orbiting, surveillance and tracking flights using predefined waypoints.

- Unplanned aerial work: UAS cannot predict their intended flight path.

4.1.4   Safety Levels

The principal regulatory bodies agree that one of the guiding principles for UAS regulation should be the equivalence with respect to manned aviation and therefore, regulatory airworthiness standards should be set to be no less demanding than those currently applied to comparable manned aircraft. This principle is known as the equivalent level of safety (ELOS) requirement and if we focus on the avoidance system, it means that the UAS must provide a method comparable to see and avoid requirements for manned aircraft. However, this concept also raises some criticism because of the difficulty of quantifying what exactly the ELOS requirement entails [2]. Conversely, another way to establish the 'required safety' is to directly specify the target level of safety (TLS) for UAS operations. In [13], both concepts are defined as follows:

- *ELOS: an approximately equal level of safety that may be determined by qualitative or quantitative means.*

- *TLS: the acceptable mean number of collisions per flight hour which could result in fatalities by modelling the end-to-end system performance.*

Numerous efforts have attempted to define the ELOS of current manned aircraft see and avoid functionality (i.e., quantifying the pilot's ability to visually detect hazardous objects or terrain and efficiently avoid them). For example, based on data from the US National Transportation Safety Board (NSTB), in [2] mid-air collision requirements are derived from fatality rates for accidents where an in-flight collision with obstacles or another aircraft occurred. According to this data, a maximum mid-air collision rate of 0.1 collisions per million flight hours is proposed for UAS. Another example is found in [1], where it is stated that SAA shall have an overall critical failure rate of no more then 0.51 per million of flight hours. This figure corresponds to the rate of general aviation mid-air collisions per year (considering a 10-year average) and was taken from the Aircraft Owners and Pilots Association (AOPA) analysis.

Conversely, the FAA proposes to establish a TLS for UAS, and in particular to further derive SAA requirements. As stated in [13], '*the TLS approach should be the most likely to succeed, relative to the others considered. This approach is a very comprehensive end-to-end analysis that is traceable and quantifies the total risk of the system*'. However, several issues should still be addressed before establishing how this value should be computed. For instance, when evaluating the collision risk for a UA flying through several airspace classes and altitudes, a decision must be made as to whether the TLS value needs to be met at every point in time or only on average across the duration of the entire flight. Moreover, it would also be reasonable to apply different TLS as a function of the intruder, this value being more stringent against cooperative aircraft than non-cooperative ones. As proposed in [13], a possible mechanism to take these considerations into account would be to apply different TLS requirements in different airspace classes.

In Figures 4.2 and 4.3, the concepts of ELOS and TLS are respectively represented by using a goal structured notation (GSN) [14]. This notation, widely used in safety case studies, consists of stating the goals that must be achieved by the system (squares), by relating them to the appropriate context and assumptions. As commented before, one of the main drawbacks of the ELOS approach is the difficulty of assessing the human perception and mid-air collision

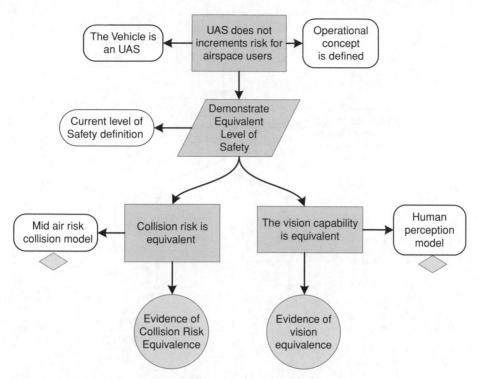

Figure 4.2   Safety representation of ELOS according to the goal structured notation

models, since several models and methodologies exist and none of them has been proved better than the others.

Nevertheless, whatever the minimum safety rate is, and however acceptable the derived level of safety is considered, it must be translated into system reliability requirements. For example, in [8] a mid-air collision risk assessment is presented, aimed at estimating the number of expected collisions per hour of flight. This study would be useful when establishing the minimum performance requirements of SAA systems in different scenarios (see also [2]).

4.2   Existing Regulations and Standards

Conventional aviation regulations have been built, taking safety as the paramount objective and embracing personnel and organizations involved in the development, production and operation of the aircraft:

- The type certification of a specific model of aircraft is usually granted thanks to its development by a certified design organization, which ensures by its quality assurance process the appropriateness of its design to the applicable certification specification.

- The airworthiness of a specific unit is usually granted by the manufacture of a type certified aircraft in a certified production organization.

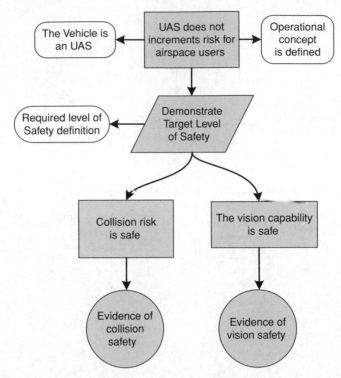

Figure 4.3   Safety representation of TLS according to the goal structured notation

- The maintenance of the aircraft airworthiness must be done in certified organizations and by personnel certified in the maintenance of a particular aircraft model.

- The safety of each flight is ensured by the airworthiness of the aircraft plus the skills of also certified pilots, which are considered the last resort in emergency situations.

This broadly accepted certification structure assumes that there is always a human pilot on-board who could take control of the aircraft at any moment and consequently aviate the aircraft. This peculiarity does not affect significantly the maintenance of the airworthiness as the maintenance organization and its trained personnel will perform their duties following procedures developed during the design of the aircraft. This consideration is also applicable to the production organization, since they will produce aircraft that comply with the type certificate.

The design of an unmanned aircraft significantly changes the way the responsibility of fulfilling the aircraft functions is ensured by the systems to be included in the type certificate of the aircraft and the responsibilities assumed by the pilots, who are no longer onboard. In fact, certification specifications that can be fulfilled at present to obtain a type certificate assume that the flight crew and all aircraft control and management systems are located onboard, which is obviously not the case for UAS. The most significant issue is when implementing the aviate and navigation functions, since the flight crew and the onboard systems are physically separated and rely on a data communications link, introducing new failure modes related to the communication links, their performance and integrity, which are not present in manned aviation. Thus, the performance of the data link, and in particular the latency of the communications, could determine the impossibility of relying on the responsibility of the ground crew, requiring a new system to be compliant with the aircraft requirements. Poor data-link integrity could also make it impossible to meet the safety requirements of the system.

Similar considerations are found with collision-avoidance regulations. In [15] it is assumed that the human pilots are the last resort to ensure the safety of the flight, especially in VFR conditions. This assumption of responsibilities by human flight crews is reflected, for example, in the absence of collision-avoidance capable electronic systems for all aircraft types. ACAS might be the exception, but it should be noted that it is not mandatory for all aircraft and, for example, is almost inexistent in general and light aviation.

These responsibilities, completely assumed by the flight crew, are embraced by the see and avoid concept, which is the last resort to avoid collisions and eliminates the need or requirement for any airborne system. The dislocation of the flight crew when operating UAS will consequently introduce new capabilities in the onboard systems in order to meet the required levels of safety.

The actual implementation of these requirements, for the different UAS categories and architectures, shall contemplate different perspectives (such as the type of airspace, the automation level of the UAS, etc.). Therefore, systems compliant with the functions aviate, navigate and mitigate (notably, the collision-avoidance sub-function) are beyond the scope of the current regulations when talking about UAS, because the requirements to be fulfilled by these systems go beyond the requirements fulfilled by the equivalent systems in conventional aviation and regulations are explicitly designed for aircraft with onboard flight crew. However, there is significant interest in enabling UAS in civil airspace, due to the large amount of civil applications that go beyond emergencies and law enforcement (such as precision farming, infrastructure monitoring, communications relays, etc.) and therefore cannot operate in segregated airspace as military operations perform currently. Then, it is expected that missing regulations will come along with several experimental flights in which the different stakeholders could evaluate the impact of UAS in future aviation scenarios.

4.2.1   Current Certification Mechanisms for UAS

At present, and as considered by the different regulation bodies, two different mechanisms allow UAS to access the civil airspace, either by obtaining a restricted type certificate, along with an airworthiness certificate, or with a special permit to fly.

Conventional aviation shall satisfy different requirements to access civil airspace. Firstly, aircraft designs shall comply with the regulations of the corresponding safety agency (for example, the European Aviation Safety Agency (EASA), or the FAA in the USA), which recognizes this safe design by a type certificate. This certificate can be granted to all designs performed by an organization previously certified with a DOA (design organization approval) or by equivalent means. Then, each aircraft produced following the approved design acquires a certificate of airworthiness when manufactured by an organization certified with a POA (production organization approval) or that is able to show an equivalent product assurance. Furthermore, this certificate of airworthiness must be renewed according to a maintenance program performed by a certified maintenance organization.

The absence of a certification specification applicable to UAS designs leads to the impossibility of obtaining a type certificate and consequently a certificate of airworthiness. When the impossibility of applying an existing certification specification occurs, a restricted type certificate can be granted only under defined and limited conditions and providing that the actual conditions of use are restricted to those in which the certification specification applicability is not compromised. For example, in the case of UAS, if operations are assumed in segregated airspace there is no need to implement the SAA capability. Thus, a restricted type certificate allows an aircraft produced accordingly to obtain a restricted certificate of airworthiness which is valid only if the aircraft is operated according to the restrictions expressed in the restricted certificate type.

In case the aircraft cannot meet the previous certification requirements, but is still capable of performing a safe flight under defined conditions, a permit to fly can also be granted. This could be the case for the majority of UAS, since not many certification specifications exist for them yet. The conditions under which an aircraft is eligible for a permit to fly may be slightly different from country to country.

United States

The Federal Aviation Administration (FAA) has specific regulations to issue experimental certificates for the following subjects (FAR §21.191): research and development; showing compliance with regulations; crew training; exhibition; air racing; market surveys; operating amateur-built aircraft; operating primary kit-built aircraft; and operating light-sport aircraft under certain restrictions.

The special interest in UAS, along with their intrinsic characteristics, motivated the FAA to develop a joint interim procedure aimed at granting access to UAS into the US national airspace [16]. The following FAA groups were involved in the development of this document: the Unmanned Aircraft Program Office (UAPO) of the FAA Aircraft Certification Service (AIR-160); the Production and Airworthiness Division of the FAA Aircraft Certification Service (AIR-200); the Flight Technologies and Procedures Division of the FAA Flight Standards Service (AFS-400); and the FAA Air Traffic Organization's Office of System Operations and Safety (AJR-3).

This procedure describes how to obtain a special permit to fly, or more precisely a certificate of approval (COA) as named by the FAA. Thus, alternative methods are proposed to the conventional compliance of FAR Part 61, which tackles certification for pilots, flight and ground instructors; and Part 91, which deals with the general operating and flight rules. However, only US governmental institutions can apply for a COA, which includes some indications on airworthiness, flight operations and personnel qualifications. Non-governmental UAS developments must obtain a restricted airworthiness certificate, as explained above.

Moreover, in the procedure it is explicitly mentioned that current onboard cameras or sensors have not yet shown enough maturity to be considered as the only mitigation means to comply with the see part of the SAA requirements. The principal milestone is the difficulty of proving that non-collaborative aircraft can be conveniently sensed and the use of external observers or equivalent means (such as chase planes) is proposed in the procedure. The observers shall be located at a distance not bigger than 1 NM in the horizontal plane and 3000 ft vertically, except for operations in class A airspace, where UAS must comply with the required equipment to operate in this airspace while proving to be safe to any other user. The use of observers is forbidden for night operations, and a special safety case is required for such operations.

The use of SAA systems must be accompanied by a suitable safety case that shows its adequacy for the intended airspace. The design of such a system shall include the following sub-functions:

1. Detect – determine the presence of aircraft or other potential hazards.

2. Track – estimate the position and velocity (state) of a single intruder based on one or more surveillance reports.

3. Evaluate – assess the collision risk based on intruder and UA states.

4. Prioritize – determine which intruder tracks have met a collision risk threshold.

5. Declare – decide that action is needed.

6. Determine action – decide on what action is required.

7. Command – communicate and/or execute the determined action.

Europe

The EASA promotes the creation of standards and regulations for safe and environmentally friendly civil aviation in Europe. So far, EASA powers are vested in the regulation 216/2008 [17], which include: airworthiness (for DOA, POA and MOA), pilots licensing, operations approval, third-country operators and permits to fly. However, it must be noted that in the particular case of UAS, EASA has no authority on unmanned aircraft with an operating mass lower than 150 kg or on aircraft explicitly designed for research, experimental or scientific purposes. In these circumstances, permits to fly for UAS must be granted at national level, by the European member state where the UAS will carry on the operations.

In August 2009, EASA published its Policy Statement Airworthiness Certification of Unmanned Aircraft Systems [18], in which the agency assumed that more experience is still needed to publish a dedicated acceptable mean of compliance (AMC) document on UAS, as done with manned aviation. The interim proposal to achieve this required experience is to use the existing CS-21, sub-part B (type certificates) modified with some guidance, especially on special conditions according to the general means [19]. These special conditions refer to:

- emergency recovery
- capability
- command and control link
- level of autonomy
- human–machine interface
- control station
- due to type of operations
- system safety assessment.

Even with the importance of the sense and avoid (or detect and avoid as designated in [18]) for ensuring the safety of on-ground personnel, SAA is not considered to be an issue exclusive to UAS. EASA considers that the type certificate obtained by following its policy statement shall be accompanied by a statement in the aircraft flight manual limiting the operations to segregated airspace, unless the mitigation measures have been accepted by the responsible authority granting access to the airspace volume in which the UAS will operate. There exist numerous conditions under which an aircraft is eligible for an EASA permit to fly, for instance flight testing of new production aircraft; flying the aircraft for authority acceptance; exhibition and air show; or for non-commercial flying activity on individual non-complex aircraft or types for which a certificate of airworthiness or restricted certificate of airworthiness is not appropriate, among others.

Other Countries: Canada and Australia

Besides Europe and the United States, other countries are also working on the development of a regulatory framework for UAS operations. It is worth briefly mentioning the cases of Canada and Australia, which have special UAS working groups for that purpose.

In December 2006, the Canadian General Aviation branch convened a joint government and industry unmanned air vehicle (UAV) working group to review existing legislation and make recommendations for a regulatory framework for UAS operations. The UAV working group published a final report in September 2007 [20], with remarks that the working group will not address sense and avoid systems or requirements. However, in March 2010, the Canadian Civil Aviation branch established a program design working group [21] aimed at developing different deliverables from 2011 to 2016. One of the deliverables should particularly address new technologies, like SAA, which will enable new safety requirements to be achieved.

On the other hand, the Australian Civil Aviation Safety Authority (CASA) is represented on two standards committees: the RTCA Committee SC-203 and the ASTM F38.01. As mentioned in [22], '*CASA is committed to the principles of harmonisation of regulations and standards with the FAA and EASA and will accept FAA and EASA UAV design standards when available*'. Australian authorities already publish UAS-specific regulations in the Civil Aviation Safety Regulation (CASR) Part 101: Unmanned aircraft and rocket operations. Concerning the SAA system, in [23] it is stated that: '*Unless the controller of a UAV is provided with sufficient visual cues to enable the acquisition and avoidance of other air traffic, UAV flights in controlled airspace will be treated as IFR flights, subject to ATC control. CASA may require a large UAV to be equipped with an SSR transponder, a collision avoidance system or forward looking television as appropriate for the type of operation.*'

4.2.2   Standardization Bodies and Safety Agencies

The acceptable means of compliance issued by the different regulatory bodies relies on standards published by different standardization bodies. Usually those organizations constitute working groups in which the different actors involved in the activity or product to be standardized are represented. There exist many of these working groups, but perhaps the most representative ones, in terms of UAS regulations and SAA requirements, are:

- EUROCAE Working Group 73 (WG-73), which is addressing the standards required for civilian UAS to fly in non-segregated airspace. This group is sub-divided into four groups:

  o SG-1: operations and sense and avoid.

  o SG-2: airworthiness and continued airworthiness.

○ SG-3: command and control, communications, spectrum and security.

○ SG-4: UAS below 150 kg for visual line of sight operations.

• RTCA Special Committee 203 (SC-203), which is developing standards for UAS aimed at helping the safe, efficient and compatible operation of UA with other vehicles, based on the premise that UAS and their operations will not have a negative impact on the existing airspace users. One of their most relevant documents is a minimum aviation system performance standard (MASPS) for SAA for UAS, which at the moment of writing this book was still not complete.

• ASTM International Committee F38 on UAS, which is devoted to standards including the design, manufacture, maintenance and operation of UAS, as well as the training and qualification of personnel. A specific standard for the design and performance of an SAA system for UAS has been published in [1]. This committee is also divided into different sub-committees:

○ F38.01 Airworthiness Standards.

○ F38.02 Operations Standards.

○ F38.03 Pilot & Maintenance Qualifications.

4.3   Sense and Avoid Requirements

In this section, we discuss the possible requirements for an SAA system for UAS, while pointing out the main issues that still need to be assessed. As already mentioned, some final requirements are not yet adopted by any regulation, but some organizations have already issued some documents. This section wraps up the available information at the moment this book was written and separates the SAA requirements into several categories.

4.3.1   General Sense Requirements

Sense functionalities in a wide approach include the detection of all the external hazards that might affect a given flight. Therefore, when operating UAS in a non-segregated airspace, it is necessary to consider that the vehicle will interact with an environment that is shared with other users. As a consequence, other aircrafts and airborne objects might be encountered and have to be detected. But, besides the detection of other flying objects, the sense systems should allow the monitoring of other hazards – for instance bad weather, wake turbulences or the proximity of terrain. Basic sense parameters which will have to be considered in the design phase of these systems are:

• The detection range of hazardous objects, which must allow the following avoidance manoeuvre to be executed with the sufficient time to result in the minimum required miss distance.

• The field of regard, being the area capable of being perceived or monitored by a sensor and which must demonstrate that the SAA system meets the right-of-way basic rules.

• Other parameters such as measurement accuracy, reliability and update rate [11].

Concerning the field of regard, right-of-way rules state that pilots must avoid all objects, with the exception of overtaking traffic, and according to [4] a horizontal azimuth angle of $\pm 110°$ off the aircraft nose is recommended for visual scanning in manned aviation, and it is expected to be demonstrated for sense systems in UA. Furthermore, [1] proposes an angle of elevation of $\pm 15°$ for UAS sense systems.

One of the main uses that the sense system will have is to avoid mid-air collisions. In that case, the UCAF (UAS collision avoidance function) must operate autonomously and independently from the ATM system or any other means of UAS separation provision [24]. Moreover, the sense system must detect cooperative and non-cooperative traffic and accommodate UAS operations in different flight modes and airspace classes. However, the system might rely in part on human intervention and the communications latency is an important factor to be assessed [9]. On the other hand, the detection of a collision threat shall be at a minimum range allowing a resolution manoeuvre resulting in a miss distance where both aircraft are *well clear*. Obviously, this minimum detection distance will depend greatly on the performances of the aircraft, such as the cruise speed, turn rate, climb or descent rates; and in the definition of the *well clear* term, which is discussed later on in this chapter. Furthermore, this detection shall be in all weather conditions that it is expected the UAV will encounter and even in case of loss of direct command, control or communications with the command ground station. Thus, it is of paramount importance to consider all these factors when designing the sense sub-system for the UAS.

The on-time detection of hazardous flying objects is a very challenging feature for the sense sub-system. Different techniques can be used in order to fulfil this objective and they are the subjects of intensive research. In [8], a classification of technologies that would be able to detect traffic is proposed, resulting in eight different categories including radar surveillance (see for instance [25]) or visual surveillance means (either by ground observers or chase planes).

Sensor technologies aim to meet or even exceed the performance of current human visual traffic detection. In 2003, NASA equipped a UA with a radar system able to detect non-cooperative targets and a traffic advisory system to detect cooperative ones. With this UA, and several surrounding traffic, some flight tests were carried out where SAA performance capabilities of the involved aircraft were varied [26, 27]. During the flight tests, the pilot's acquisition capabilities were also assessed, in order to be compared with the UAS sense capabilities. As reported in [8], only the traffic advisory system was sufficient for all encounter scenarios. The radar had a limited range of 4 miles detecting targets, being too late to perform appropriate avoidance manoeuvres. On the other hand, the human pilot's perception was between 1 and 1.5 nm. Further research showed that the human eye was inadequate to detect and prevent collisions in several situations and even limited sensors performed better than the human eye [26].

Obviously, the difficulty of detecting other flying objects will depend on the nature of these objects themselves. Nevertheless, not all possible hazardous objects are present in all situations and therefore, if UAS operations are restricted to a certain type of conditions (such as altitudes or airspace classes), the sense requirements will depend on the type of objects that the UA might encounter during its operation. In this context, a definition of the attributes of these potential threats becomes extremely import when developing a sense system. In the work done in [24], an exhaustive analysis was performed typifying all possible flying objects that may represent a threat of collision. Thence, 17 different categories are proposed, ranging from fauna, parachutists, kites and radio-controlled model aircraft to all types and sizes of aircraft. For each type of object it is explained under what conditions these objects may not be encountered, for example above certain altitudes, weather conditions or airspace classes. These categories are:

- **F (Fauna):** birds the size of a goose or larger, which do not generally fly in IMC or above 1000 ft AGL (above ground level). However, migrating birds can be encountered higher than this, typically in the range 5000 ft to 7000 ft AGL, often at specific times of year and in specific locations. Generally, the greater the height above the ground the less likely it is that birds will be encountered.

- **K (Kites and tethered balloons):** both the object itself and the cable connecting them to the ground. In general, operations above 400 ft should be notified by NOTAM.

- **R (Radio-controlled model aircraft operated by hobbyists):** generally operated in VMC below 400 ft AGL and within line of sight of the operator (typically 500 m). Operation above 400 ft should also be notified by NOTAM.

- **B (Hot air balloons):** which do not operate in IMC.

- **D (Dirigible airships).**

- **G (Gliders):** which do not operate in IMC.

- **P (Parachutists):** which are not usually present in IMC. Their activity is usually notified by NOTAM or known by the ATS.

- **S (Powered air sports):** such as very light aircraft, ultra-lights, motor gliders, motor paragliders, etc. Do not operate in IMC.

- **A (Unpowered air sports):** such as hang gliders, paragliders, etc. Do not operate in IMC.

- **H (Helicopters):** considering both civil and military.

- **L (Light aircraft):** such as non-pressurized general aviation.

- **Q (Pressurized general aviation with a maximum take-off mass (MTOM) less than 5700 kg).**

- **M (Military fighters and high-performance jets).**

- **N (Pressurized passenger aircraft not required to carry ACAS).**

- **T (Pressurized passenger aircraft required to carry ACAS).**

- **C (Cargo aircraft or military air transport):** generally with MTOM over 5700 kg and thus, expected to be ACAS equipped.

- **U (Unmanned aircraft):** a wide-ranging group covering a variety of sizes, airframe designs and capabilities.

All previous categories in turn can be grouped into five different categories of flying objects depending on their level of cooperation and capability to avoid a mid-air collision (see Table 4.3).

Categories 1 and 2 are the most challenging ones as the objects are non-cooperative, meaning that active sensors will be required to detect them. Objects of category 2 are able to avoid collisions in VMC and therefore, mitigation actions such as improving the UA visibility

Table 4.3 Categorization of flying threatening objects as a function of their cooperativeness and avoidance capabilities

Category	Cooperative	Can initiate avoiding action?	Category of objects
1	No	No	F, K, B, P, A, D
2	No	Yes in VMC	R, G, S, H, L, U
3	Yes	No	D
4	Yes	Yes in VMC and with ATC intervention in IMC	H, L, Q, N, T, C, M, U
5	Yes	Yes in VMC and with ATC intervention in IMC and in any situation if the intruder is equipped with a transponder	T, C, M, U

may be foreseen. Categories 3, 4 and 5 are cooperative and their detection can be performed more easily than with the others. On the other hand, an interesting factor to take into account is that the sense system shall detect intruders, but also identify them in order to comply with the right-of-way rules and decide which aircraft has the priority.

4.3.2   General Avoidance Requirements

After a collision threat has been conveniently sensed, the UA must select the appropriate avoidance manoeuvre and execute it. This manoeuvre must be compatible with the performance of the aircraft and remain below its structural and aerodynamic limits. These avoidance or resolution manoeuvres may include one or more of the following changes in the flight trajectory: airspeed, altitude or heading. If, as a consequence of an avoidance manoeuvre, the flight path deviates from an ATC clearance or instruction, it must be notified as soon as possible. Moreover, after the conflict is solved, subsequent manoeuvres must return the aircraft to the original flight plan or to a newly assigned flight path, while being compliant with the right-of-way rules.

The most basic requirement for the avoidance manoeuvre is to perform it in such a way that the distance from the intruder aircraft or object is equal to or greater than a minimum required miss distance. Current manned regulations state that the aircraft must remain *well clear* from the intruder and no explicit distances are given (see for example FAR §91.113). It is generally and implicitly understood that the minimum miss distance should be at least 500 ft in all directions [28]. However as reported in [29], the industry itself regards 500 ft of lateral separation as the worst-case minimum distance for SAA. Quoting this document, '*the application of 500 ft. horizontal separation could generate a heightened sense of collision risk [and therefore, it is proposed] an increase in horizontal separation to 0.5 NM, [which] would reduce this perception and also the collision risk itself. [. . .] These minima would only apply away from aerodromes.*' Yet, in [13], the term *well clear* is considered at separation level and not at avoidance level, since it is defined as the state by which two aircraft are separated in such a way that they do not initiate a collision avoidance manoeuvre. Therefore, according to this definition, this *well clear* boundary would vary as a function of the UA and intruder performances, conflict geometry, closure rates and relative accelerations.

Besides the actual value of this minimum miss distance or boundary, special consideration should be given to collaborative aircraft which will be equipped with ACAS, since the avoidance sub-system safety analysis must show compatibility with existing ACAS-II manoeuvres onboard some manned aircraft. In this context, coordinated manoeuvres can range from complex full 4D coordinated manoeuvres to only basic heading or altitude changes in the horizontal and vertical planes respectively. Carriage requirements for ACAS-II equipment are addressed in [30], where it is stated that '*all turbine-engined aeroplanes of a maximum certificated take-off mass in excess of 5700 kg or authorized to carry more than 19 passengers shall be equipped with an airborne collision avoidance system (ACAS-II).*' The same document also recommends equipping all aircraft with such a system, while flight crew procedures for the operation of ACAS are found in [31] and procedures regarding the provision of air traffic services are described in [32]. For example, in [31] it is explained how pilots should react to ACAS advisories and some guidelines in the training of pilots are also given. Conversely [32] explains, for instance, how air traffic controllers should deal with ACAS equipped aircraft in case they deviate from the least clearance due to a resolution advisory. In both documents, phraseology in the operation of ACAS is also described.

RTCA minimum operational performance standards (MOPS) for TCAS-II are found in [33] and could be applied to UAS to some extent. For instance, and as [11] already points out, TCAS-II assumes typical transport category aircraft performance for collision avoidance and resolution advisories (RAs) algorithms, while many UAS may not be capable of the same performance characteristics. Moreover, RAs are executed by pilots in manned aircraft and if RAs are executed autonomously by a UA system, this increases the safety requirements on the system. Conversely, if a UA operator executes a RA, issues of data-link latency and reliability must be addressed.

Unusual UAS performance (compared with transport category aircraft) must also be assessed from the ATC point of view, since current ATC practices and training are based on current existing manned aircraft. For example, some UA are able to fly at high altitudes comparable with current manned commercial aviation, but their rate of climb is dramatically lower than modern airliners. Therefore, UA performances will have to be included in ATC handbooks in order to be able to accommodate UA and provide safe and efficient separation and traffic information services. Moreover, it should be noted that foreseen UAS SAA systems will also support self-separation functions and consequently, some responsibility could eventually be shifted from ATC to UAS (as is also foreseen in manned aircraft [34]).

Concerning terrain avoidance, some systems exist in manned aviation such as the terrain avoidance warning system (TAWS) or the ground proximity warning system (GPWS). However, these systems are advisory in nature and as we observed with the resolution advisories of the TCAS-II, the onboard pilot has the ultimate responsibility to execute the avoidance manoeuvre. Therefore, existing standards will also need to be updated in order to address remote pilot operations and/or UAS automation to avoid terrain [11].

Finally, avoidance means must also be designed to comply with visibility and cloud clearance criteria (with specific requirements depending mainly on the airspace class) for the UA to be seen by other aircraft and therefore comply with the flight rules. Besides SAA requirements, severe weather could result in damage to the UAS and affect its airworthiness and must therefore be detected and avoided.

Summing up, avoidance design parameters should take into consideration several factors, such as weather, terrain and flying objects. Traffic avoidance design will depend mainly on the minimum miss distance from the intruder aircraft, actual aircraft performances and limitations, a correct interpretation and implementation of the right-of-way rules, the collision avoidance capabilities of the intruder and the compatibility with the ATC clearances.

The next sub-sections focus on the particularities that different airspace classes, altitudes of operation and weather conditions, UAS communications data links and automation levels may have on the requirements for a safe design of SAA system.

4.3.3   Possible SAA Requirements as a Function of the Airspace Class

The type of airspace where UAS operations will be carried out will mainly determine the level of cooperativeness of the other traffic and the availability of ATC for assuring separation or enhancing the situational awareness with respect to other traffic. It is possible to aggregate the types of airspace into two large categories, as proposed in [24]: those airspaces where all aircraft are cooperative; and the remaining ones where some of the aircraft may not be cooperative. As mentioned in Section 4.1.2, the specific requirements for each airspace class may differ slightly from one country to another. Yet, in airspace classes A to D it is in general required to operate with a transponder. Moreover, it is also quite usual to mandate the use of a transponder above a certain altitude regardless of the airspace class. For example, in the USA transponders are required in airspace classes A, B, C and E above FL100, although some aircraft are exempted (FAR §91.215). The RTCA MOPS for transponders are found in [35], which specifies some requirements for the flight crew control and monitoring of the operation of the transponder that could be sufficient for the application in UAS, while other requirements may be sufficient for UAS.

The signal emitted by aircraft transponders is received by secondary surveillance radars (SSR) and the derived aircraft positions are enhanced with the information encoded in those signals. Thus, a mode A transponder transmits just a five-digit identifier, while a mode C transponder also transmits the barometric altitude of the aircraft. The newer mode S transponders [36] have the capability to transmit even more information, such as the position of the aircraft if an ADS-B system is present. All this information is presented in the ATC positions.

Besides ATC stations, ADS-B allows aircraft and ground vehicles to send or receive surveillance information, including an identifier, latitude, longitude and altitude. Therefore, the flight crew situational awareness is significantly improved and some separation responsibilities can eventually be shifted from the ATC to the flight crew. However, ADS-B is not currently mandated and different data-link technologies exist (other than mode S transponders) and are standardized in documents [37, 38], while the data-link-independent standard for ADS-B systems is published in [39].

Besides SSR detection, transponders are the base technology for current implementations of the TCAS. Obviously, TCAS (or more generally ACAS) equipped aircraft will be easier to detect and, if the UA is equipped with such a system, the generated traffic alerts would increase the situational awareness of the UAS flight crew and improve the sense capabilities. The TCAS-I standard is published in [40], while relaxed requirements can be found in [41] if only the traffic advisory airborne equipment is implemented. This could reduce the cost of such a system with respect to a TCAS-I and could be useful for certain UAS implementations. As explained before, TCAS-II implements resolution advisories too, but the way these derive into effective avoidance manoeuvres for UAS still remains an open issue.

It is worth mentioning that UAS usually perform a type of flight which is different from the one performed by the majority of commercial aviation. In the majority of cases, UAS will be used to perform missions that might include holding, scans and other non-conventional procedures whilst commercial aviation mainly operates a point-to-point flight, the carriage of people and goods being its main purpose. Therefore, commercial aviation uses airways to fly from one point to another. This leads to a situation where the probability of a mid-air collision is higher flying on those airways and in their vicinity than flying away from them. Therefore, as computed in [8], the mid-air collision risk outside the main congested airways and areas might be low enough to allow a reduction in the performances needed to detect and avoid other traffic.

4.3.4   Possible SAA Requirements as a Function of the Flight Altitude and Visibility Conditions

Some systems operate in a given range of altitudes, and if the UA is flying outside these altitudes it is not likely to encounter them. In a similar way, the weather conditions (VMC or IMC) will affect the capabilities of some sensors, for example those based on vision, but will also affect the type of hazardous objects that might be encountered (i.e., with bad visibility the UA will not collide with a glider as they do not operate in IMC conditions). These dependencies are analysed in [24] and summarized in Table 4.4, where for each category of objects the likelihood of finding them in certain altitudes and meteorological conditions is given.

As mentioned before, transponders are generally mandated for aircraft evolving above FL100 and that is why in this table a clear separation as a function of this altitude appears. Besides the fact that this requirement is not enforced in all countries, it should also be noted that even in those countries where it is applied, some exemptions may apply (for example, gliders flying above FL100 and not equipped with a transponder). On the other hand, below FL100 aircraft are usually limited to speeds below 250 kt in the majority of airspace classes (depending on the country, classes A and B are usually exempt from this speed limitation).

Table 4.4 Category of objects that may be found as a function of the flying altitude, meteorological conditions and traffic environment (from [24])

Object Category		Unknown traffic environment Below FL100		Known traffic environment Below FL100		Above FL100	
		VMC	IMC	VMC	IMC	VMC	IMC
Non-cooperative objects	1	√	√*	√	√*	×	×
	2	√	×	×	×	×	×
Cooperative objects	3	√	√	√	√	×	×
	4	√	√	√	√	√	√
	5	√	√	√	√	√	√

√ Category of objects that the UA might encounter.
× Category of objects that are unlikely to be encountered by the UA.
√* Tethered objects below 500 ft AGL are the only category of objects for this category.

Therefore, Table 4.4 shows the most common situations providing that some exemptions may exist as a function of the national regulations. Moreover, it is worth mentioning that even if some objects are not likely to be found above a certain altitude (such as fauna or light aircraft), the UAS may encounter them during its climb and descent phases. Nevertheless, some possible solutions allowing a less demanding SAA system would be to temporarily segregate some airspace to fit these climb and descent phases; use chase aircraft following the UA during these phases; climb/descend in a controlled airspace class with a completely known environment for the ATC, while avoiding overflying populated areas to reduce the risk of a crash due to a potential collision with fauna, etc.

4.3.5   Possible SAA Requirements as a Function of the Type of Communications Relay

Latency of the communications with the ground control could determine the performance of the systems onboard. Since nobody questions the presence of a human controlling the UAS, the distance between pilot and aircraft could introduce delays that may threaten certain SAA architectures or solutions. This latency includes: communication delays, SAA scan rates, onboard and ground processing times of the different algorithms involved, pilot-in-the-loop reaction times and coordination with ATC. Two main categories exist when talking about command, control and telemetry communications: those which are at line of sight with the UAS and those beyond line of sight.

Line of Sight Communications

Line of sight operations have no visual obstacles between the flight crew and the UA and the range between them can vary from a few metres to tens of nautical miles. For short distances (up to a few miles), the visual contact with the UA that can establish the flight crew on ground could even justify the absence of onboard means to sense the traffic. For greater distances, the visual performance of the flight crew along with the reduction of the spatial situational awareness will not be able to fulfil the sense requirements and other support systems will be required. Moreover, the orography is an important factor to consider when operating in line of sight conditions, since some traffic could be hidden behind the terrain. Therefore, not only is it required to have visibility of the UA, but also of the surrounding airspace where threatening traffic may exist.

Small latencies in line of sight conditions could even allow a direct video link from the UA to the flight crew. This video link and the assumption of responsibility by the flight crew when assessing the situation, elaborating the evasive actions and performing them simplifies the whole SAA system. However, the data link itself remains a critical component and the exigencies over real-time performances would be very demanding.

Beyond Line of Sight Communications

An aircraft beyond line of sight impedes the assumption of sensing responsibility by the flight crew exclusively. It is worth mentioning that these kinds of operation do not always imply a far location from the ground control station. For example, the flight crew could be located on one side of a building controlling a UA performing a perimeter surveillance of the same building. Since there is not a straight obstacle-free line between the UA and the flight crew for the entire mission, this operation would be considered beyond line of sight. Different communication strategies are foreseen for these types of operation and are summarized as follows:

- **Direct radio frequency (RF) communications.** Even if the UA is beyond line of sight, RF communications can be successfully implemented under some conditions. These architectures provide fast links and latency is not deemed a big issue for SAA applications. However, the frequency employed for these communications could suffer from insufficient bandwidth, especially if video is transmitted aimed at implementing the sense function on the ground station.

- **Terrestrial networks.** This architecture externalizes part of the communications to a service provider that already owns a dedicated network over a certain area. With this architecture, the bandwidth limitations depend on the actual means of the provider and sense and avoid video architecture will be conditioned to this available bandwidth, the latency of the data transmission and the stability of the communications. Thus, depending on the quality of service (QoS) guaranteed by the provider, a strategy for SAA including video transmission could be considered or discarded as a function of the real-time performance and integrity of the network.

- **Satellite communications.** These kinds of communication introduce large latencies because of the long distances that communication signals must travel. These delays mean the SAA functionality to be performed entirely onboard the UA.

4.3.6   Possible SAA Requirements as a Function of the Automation Level of the UAS

As commented before, the degree of automation of the UAS is also a key factor in the SAA capability, since highly automated platforms will need more reliable sense sub-systems than those with a high contribution of the human crew and, therefore, closer to manned aviation see and avoid performance.

Therefore, the level of autonomy envisaged for the UAS could notably simplify the requirements for the SAA system. During the phase of the allocation of aircraft functions to systems [42], the abstract functionality of the aircraft is divided between the flight crew and the automated systems supporting each of the functionalities. As a function of the level of responsibility assumed by the flight crew, we can establish four levels of automation:

- Radio-controlled aircraft: in this situation, the pilot directly commands the attitude of the aircraft by controlling the flight surfaces through a dedicated radio control system. As the aircraft itself has no means to autonomously modify its trajectory, all avoidance manoeuvres will be performed exclusively by the human pilot and therefore all sense information must be available to this pilot in real time.

- Pilot in line: this architecture allows the aircraft to follow programmed flight plans but allowing the pilot to take control of the aircraft and aviate it at any time. Since the flight crew retains the capability of assuming control of the trajectory, the responsibility of the SAA functions can be shared by systems and humans.

- Pilot on line: here the pilot has the capability to take control of the UA but only to navigate it, the aviation function being performed autonomously. The difference between allowing the flight crew to directly aviate or navigate the aircraft may seem subtle, but has an enormous implication in SAA systems. Aviate is defined as modifying the trajectory of the aircraft by controlling its attitude in real time. Whereas navigate means to give guidance inputs to the aircraft (in the form of headings or even waypoints, for example) aimed at also modifying the flight path. Therefore, the time scale for pilot on line operations is significantly larger than in previous categories and as a consequence it is likely that SAA systems would be built onboard and the UA would assume the entire responsibility for sensing and performing avoiding manoeuvres.

- Full autonomous UAS: full autonomy shall be understood as the capacity of the UA to achieve its entire mission without considering any intervention of the human flight crew. Obviously, with this architecture the SAA must be performed exclusively by onboard means. Nevertheless, these kinds of operation are still not contemplated by any regulatory body in the world.

4.4 Human Factors and Situational Awareness Considerations

Although it might be technically feasible to build up a SAA system which operates completely autonomously and without reporting traffic and decisions taken to the UAS flight crew, it would be advisable for the pilot in command to have access to any traffic information sensed and the separation or collision avoidance manoeuvres carried out by the UA. This would help to validate the correct operation of the SAA system, enhance the situational awareness of the flight crew and allow them to report traffic to ATS, if necessary. This means that some human–machine interfaces (HMIs) should be developed to present all the necessary information to flight crew.

As a general statement in aviation, the more information the flight crew have access to, the higher the situational awareness is, and the 'thicker' the different safety layers are. However, presenting more information to the operator can also lead to problems such as cognitive overload. Perhaps using contextual information is one of the solutions to cope with this problem, but no real experience exists so far with these kinds of HMI, since most of the analyses are currently limited to UAS operating under military conditions. Therefore, the definition of these interfaces, along with all the human factors derived from the involvement of the flight crew in the SAA process, is also the subject of intense ongoing research and standardization processes.

In manned aviation, we already find some interfaces that enhance the see and avoid capabilities of the pilots. At collision avoidance level, ACAS systems must notify aircraft crew about possible conflicts. As mentioned earlier, the most basic ACAS systems trigger traffic alerts in case of potential conflict with another cooperative aircraft (such as the TCAS-I), while others can also provide the pilots with resolution advisories. These systems interface with the cockpit indicators or displays, being the navigation display (ND) of modern aircraft used for showing the surrounding traffic; and the primary flight display (PFD) for traffic alerts and resolution advisories indications. Current ASAS developments and prototypes propose the use of a cockpit display traffic information (CDTI), which in some cases could be implemented on upgraded versions of current navigation display concepts. A CDTI is also meant to present surveillance information about the surrounding traffic, but with a larger scope than ACAS implementations, allowing in this way to perform separation functions onboard. Thus, the information presented may include relative positions of other aircraft gathered from ACAS, ADS-B systems, traffic information services (TIS) and, in short, all traffic information sensed by the UAS SAA system(s).

Technically speaking, the 'cockpit' does not exist in a UAS, as the onboard location where flight control is conducted by flight crew. Yet, similar displays and with the same philosophy as CDTIs will certainly be found in the UAS control ground station. The RCTA document DO-243 [43] contains guidance material for the implementation of a CDTI and a few basic features and capabilities are provided (for example, formats for surrounding aircraft identification, closure rates, ground speed information, ground track indicator, target selection, traffic alert, etc.). Even if this standard refers to the cockpit of manned aircraft, it would easily be applicable to UAS ground stations since the focus is given on the presentation to the flight crew. On the other hand, [44] presents some ASAS applications by using CDTI interfaces. Some of them would also be directly applicable to UAS, while others would need some minor or major modifications (like, for example, for all the applications aimed at improving situational awareness of the flight crew when performing visual self-separation operations). Some high-level definitions and specifications on ASAS applications are given in the RTCA MASPS Do-289 [45], while document [46] focuses on ADS-B specific applications. On the other hand, TIS-B messages are standardized in [47] and interchange standards for terrain, obstacle and aerodrome data are found in [48].

As already observed in [11], all the above-mentioned standards and discussions are general enough to apply to most UAS applications and these documents would undoubtedly be an excellent starting point for UAS regulations and standards. Yet, there is still the need to particularize for UAS, especially when dealing with critical issues such as signal quality, data latency, overall integrity and interaction with the crew. For example, current experimental UAS platforms only address the way in which the surrounding traffic is displayed to the UAS crew. However, they do not address critical issues like how to report the best reaction strategy to the crew; or how to provide support to the potential instructions that the ATC may deliver. All these elements are still under research, with little or no experience available.

4.5 Conclusions

In conventional aviation, see and avoid is the capacity of the human crew to detect other threatening objects or terrain and perform evasive actions to avoid a loss of separation and collisions as a last resort. In fact, all current regulations in aviation assume the presence of a human pilot onboard who could take control of the aircraft at any moment. UAS try to go beyond this limitation by removing the human from the aircraft, which somehow remains on the loop. The equivalent sense and avoid for UAS is one of the principal pitfalls in their way of performing normal operations in civil and non-segregated airspace. However, there is no regulation about SAA yet, since this functionality is specific to UAS, whose operations are at present not fully regulated in non-segregated airspace. Nevertheless, the big interest in UAS technologies demonstrated by military agencies in recent years, as well as the amount of applications in which such a technology seems to be beneficial, have pushed the regulatory bodies to start the regulation process for UAS. Yet, this process is very complex, since it needs agreement between actors with possible conflicting interests; for example, private aviation willing to fly with the minimum possible required avionic equipment versus highly technological and automated UAS operations, companies offering aerial services with UAS versus ATC conventional practices, etc.

In this chapter an overview of existing regulations and their limitations regarding SAA has been given. The principal factors that would determine future SAA requirements have been identified: class of airspace, altitude and visibility conditions, type of UA communication relay and UAS automation level. From this analysis, a significant paradox arises when regarding the sense capability: large UA will in general be required to detect fewer types of object because of their usual altitude of operation, airspace class and visibility conditions. This is a direct implication of the fact that large platforms usually perform missions at high altitudes where terrain and fauna are not an issue and other flying aircraft are all cooperative and known by the ATC. On the other hand, very small UAS will definitely evolve in a very challenging environment, with non-cooperative flying aircraft, fauna and ground obstacles. However, some collisions might not be a real threat for humans since these platforms have a really small weight and kinetic energy. Therefore, the risk of fatalities in some operational scenarios (such as in line of sight and away from populated areas) would be well below the required safety levels. Then, since the UAS spectrum covers a great variety of UA sizes, airframe designs, capabilities and mission particularities, the particular requirements for their SAA system will be strongly related to all these variables.

Furthermore, some issues around SAA systems still linger. For example, there exists the need to define the minimum requirements on minimum detection distances/times and aircraft separation margins; along with minimum avoidance distances/margins and efficient deconfliction algorithms to replace pilot action. The main reason for this is the non-specificity of the 'well clear' term which sets the basis for current manned aviation regulations, allowing for a pilot's subjective assessment on their actions but not being valid for designing an engineering solution meeting the SAA requirements. Moreover, collision avoidance must be regarded as a last resort capability to prevent an imminent collision and consequently a separation function is also required. Again, the minimum separation values are not always objective and are difficult to quantify with present regulations. On the other hand, cooperativeness in SAA systems should be addressed when considering UAS since not all airspace users are (or will be) equipped with transponders, ADS-B based systems or ACAS. Moreover, even current ACAS implementations will not be valid for some UAS, since they assume that avoidance manoeuvres could be performed with performances typical of transport category aircraft.

It seems obvious that SAA functions will be necessary for future UAS operations in civil non-segregated airspace, being one of the most important systems since SAA malfunction will represent a flight critical emergency. However, SAA requirements cannot be the same for all UAS, because of the great variability of platforms, systems and missions. It is true that UAS integration in civil airspace should not incur any cost to current airspace users [49], but it is also true that the UAS should pose no greater risk to persons or property than other current airspace users and therefore, requirements and regulations will be expected to adapt to the specificities and particularities of this promising and exiting new chapter in the history of aviation.

Acknowledgments

This work has been partially funded by Ministry of Science and Education of Spain under contract CICYT TIN 2010-18989. This work has been also co-financed by the European Organization for the Safety of Air Navigation (EUROCONTROL) under its CARE INO III programme. The content of the work does not necessarily reflect the official position of EUROCONTROL on the matter.

References

1. ASTM International, 'Standard Specification for Design and Performance of an Airborne Sense-and-Avoid System', American Society of Testing & Materials, Document F2411-07, February 2007.

2. Dalamagkidis, K., Valavanis, K.P., Piegl, L.A., and Tzafestas, S.G. (eds), *On Integrating Unmanned Aircraft Systems into the National Airspace System: Issues, Challenges, Operational Restrictions, Certification and Recommendations*, Vol. 26 of International Series on Intelligent Systems, Control, and Automation: Science and Engineering, Springer-Verlag, Berlin, 2009.

3. ICAO, *Annex 10 to the convention on international civil aviation – Aeronautical Telecommunications. Volume IV – Surveillance radar and collision avoidance systems*, 2nd edn, International Civil Aviation Organisation, Montreal (Canada), 1998.

4. ICAO, *Annex 2 to the convention on international civil aviation – Rules of the air*, 9th edn, International Civil Aviation Organisation, Montreal (Canada), 1990.

5. ICAO, *Annex 11 to the convention on international civil aviation – Air traffic services*, 13th edn, International Civil Aviation Organisation, Montreal (Canada), 2001.

6. Eurocontrol, 'Implementation of the Airspace Strategy Task Force A (IAS-TFA)', available at http://www.eurocontrol.int/airspace/public/standard_page/17111_IASTFA.html (last checked January 2011).

7. UK Civil Aviation Authority (CAA), *CAP 722 Unmanned Aircraft System Operations in UK Airspace – Guidance*, 2010.

8. Weibel, R.E. and Hansman, R.J., 'Safety Considerations for Operation of Unmanned Aerial Vehicles in the National Airspace System', Technical report, MIT International Center for Air Transportation, March 2005.

9. RTCA, 'Operational Services and Environmental Definition (OSED) for Unmanned Aircraft Systems (UAS)', Radio Technical Commission for Aeronautics, Document Do-320, Washington, DC, June 2010.

10. NASA, 'Potential use of unmanned aircraft systems (UAS) for NASA science missions', National Aeronautics & Space Administration, 2006.

11. RTCA, 'Guidance material and considerations for unmanned aircraft systems', Radio Technical Commission for Aeronautics, Document Do-304, Washington, DC, March 2007.

12. UAVNET, 'European civil unmanned air vehicle roadmap, volumes 1 and 2', available at http://www.uavnet.com, 2005.

13. FAA, 'Sense and Avoid (SAA) for Unmanned Aircraft Systems (UAS)', Federal Aviation Administration sponsored Sense and Avoid Workshop, Final report, October 2009.

14. Weaver, R.A. and Kelly, T.P., 'The goal structuring notation – a safety argument notation', Dependable Systems and Networks, Proceedings of Workshop on Assurance Cases, July 2004.

15. Hayhurst, K.J., Maddalon, J.M., Miner, P.S., Szatkowski, G.N., Ulrey, M.L., DeWaltCary, M.P., and Spitzer, R., 'Preliminary considerations for classifying hazards of unmanned aircraft systems', February 2007.

16. FAA, 'Interim Operational Approval Guidance,08-01: Unmanned Aircraft Systems Operations in the U.S. National Airspace System', Federal Aviation Administration, Aviation Safety Unmanned Aircraft Program Office, AIR-160, March 2008.

17. European Parliament, 'Regulation (EC) No. 216/2008 of the European Parliament and the Council of 20 February 2008 on common rules in the field of civil aviation and establishing a European Aviation

Safety Agency and repealing Council Directive 91/670/EEC, Regulation (EC) No. 1592/2002 and Directive 2004/36/EC', *Official Journal of the European Union*, February 19, 2008.

18. EASA, 'Policy Statement Airworthiness Certification of Unmanned Aircraft Systems (UAS)', European Aviation Safety Agency, August 2009.

19. EASA, 'Acceptable means of compliance and guidance material to Part 21', decision no. 2003/1/RM, October 2003.

20. Transport Canada Civil Aviation, 'Unmanned Air Vehicle (UAV) Working Group Final Report', September 2007, available at: http://www.tc.gc.ca/eng/civilaviation/standards/general-recavi-uavworkinggroup-2266.htm.

21. Transport Canada Civil Aviation, 'Unmanned Air Vehicle (UAV) Systems Program Design Working Group', March 2010, available at: http://www.h-a-c.ca/UAV_Terms_of_Reference_2010.pdf.

22. Carr, G., 'Unmanned aircraft CASA regulations', Civil Aviation Safety Authority (CASA), Australia, on-line presentation available at: http://www.uatar.com/workinggroups.html.

23. CASA, 'Unmanned aircraft and rockets. Unmanned aerial vehicle (UAV) operations, design specification, maintenance and training of human resources', Civil Aviation Safety Authority, Advisory Circular 101-01(0), Australia, July 2002.

24. Eurocontrol, *Unmanned Aircraft Systems – ATM Collision Avoidance Requirements*, Edition 1.3, May 2010.

25. Wilson, M., 'A mobile aircraft tracking system in support of unmanned air vehicle operations', Proceedings of the 27th Congress of the International Council of the Aeronautical Sciences, ICAS, Nice, France, September 2010.

26. Wolfe, R., 'NASA ERAST non-cooperative DSA flight test', Proceedings of the AUVSI Unmanned Systems Conference, Baltimore, MD, July 2003.

27. Schaeffer, R.J., 'A standards-based approach to sense-and-avoid technology', AIAA 3rd 'Unmanned Unlimited' Technical Conference, Workshop and Exhibit, Chicago, IL, September 2004, Paper AIAA 2004-6420.

28. FAA Order 8700.1, Change 3, Chapter 169, §5A.

29. Eurocontrol, 'Eurocontrol specifications for the use of military unmanned aerial vehicles as operational air traffic outside segregated airspace', Document SPEC-0102, July 2007.

30. ICAO, 'Annex 6 to the Convention on International Civil Aviation – Operation of Aircraft, Part I – International Commercial Air Transport – Aeroplanes', 8th edn, International Civil Aviation Organisation, Montreal, Canada, 2001.

31. ICAO, 'Procedures for Air Navigation Services. Aircraft Operations, Volume I – Flight Procedures', 5th edn, International Civil Aviation Organisation, Montreal, Canada, 2006, Document 8168.

32. ICAO, 'Procedures for Air Navigation Services. Air Traffic Management', 14th edn, International Civil Aviation Organisation, Montreal, Canada, 2001, Document 4444.

33. RTCA, 'MOPS for traffic alert and collision avoidance system III (TCAS II) airborne equipment', Radio Technical Commission for Aeronautics, Washington, DC, December 1997, Document Do-185A.

34. Eurocontrol, 'Review of ASAS applications studied in Europe', Technical Report, CARE/ASAS action, CARE/ASAS activity 4, February 2002, available at: http://www.eurocontrol.int/care-asas/gallery/content/public/docs/act4/care-asas-a4-02-037.pdf.

35. RTCA, 'Minimum operational characteristics – airport ATC transponder systems', Radio Technical Commission for Aeronautics, Washington, DC, October 2008, Document Do-144A.

36. RTCA, 'Minimum operational performance standards (MOPS) for air traffic control radar beacon system/mode select (ATCRBS/mode S) airborne equipment', Radio Technical Commission for Aeronautics, Washington, DC, February 2008, Document Do-181D.

37. RTCA, 'MOPS for 1090 MHz extended squitter automatic dependent surveillance – broadcast (ADS-B) and traffic information services – broadcast (TIS-B)', Radio Technical Commission for Aeronautics, Washington, DC, February 2009, Document Do-260B.

38. RTCA, 'MOPS for universal access transceiver (UAT) automatic dependent surveillance – broadcast (ADS-B)', Radio Technical Commission for Aeronautics, Washington, DC, February 2009, Document Do-282B.

39. RTCA, 'Minimum aviation system performance standards for (MASPS) for automatic dependent surveillance broadcast (ADS-B)', Radio Technical Commission for Aeronautics, Washington, DC, June 2002, Document Do-242A.

40. RTCA, 'MOPS for an active traffic alert and collision avoidance system I (Active TCAS I)', Radio Technical Commission for Aeronautics, Washington, DC, December 1994, Document Do-197A.

41. FAA, 'Technical Standard Order (TSO) C147, Traffic advisory system (TAS) airborne equipment', Federal Aviation Administration, Department of Transportation, Aircraft Certification Service, Washington, DC, April 1998.

42. SAE, *Guidelines for Development of Civil Aircraft and Systems*, Society of Automotive Engineers, Standard ARP4754-A, December 2010.

43. RTCA, 'Guidance for initial implementation of cockpit display of traffic information', Radio Technical Commission for Aeronautics, Washington, DC, February 1998, Document Do-243.

44. RTCA, 'Applications descriptions for initial cockpit display of traffic information (CDTI) applications', Radio Technical Commission for Aeronautics, Washington, DC, September 2000, Document Do-259.

45. RTCA, 'Minimum aviation system performance standards (MASPS) for aircraft surveillance applications (ASA)', Radio Technical Commission for Aeronautics, Washington, DC, September 2003, Document Do-289.

46. RTCA, 'Development and implementation planning guide for automatic dependent surveillance broadcast (ADS-B) applications', Radio Technical Commission for Aeronautics, Washington, DC, June 1999, Document Do-249.

47. RTCA, 'Minimum aviation system performance standards (MASPS) for traffic information service – broadcast (TIS-B)', Radio Technical Commission for Aeronautics, Washington, DC, October 2007, Document Do-286B.

48. RTCA, 'Interchange standards for terrain, obstacle, and aerodrome mapping', Radio Technical Commission for Aeronautics, Washington, DC, February 2009, Document Do-291A.

49. Gonzalez, L.-F., *Australian Research Centre of Aerospace Automation. Unmanned Aircraft Systems, the Global Perspective 2009/2008*, 2007, pp. 17–18.

5

Human Factors in UAV

Marie Cahillane[1], Chris Baber[2] and Caroline Morin[1]
[1]*Cranfield University, Shrivenham, UK*
[2]*University of Birmingham, Birmingham, UK*

5.1   Introduction

Several human factors in human–unmanned vehicle interaction are considered here through a synthesis of existing research evidence in the military domain. The human factor issues covered include the potential for the application of multimodal displays in the control and monitoring of unmanned vehicles (UVs) and the implementation of automation in UVs. Although unmanned aerial vehicles (UAVs) are the focus of this book, this chapter reviews research evidence involving the supervisory control of unmanned ground vehicles (UGVs), as the results are relevant to the control of UAVs. This chapter also aims to highlight how the effectiveness of support strategies and technologies on human–UV interaction and performance is mediated by the capabilities of the human operator.

Allowing operators to remotely control complex systems has a number of obvious benefits, particularly in terms of operator safety and mission effectiveness for UVs. For UVs, remote interaction is not simply a defining feature of the technology but also a critical aspect of operations. In a military setting, UAVs can provide 'eyes-on' capability to enhance commander situation awareness in environments which might be risky or uncertain. For example, in a 'hasty search' the commander might wish to deploy UAVs to provide an 'over-the-hill' view of the terrain prior to committing further resource to an activity [1]. In addition, advances in image-processing and data-communication capabilities allow the UAV to augment the operators' view of the environment from the multiple perspectives using a range of sensors. Thus, augmentation is not simply a matter of aiding *one* human activity, but can involve

Sense and Avoid in UAS: Research and Applications, First Edition. Edited by Plamen Angelov.
© 2012 John Wiley & Sons, Ltd. Published 2012 by John Wiley & Sons, Ltd.

providing a means of enhancing multiple capabilities (for example, through the use of thermal imaging to reveal combatants in hiding).

In parallel with improved imaging and sensing capabilities of UAVs comes the increasing autonomy of UAVs [2]. Thus, for example, UAVs can fly with little direct intervention; rather, the operator defines waypoints to which the vehicle routes itself. This leaves the operator free to concentrate on the control and analysis of the payload, i.e., onboard sensors. While UAVs are increasing in their ability to fly autonomously, it is likely that human-in-the-loop operation will remain significant. There are several reasons why there might need to be a human in the control loop, particularly for military UAVs. The first relates to accountability and responsibility for offensive action taken by the UAV; it remains imperative that any decisions to attack a target are made by a human operator. The second reason why there might need to be a human in the loop relates to the interpretation of imagery of potential targets (and the need to decide whether to reroute to take advantage of 'opportunistic' targets); even with excellent image-processing capabilities, the UAV might require assistance in dealing with ambiguity or changes in mission plan. The third reason relates to immediate change in UAV operation, either as the result of malfunction or of hostile intervention. These reasons imply a hierarchy of intervention, from high-level goals and accountability to low-level operation and monitoring. What is also apparent from this brief discussion is that the 'control loop' extends well beyond piloting the UAV. Schulte et al. [3] argue that there is a need to shift design from the conventional view of supervisory control (in which the human operator monitors and manages the behaviour of an autonomous system) to the development of cognitive and cooperative automation (in which the autonomous system works with its human operators to share situation awareness (SA), goals and plans). A common observation is that the tasks that people perform change as technology becomes more autonomous, i.e., from physical to cognitive tasks [4]. This shift in human activity, from physical to cognitive work, is illustrated in Figure 5.1.

This shift in human activity raises the question of what might constitute the role of the human operator in UAV operations. Alexander et al. [5] identify seven primary roles for humans in UAV operations (although they acknowledge that there may well be far more roles

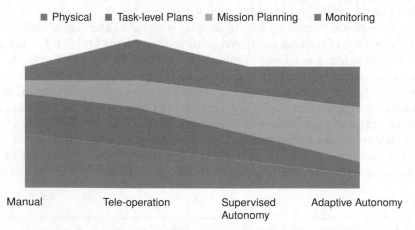

Figure 5.1   Relative contribution of human activity to increasing levels of automation

possible). These roles relate to the earlier discussion of the need to retain human-in-the-loop interaction with UAV, and are:

- Providing the capability to intervene if operations become risky, i.e., a safety function.

- Performing those tasks that cannot yet be automated.

- Providing the 'general intelligence' for the system.

- Providing liaison between the UAV system and other systems.

- Acting as a peer or equal partner in the human–UAV partnership.

- Repairing and maintaining the UAV.

- Retrieving and rescuing damaged UAVs.

It is worth noting that these roles are often (but not always) distributed across different members of the UAV crew and can require different skill sets. Furthermore, the last role, in particular, introduces some interesting issues for human factors because it moves the human from remote operator into the risky area in which the UAV is operating. For example, Johnson [6] describes an incident in which a British officer was killed attempting to retrieve a downed UAV.

As the level of autonomy increases, so the role of the human operator becomes less one of direct control and more one of monitoring and supervision. In other words, the human operator can potentially become removed from the direct 'control loop' and given tasks of monitoring or a disjointed collection of tasks that cannot be automated. The issue of human-in-the-loop has been discussed above, but failure to consider the ramifications of designing to explicitly remove the operator has the potential to lead to a state of affairs that has been termed the 'irony' of automation [7]. This raises two issues. First, humans can be quite poor at monitoring, particularly if this involves detecting events with low probability of occurrence [8]. As discussed later in this chapter, this performance can be further impaired if the information displayed to the operator makes event detection difficult to perform. Second, the role of humans in automated systems is often intended to be the last line of defence, i.e., to intervene when the system malfunctions or to take over control when the system is unable to perform the task itself. If the user interface does not easily support this action, then human operators can either fail to recognize the need to intervene or can make mistakes when they do intervene; if the human has been removed from the control loop, and has not received sufficient information on system status, then it can be difficult to correctly determine system state in order to intervene appropriately. A report from the Defense Science Study Board [9] suggested that 17% of UAV accidents were due to human error, and most of these occur at take-off or landing. In order to reduce such 'human errors', some larger UAVs, such as Global Hawk, use automated take-off and landing systems. However, our discussions with subject matter experts suggest there continue to be cases of medium-sized UAVs (such as the Desert Hawk flown by UK forces) crashing on landing. One explanation of these accidents is that the configuration of cameras on these UAVs is such that they support the primary reconnaissance task but, because the cameras are mounted to support a view of the ground while in flight, they do not readily support the forward view required to effectively land the UAV. The situation could be remedied by the inclusion of additional cameras, but obviously this adds to the weight of the payload and the need to manage additional video-streams. Another remedy would be to increase the autonomy of these UAVs to take off and land themselves, but this has cost implications for what are supposed to be small and relatively cheap vehicles. This raises the question of how operators remotely, or teleoperate, UAVs.

5.2   Teleoperation of UAVs

According to Wickens [10], there are three categories of automation. Each category has different implications for the role of the human operator.

(i) *Substitution*. The first category concerns operations that the human cannot perform due to inherent limitations, either in terms of physical or cognitive ability. An example of this type of automation relating to teleoperation would be the control of a high-speed missile or a modern fighter aircraft, where the time delay in operator control could cause instability to the flight dynamics of the system. Thus, some teleoperated systems require automation in order to maintain stable and consistent performance, where human intervention could cause instability and inconsistency.

(ii) *Addition*. The second category of automation concerns systems which perform functions that are demanding or intrusive for human performance. In other words, performance of these functions might cause undue increases in workload or might lead to disruption of primary task activity. An example of such automation is the ground proximity warning system (GPWS), which provides an auditory warning when a combination of measures indicates that the air vehicle is unacceptably close to the ground [11]. In this example, the automation continually monitors specific variables and presents an alert to the operator when the values exceed a defined threshold.

(iii) *Augmentation*. The third example of automation can be considered as a form of augmentative technology. In other words, automation is used to complement and support human activity, particularly where the human is fallible or limited. One example of this is the use of automation to reduce clutter on a display, for example by merging tracks on a radar screen on the basis of intelligent combination of data from different sources. This could prove beneficial in teleoperation, particularly if several UVs are to be tracked and managed. However, this can lead to problems as the data fusion could reduce system transparency and remove the operator from the control loop [12].

In his review of human–machine systems built between the mid-1940s and mid-1980s, Sheridan [13] proposed four trends. The first trend was the removal of the human operator either 'up' (super-) or 'away' (tele-) from direct operation of the system under control. This meant significant changes in the form of control the operator could exercise over the system. The most obvious consequence of these changes was the reduction in direct physical and local contact with the system. This suggests the issue of engagement between human and system as a central theme. The question of engagement with teleoperated systems raises specific human factor questions about the operation of the system and about the type of feedback that the human operator can receive. This in turn leads to questions about the design of the control loop and the operator's position in that control loop. For example, a teleoperated system could be capable of fully autonomous flight, leaving the operator with the task of monitoring performance. The notion of a fully autonomous system introduces Sheridan's concept of super-operated systems, where the operator is effectively a reporting-line superior to the UAV – a manager who requires achievement of a specific goal without dictating means of achievement.

The second trend identified by Sheridan was the need to maintain systems with multiple decision-makers, each with partial views of the problem to be solved and the environment in which they were performing. This raises questions about teamwork and collaborative working. Thus, one needs to consider the role of teleoperation within broader military systems, particularly in terms of communications [14]. From this approach, one ought to consider the role of the UAV within a larger system, for example, if the UAV is being used to gather intelligence, how will this intelligence be communicated from the operator to relevant commanders? This might require consideration not only of the workstation being used to operate/monitor the UAV, but also the means by which information can be compiled and relayed to other parties within the system.

From a different perspective, if the role of the operator is reduced to monitoring and checking (possibly leading to loss of appreciation of system dynamics), operational problems may be exacerbated by the operator's degradation of skill or knowledge. While it might be too early to consider possible socio-technical system design issues surrounding teleoperation systems, it is proposed that some consideration might be beneficial. A point to note here, however, is that allocation of function often assumes a sharing of workload. Dekker and Wright [15] make the important point that not only will tasks be allocated to change workload levels, but that the very sharing of tasks will significantly transform the work being performed (by both the machine and the human). In a review of 'adaptive aiding' systems, Andes [16] argues that specific human factor issues requiring attention include: the need to communicate the current aiding mode to the operator, the use of operator preference for aiding (as evidenced by the studies reviewed above), and human performance effects of adding or removing aiding.

The third trend identified by Sheridan relates to the shift in operator skill from direct (and observable) physical control to indirect (unobservable) cognitive activity, often related to monitoring and prediction of system status. This relates, in part, to the question of engagement raised above. However, this also relates to the issue of mental models held by operators and the provision of information from the system itself.

This chapter has already noted the shift from physical to cognitive interaction between human operators and UAVs, leading to a tendency towards supervisory control. Ruff *et al.* [17] compared two types of supervisory control of UAVs – management by consent (where the operator must confirm automation decisions before the system can act) and management by exception (where the system acts unless prevented by the human operator). While management by consent tended to result in superior performance on the experimental tasks, it also (not surprisingly) resulted in higher levels of workload, especially when the number of UAVs under control increased from 1 to 4. In a later study, Ruff *et al.* [18] found less of an effect of management by consent. Indeed, the latter study suggested that participants were reluctant to use any form of automated decision support.

5.3   Control of Multiple Unmanned Vehicles

In their analysis of operator performance, Cummings and Guerlain [19] showed that it was possible for an operator to effectively monitor the flight of up to 16 missiles, with more than this amount leading to an observed degradation in performance. They compared this with the limit of 17 aircraft for air traffic controllers to manage. In a modelling study, Miller [20] suggested a limit of 13 UAVs to monitor. However, it should be noted that monitoring the movement of vehicles in airspace was only part of a UAV operator's set of tasks. When the number of control tasks increases, or with the demands of monitoring multiple information sources, it is likely that the number of UAVs will be much smaller. Indeed, Galster *et al.* [21] showed that operators could manage 4, 6 or 8 UAVs equally well unless the number of targets to monitor also increased. Taylor [22] suggested that 4 UAVs per operator is a typical design aim for future systems. Cummings *et al.* [23] presented a meta-review of UAVs with different levels of automation, and suggested that studies converge on 4–5 vehicles when control and decision-making were primarily performed by the operator (but 8–12 when UAVs had greater autonomy). Their model of operator performance suggested that control of 1 UAV is superior to 2, 3 or 4 (which result in similar performance), with degradation in performance when controlling 5 or more UAVs. Liu *et al.* [24] found significant impairment in performance on secondary tasks (response to warnings or status indicators) when operators controlled 4 UAVs, in comparison with controlling 1 or 2 UAVs. Research therefore clearly demonstrates that the level of automation applied to unmanned vehicles needs to be considered when determining the number of autonomous vehicles an operator can control effectively, and there is growing consensus that operators struggle to monitor more than 4 UAVs.

5.4   Task-Switching

The concurrent management and control of multiple UAVs by human operators will lead to prolonged periods of divided attention, where the operator is required to switch attention and control between UAVs. This scenario is further complicated by additional secondary tasks that will be required, such as the operator communicating with other team members. A consequence of switching attention in a multitask setting is the difficulty experienced by an operator in reverting their attentional focus back to the primary task at hand, such as controlling unmanned assets in a timely manner affording the appropriate identification, detection and response during critical situations [25]. Whether or not an individual can effectively conduct two or more tasks at once has been the subject of much basic psychological research. Basic research in dual-task interference has highlighted that people have difficulty in dual-task scenarios, despite a common misconception that they can perform tasks simultaneously. This is found to be the case, even with tasks that are relatively simple compared to the complex dynamic conditions within a military context (e.g. [26, 27]). Laboratory research on the consequences of task-switching highlights that the responses of operators can be much slower and more error-prone after they have switched attention to a different task (e.g. [28–30]).

Squire *et al.* [31] investigated the effects of the type of interface adopted (options available to the operator), task-switching and strategy-switching (offensive or defensive) on response time during the simulated game, RoboFlag. Response time was found to be slower by several seconds when the operator had to switch between tasks, in particular when automation was involved. Similarly, switching between offensive and defensive strategies was demonstrated to slow response time by several seconds. Therefore the task-switching effect is also seen when operators managing multiple UVs switch between different strategies. However, when operators used a flexible delegation interface, which allowed operators to choose between a fixed sequence of automated actions or selectable waypoint-to-waypoint movement, response time decreased even when task or strategy switching was required. This advantage was ascribed to operators recognizing conditions where the automation was weak and thus needed to be overridden by tasking the unmanned vehicles in a different way.

Chadwick [32] examined operator performance when required to manage 1, 2 or 4 semi-autonomous UGVs concurrently. Their performance on monitoring, attending to targets and responding to cued decision requests and detecting contextual errors was assessed. Contextual errors came about when the UGV was operating correctly but inappropriately given the context of the situation. One example is when the navigation system of a UGV fails and the operator is required to redirect the UGV to the correct route. The UGV was unable to recognize contextual errors and couldn't alert the operator and so it was down to the operator to recognize and respond to it. The tasks participants were asked to conduct included the detection and redirection of navigation errors as well as attending to targets. An attentional limitation was evident when operators were required to detect contextual errors, the detection of which was found to be very difficult when control of multiple UGVs was expected. When 1 UGV was controlled, contextual errors were spotted rapidly (within 10 seconds) whereas when 4 UGVs were controlled, the response to these errors slowed to nearly 2 minutes. Chadwick argued that having to scan the video streams from 4 UGVs inhibited operators from focusing attention on each display long enough to understand what was going on in each display.

Task-switching has been found to have an impact on situation awareness (SA). For example, when operators are required to switch attention from a primary task (for example, the supervisory control of UAVs) to an intermittent secondary task (for example, a communication task) SA is reduced when they switch their attention back to the primary task [33, 34]. There is also evidence to suggest that task-switching may result in change blindness, a perceptual phenomenon which refers to an individual's inability to detect a change in their environment. This perceptual effect in turn may have an affect on SA. The impact of this phenomenon within a supervisory control task was investigated by Parasuraman *et al.* [35]. The study involved operators monitoring a UAV and a UGV video feed in a reconnaissance tasking environment. The operators were asked to perform four tasks, of which target detection and route planning were the primary tasks. A change detection task and a verbal communication task were used as secondary tasks to evaluate SA. These latter two tasks interrupted the two former (primary) tasks. The routes for the UAV and UGV were programmed and so participants only had to control a UAV if it needed to navigate around an obstruction. For the change detection tasks, participants were asked to indicate every time a target icon, that they had previously detected, had unexpectedly changed position on a map grid. Half of these changes took place when participants were attending to the UAV monitoring task and the other half occurred during a transient event, when the UGV stopped and its status bar flashed. The results demonstrated the low accuracy of participants at detecting changes in the position of target icons, in particular during the transient events. Parasuraman *et al.*'s results indicate that most instances of change blindness occurred in the presence of a distractor (a transient event). However, change blindness was also observed when participants switched their attention from monitoring the UAV to monitoring the UGV.

In the wake of the current military need to streamline operations and reduce staff, research has been conducted on designing systems that will allow a shift in the control of a single UAV by multiple operators to a single operator. Similarly, there is a drive towards single operators controlling multiple vehicles, which applies to land, air and underwater vehicles [23]. This will require UAVs to become more autonomous and the single operator would be expected to attend to high-level supervisory control tasks such as monitoring mission timelines and responding to mission events and problems as they emerge [36, 37]. Each UV carries out its own set plan, and as a result a single operator may experience high workload when critical tasks for more than one UV require their attention concurrently. This situation can result in processing 'bottlenecks'. The use of an automated agent to signal the likelihood of processing bottlenecks occurring has been shown to aid the operator by affording them the opportunity to prepare a course of action to lessen the effects of any bottleneck, if necessary [36]. For example, Cummings *et al.* [37] conducted a decision support experiment that examined the effectiveness of an intelligent schedule management support tool that provided different levels/types of automated decision support and alerted operators to potential scheduling conflicts for 4 UAVs. The effectiveness of a timeline display for each UAV which included an intelligent configurable display was examined. The configurable display, presenting potential scheduling conflicts, was called 'Star Visualization' (StarVis). A configurable display was a single geometrical form that mapped multiple variables onto it and changes in these individual variables resulted in form variation. The variables incorporated within the display were, for example, the type of schedule problem (late arrival or time-on-targets conflict) and the number of targets involved in a specific problem type and their relative priorities (low, medium or high). The configurable display featured a grey rectangle, which represents the ideal state. As problems were detected by the automation, grey triangles appeared on the display for one or more target problems.

The design of the configurable display was such that the emerging features (grey triangles) identified potential scheduling conflicts and these features influenced direct perception/action. This provided operators with the capability to use more effective perceptual processes as opposed to adopting cognitively demanding processes that require, for example, memory. This configuration also allowed operators to denote whether or not scheduling conflicts were present not only for one, but for all the UAVs at a glance.

The StarVis not only displayed current scheduling conflicts but also provided operators with the opportunity to view the effect of an accepted time-on-target delay request, for example, before they performed any action. This aided the decision-making process. Two different implementations of the configurable display were examined, the local StarVis and the global StarVis. In the local StarVis version, each UAV timeline only displayed problems and how time-on-target delay requests affected that specific UAV's schedule. Conversely, the global StarVis only displayed how a time-on-target delay request for a single UAV would impact on all UAVs. Cummings *et al.* [37] found that operators presented with the local StarVis version of the decision support aid performed better than those provided with the global StarVis version and no visualization for support. Moreover, when future problems for all 4 UAVs were presented in the global context, the performance of operators fell to a level that was equivalent to performance found when no decision support was provided. The fact that the global display did not aid performance was explained with reference to the fact that the global visualization presented information that was not vital in the decision-making process. In addition, operators had to look at all UAV displays to see the consequences of a decision. The local StarVis presented 'just enough' information to the operator to allow for efficient decision-making, a characteristic that would be effective in a time-critical command and control environment. Cummings *et al.* [37] suggest that these findings highlight the difficulties surrounding the design of automated decision support tools, particularly with respect to how information is mapped and presented on a display.

Research has indicated ways in which UV systems can be designed so as to support an operator's performance on high-level supervisory control tasks for multiple UVs by enabling them to manage the workload across UV systems effectively. Work with delegation interfaces showed that providing an operator flexibility in their decision-making on how to maintain awareness of the state of a UV or how to task it reduced the negative effects of task-switching [31]. Cummings *et al.* [37] demonstrated how decision support tools can help operators perform high-level supervisory control tasks for multiple UVs by helping them manage the workload effectively. Moreover, they recommend that decision support displays be designed to leverage the ability of operators to not only notice and identify the nature of a problem but also to help operators solve problems. Thus, any decision support tool adopted must present useful solutions to emerging critical events as opposed to only displaying visualizations of potential critical events requiring attention.

5.5 Multimodal Interaction with Unmanned Vehicles

There have been several studies of the potential application of multimodal display in the control and monitoring of UAVs (as the following review illustrates) but less work to date on the potential for multimodal control of payload. In such systems, feedback is typically visual and audio, although there is growing interest in uses of haptic feedback. Multimodal display appeared to both reduce UAV operator workload and provided access to multiple streams of information [38–41]. Auditory presentation of *some* information can be combined with ongoing visual tasks [42], and these improvements can be particularly important when dealing with multiple UAVs, providing they do not interfere with other auditory warnings [43]. However, combining the control of a UV with other tasks can impair performance on target detection [38, 44] and reduce SA [45]. Chen [46] reported studies in which aided target recognition (AiTR) significantly enhanced operators' ability to manage the concurrent performance of effector and vehicle control, in comparison with performing these tasks with no support.

Draper *et al.* [47] compared speech and manual data entry when participants had to manually control a UAV, and found speech yielded less interference with the manual control task than manual data entry. Chen *et al.* [48] showed that target detection was significantly impaired when participants had to combine search with control of the vehicle, in comparison with a condition in which the vehicle was semi-autonomous. Baber *et al.* [49] looked at the use of multimodal human–computer interaction for the combined tasks of managing the payload of an autonomous (simulated) UV and analysing the display from multiple UVs. Speech was the preferred mode of choice when issuing target categorization, whereas manual control was preferred when issuing payload commands. Speech combined with gamepad control of UVs led to greater performance on a secondary task. Performance on secondary tasks degraded when participants were required to control 5 UVs (supporting work cited earlier in this chapter).

The support for the benefits of multimodal interaction with UAVs (for both display of information and control of payload) not only speaks of the potential for new user interface platforms but also emphasizes the demands on operator attention. A common assumption, in studies of multimodal interfaces, is that introducing a separate modality helps the operator to divide attention between different task demands. The fact that the studies show benefits of the additional modality provides an indication of the nature of the demands on operator attention. Such demands are not simply 'data limited' (i.e., demands that could be resolved by modifying the quality of the data presented to the operator, in terms of display resolution or layout) but also 'resource limited' (i.e., imply that the operator could run out of attentional 'resource' and become overloaded by the task demands).

5.6   Adaptive Automation

The supervision of multiple UVs in the future is likely to increase the cognitive demands placed on the human operator and as such timely decision-making will need to be supported by automation. The introduction of decision aids into the system is likely to increase the time taken for tactical decisions to be made. These decision aids will be mandated owing to the high cognitive workload associated with managing several UVs [19]. It is important to consider the human–automation interaction in terms of how information gathering and decision support aids should be automated and at what level (from low (fully manual) to high (fully autonomous) automation). Consideration of the type of automation required is also paramount. The type and level of automation can be changed during the operation of a system, and these systems are referred to as adaptive or adaptable systems. Parasuraman *et al.* [50] argued that decision support aids should be set at a moderate level of automation, whereas information gathering and analysis functions can be at higher levels of automation, if required. However, the human operator is ultimately responsible for the actions of a UV system and thus even highly autonomous assets require some level of human supervision.

A level of human control is especially important when operating in high-risk environments, such as military tasks involving the management of lethal assets where, due to unexpected events that cannot be supported by automation, the operator would require the capability to override and take control. Parasuraman *et al.* therefore proposed moderate levels of automation of decision support functions because highly reliable decision algorithms cannot be assured and are thus coined 'imperfect (less than 100%) automation'. In a study by Crocoll and Coury [51], participants were given an air-defence targeting task that required identification and engagement. The task incorporated imperfect automation where participants received either status information about a target or decision automation which provided recommendations regarding the identification of a target. Decision automation was found to have a greater negative impact on performance than information automation. Crocoll and Coury argued that when provided with only information concerning the status of a target, participants use this information to make their own decisions. This is because information automation is not biased towards any decision in particular, presenting the operator with the raw data from which they can generate alternative choices and thus lessen the effects of imperfect automation. Furthermore, the costs of imperfect decision support are observed for various degrees of automation [52]. Moderate levels allow the human operator to be involved in the decision-making process and ultimately the decision on an action is theirs. Moreover, automation in systems has not always been found to be successful, with costs to performance stemming from human interaction with automated systems which have involved unbalanced cognitive load, overreliance and mistrust [53]. Therefore, the analysis of an appropriate level of automation must consider the impact of imperfect automation, such as false alarms and incorrect information, on human–system interaction [50].

Context-sensitive adaptive automation has been shown to mitigate the issue of skills fade, reduced SA and operator overreliance arising from static (inflexible) automation introduced into systems [54–56]. In contrast to decision aids or alerts implemented in static automation, those presented using adaptive automation are not fixed at the design stage. Rather, their delivery is dictated by the context of the operational environment. The delivery of this adaptive automation by the system is based on operator performance, the physiological state of the operator or critical mission events [57].

The flexible nature of adaptive automation allows it to link with the tactics and doctrine used during mission planning. To illustrate, adaptive automation based on critical events in an aircraft air defence system would initiate automation only in the presence of specific tactical environmental events. In the absence of these critical events the automation is not initiated. To our knowledge a small number of empirical studies on adaptive automation to aid human management of UVs have been conducted (for example, [35, 58, 59]. Parasuraman *et al.* [35] presented adaptive support to participants who had to manage multiple UVs. This adaptive automated support was based on the real-time assessment of their change detection accuracy. Parasuraman *et al.* [35] compared the effects of manual performance, static (model-based) automation and adaptive automation on aspects of task performance including change detection, SA and workload in managing multiple UVs under two levels of communications load. Static (model-based) automation was invoked at specific points in time during the task based on the prediction of the model that human performance was likely to be poor at that stage. In contrast, adaptive automation was performance-based and only invoked if the performance of the operator was below a specified criterion level. The results demonstrated that static and adaptive automation resulted in an increase in change detection accuracy and SA and a reduction in workload compared to manual performance. Adaptive automation also resulted in a further increase in change detection accuracy and an associated reduction in workload in comparison to performance in the static condition. Participants also performed better on the communications task, providing further evidence that adaptive automation acts to reduce workload. Participants using static automation responded more accurately to communications than did those using manual performance. Parasuraman *et al.*'s findings demonstrate that adaptive automation leads to the availability of more attentional resources. Moreover, its context-sensitive nature supports human performance, in this case, when the change detection performance of participants is reduced indicating reduced perceptual awareness of the evolving mission events.

Parasuraman *et al.* [50] point out issues with adaptive automation. First, system unpredictability may impact on operator performance. Second, systems designed to reduce cognitive load may actually work to increase it. User acceptance is a problem raised in systems that implement adaptive automation, where the decision to invoke automation or pass control back to the operator is made by the system. Operators who see themselves as having manual control expertise may not be willing to comply with the authority of a system. These potential limitations highlight the issue of the efficacy of adaptive automation when compared to adaptable systems, where the human operator decides when and at which point in time to automate. However, providing the human operator with the capability to make decisions on automation may also act to increase workload. Therefore, Miller and Parasuraman [60] argue for a trade-off between increased workload versus increased unpredictability where automation is started by the system or by the human operator.

5.7   Automation and Multitasking

Several studies have investigated the influence of different factors on the ability of individuals to concurrently manage or operate unmanned vehicles whilst performing other primary tasks. Mitchell [61] carried out a workload analysis, using the Improved Performance Research Integration Tool (IMPRINT), on crewmembers of the Mounted Combat System (MCS). The MCS is a next generation tank, which forms part of the US Army Future Combat System (FCS).

The FCS vision for the MCS is that it will be operated by three soldiers (vehicle commander, gunner and driver), one of which will be required to concurrently operate the UGV. As no operator is dedicated to the operation and control of the platoon's UGV, Mitchell modelled the workload of each crewmember in order to examine how the UGV could be most effectively used and who would be most appropriate to operate it. The gunner was observed to experience the least instances of work overload and was able to effectively assume control of the UGV and conduct the secondary tasks associated with its operation. However, scenarios that required the teleoperation of the UGV led to consistent instances of cognitive overload, rendering concurrent performance of the primary tasks of detecting targets more difficult. Mitchell's IMPRINT analysis also identified instances where the gunner ceased to perform their primary tasks of detecting and engaging targets in order to conduct UGV control tasks, which could have serious consequences during a military operation. This is supported by research demonstrating that target detection is lower when combined with teleoperation of a UGV in comparison to when the UGV is semi-autonomous [48]. UGV operation requires more attention, through more manual operation (teleoperation) and/or manipulation (for example, via an interface to label targets on a map) than simply monitoring information on a display. Moreover, situation awareness of UGV operators has been observed to be better when the UGV has a higher level of automation [45].

The level of automation and consideration of the cognitive capabilities of the human, in particular the demands placed on attention, is of critical importance. In general, lower levels of automation result in a higher workload whereas higher levels of automation produce a lower workload [62]. As discussed, research on adaptive systems indicates moderate levels of workload will produce optimal human performance. Unfortunately, automation has not always led to enhancement in system performance, which is mainly due to problems in using automated systems experienced by human operators. Examples of human–automation interaction problems stemming from a level of automation which is too low are: cognitive overload in time-critical tasks, fatigue and complacency and an increase in human interdependency and decision biases [63, 64]. In contrast, problems with human–automation interaction resulting from a level of automation which is too high are: an increase in the time taken to identify and diagnose failures and commence manual take-over when necessary, cognitive and/or manual skill degradation and a reduction in SA [63]. Thus, given the consequences of introducing an inappropriate level of automation into a system, automation should only be utilized when there is a requirement for its introduction.

Simulation experiments have been conducted in order to validate the results of Mitchell's [61] IMPRINT analysis and to assess the effect of AiTR for the combined job of gunner and UGV operator. Chen and Terrence [65] found that assisting the gunnery task using AiTR information significantly enhanced performance. Participants in this multitasking scenario not only had to detect and engage hostile targets but also neutral targets, which were not cued by the AiTR system. Significantly fewer neutral targets were detected when participants were required to teleoperate the UGV or when AiTR was employed to aid the gunnery tasks. This was taken to indicate that some visual attention had been removed from the primary gunnery task. A plausible explanation put forward by Chen and Terrence is that cueing information delivered by the AiTR augmented the participants' ability to switch between the primary gunnery task and secondary task of controlling the UGV. Whilst enhancing task-switching capability, AiTR assistance had a negative impact on the detection of the neutral targets. In addition to the enhancement of gunnery (primary) task performance aided by AiTR, participants' concurrent (secondary) task performance was observed to improve when the gunnery task was aided by AiTR. This was found for both UGV control and communication tasks. The work of Chen and Terrence demonstrated how reliable automation can improve performance on an automated primary task and a simultaneous secondary task.

Chen and Joyner [66] also conducted a simulation experiment to examine both performance and cognitive workload of an operator who performed gunner and UGV tasks simultaneously. They found that performance on the gunnery task degraded significantly when participants had to concurrently monitor, control or teleoperate an unmanned asset relative to their performance in the baseline control condition featuring only the gunnery task. Furthermore, the degradation in gunnery task performance was a function of the degree of control of the UV, such that the lowest level of performance was observed when the gunner had to teleoperate the unmanned asset concurrently. Looking at the concurrent task of operating the UV, performance was worst when the UGV that participants had to control was semi-autonomous. This finding was interpreted as indicative of an increased reliance on automation in a complex high-load multitasking scenario and of a failure to detect more targets that were not cued by automation. In contrast to Chen and Terrence [65], the semi-autonomous system in Chen and Joyner's simulation experiment was imperfectly reliable and thus seems more representative of real-world environments where AiTR systems are never perfectly reliable.

Research has also examined how unreliable AiTR systems can moderate task performance of operators. There are two types of alert in unreliable AiTR systems, those that deliver false alarms (signal the presence of a target when no target is present) and those that present misses (fail to alert an operator to the presence of a target). Research has demonstrated that the performance of participants on an automated primary task, such as monitoring system failures, is degraded when the false alarm rate is high [67]. This illustrates how high false alarm rates reduce operator compliance in false alarm prone (FAP) AiTR systems and as a consequence operators take fewer actions based on the alerts presented. High miss rates, however, impair performance on a concurrent task more than on an automated primary task as participants have to focus more visual attention to monitor the primary task. As a result, the reliance on automation is reduced. That is, there is a reduction in the assumption that an automated system is working properly and the failure to take precautionary actions when no alerts are emitted [67, 68]. However, research has refuted the idea that FAP and miss prone (MP) AiTR systems have independent effects on primary and concurrent task performance and how personnel use AiTR systems. A greater degradation in performance on automated tasks has been observed in FAP systems relative to that found in MP systems [69]. Nevertheless, performance on a concurrent robotics task was found to be affected equally negatively by both FAP and MP alerts. Both operator compliance and reliance was moderated by FAP systems, whereas automation producing high miss rates was found to only influence the reliance of the operator on the AiTR system [69].

5.8 Individual Differences

5.8.1 Attentional Control and Automation

Basic research has demonstrated that performance in multitasking environments is mediated by individual differences such that some individuals are less susceptible to performance impairment during multitask scenarios. For example, Derryberry and Read [70] conducted an experiment which looked at anxiety-related attentional biases and how these are regulated by attentional control during a spatial orienting task. They demonstrated that individuals with better attentional control could allocate their attentional resources more effectively and efficiently and were found to be better at resisting interference in a spatial orienting task.

Chen and Joyner [66] examined whether operators with higher perceived attentional control (PAC) could perform a gunnery and UGV control task better relative to operators exhibiting lower attentional allocation skills when they also had to perform a concurrent intermittent communication task. This concurrent task simulated the communication that would go on between the gunner and other tank crew members. PAC was measured using The Attention Control Questionnaire [70], which consists of 21 items that measure attentional control in terms of attention focus and shifting. Derryberry and Read [70] reported that factor analysis has confirmed the scale measures the general capacity for attentional control and has revealed correlated sub-factors associated with attention focus and attention shifting. An example of a sub-factor linked to attention focus is, 'My concentration is good even if there is music in the room around me' [70, p. 226]. An example of an item measuring attention shifting is, 'It is easy for me to read or write while I'm also talking on the phone' [70, p. 226].

Chen and Joyner's [66] findings provided partial support for the notion that operators who have higher attentional control are better able to allocate their attention between tasks. Operators reporting higher PAC performed the concurrent (communication) task more effectively in comparison to lower PAC operators, in particular when the UGV control tasks required more manipulation and attention. However, there was no significant difference between low and high PAC individuals in terms of performance on the gunnery and UGV control tasks. Chen and Joyner argue that the operators channelled most of their attentional resources into the gunnery and UGV control tasks (even more so for the UGV teleoperation) and that only operators with higher PAC could conduct the communication task more effectively. During the single gunnery task (baseline) and the UGV monitoring task, the high PAC and low PAC individuals displayed an equivalent level of performance for the communication task. Thus it seems that monitoring the video feed from a UGV left operators with sufficient visual attentional resources to perform the communication task.

Chen and Terrence [71] investigated the effect of unreliable automated cues in AiTR systems on gunners' concurrent performance of gunnery (hostile and neutral target detection), UGV operation (monitor, semi-autonomous, teleoperation) and communication tasks. Moreover, Chen and Terrence examined whether participants with different attentional control skills react differently to FAP and MP AiTR systems. In other words, whether the reaction of low PAC and high PAC participants to automated cues delivered by a gunnery station differed depending on whether the system was FAP or MP. Following the methodology employed by Chen and Joyner [66], Chen and Terrence simulated a generic tank crew station setting and incorporated tactile and visual alerts to provide directional cueing for target detection in the gunnery task. The directional cueing for target detection was based on a simulated AiTR capability. The detection of hostile targets for the gunnery task was better when operators had to monitor the video feed of a UGV in comparison to when they had to manage a semi-autonomous UGV or teleoperate a UGV. This result is consistent with Chen and Joyner's findings and further supports the notion that operators have a lower cognitive and visual workload when required to only monitor the UGV video feed, affording more cognitive resources to be allocated to the gunnery task. Further, Chen and Terrence observed a significant interaction between type of AiTR automation system and the PAC of participants for hostile target detection. Individuals found to have high PAC did not comply with FAP alerts and did not rely on automation, detecting more targets than were cued when presented with MP alerts. This is in line with the idea that operator's compliance with and reliance on AiTR systems are independent constructs and are affected differently by false alarms and misses (e.g. [67]). The picture was rather different for those with low PAC. In the FAP condition, low PAC operators demonstrated a strong compliance with the alerts and as a consequence this led to good target detection performance. In contrast, with MP automation, low PAC operators relied too much on the AiTR which resulted in very poor performance. As workload became heavier (e.g. with more manual manipulation of the UGV), low PAC operators became increasingly reliant on automation, whereas operators with strong attention shifting skills retained a relatively stable level of reliance during the different experimental conditions.

Considering neutral target (not cued by the AiTR system) detection during the gunnery task, Chen and Terrence [71] found that when gunners had to teleoperate a UGV their detection was much poorer in comparison to when gunners had to manage a semi-autonomous UGV. This is consistent with Chen and Joyner's [66] finding, that operators allocated much less attentional resources to the gunnery task when the concurrent UV task required manual manipulation. Operators with low attention allocation skills performed at an equivalent level, independent of the AiTR system they were exposed to. However, operators with higher PAC displayed greater target detection performance when the AiTR system was MP, indicating that individuals with high PAC allocated more attentional resources to the gunnery task because they relied less on the MP cues for target detection.

For the concurrent robotics task, the highest level of performance observed by Chen and Terrence [71] was when the operator had only to monitor a video feed from a UGV. The type of AiTR received had no effect on performance in this condition. As observed for hostile and neutral target detection during the gunnery task, a greater adverse effect of MP cueing was observed when the concurrent robotics task became more challenging. When teleoperation of a UGV was required, MP cueing produced a larger performance decrement than did FAP. A significant interaction between type of AiTR and PAC was also found. Low PAC individuals' performance was worse when presented with a MP AiTR system; however, FAP alerts received for the gunnery task improved their concurrent task performance. In contrast, high PAC individuals were less reliant on MP cues and thus demonstrated better concurrent task performance. Also, high PAC operators complied less with FAP cues, although this did not result in improved performance. Performance for the tertiary communication tasks was also moderated by the complexity of the robotics task. Better communication performance was demonstrated when operators were required to monitor the video feed of a UGV than when a UGV was teleoperated, as observed by Chen and Joyner [66]. Chen and Terrence [71] suggest the information-encoding processes required for manipulating a UGV during teleoperation demand more attention and are more susceptible to the impact of competing requirements in multitask scenarios.

It seems that reliance on automation during multitask scenarios is influenced by PAC. Only operators with a low attention shifting ability seemed to rely on AiTR in a heavy workload multitasking environment [67, 71]. MP alerts seemed to be more detrimental to the performance of low PAC operators than FAP alerts [71], whereas FAP cues impaired performance on both automated (gunnery) and concurrent tasks in those with high PAC more than MP alerts [69, 71]. Low PAC operators seemed to trust automation more than high PAC operators and found performing multiple tasks simultaneously more difficult, leading to an overreliance on automation when available. High PAC individuals, however, displayed reduced reliance on MP automation and seemed to have greater self-confidence in their ability to multitask in a complex environment. These findings therefore suggest that PAC can mediate the link between self-confidence and degree of reliance.

5.8.2 Spatial Ability

There is a growing body of research that discusses the influence of spatial ability (SpA) in the context of unmanned system performance and operations. Spatial ability can be divided into two sub-components: spatial visualization and spatial orientation (e.g. [72, 73]). Ekstrom *et al.* [72] defined spatial visualization as the 'ability to manipulate or transform the image of spatial patterns into other arrangements', and spatial orientation as the 'ability to perceive spatial patterns or to maintain orientation with respect to objects in space'. Previous research has shown that these two sub-components are distinct [74, 75].

Spatial ability has been found to be a significant factor in military mission effectiveness [76], visual display domains [77], virtual environment navigation [78], learning to use a medical teleoperation device [79], target search task [48, 65, 66, 71] and robotics task performance [80].

Lathan and Tracey [81] found that people with higher SpA completed a teleoperation task through a maze faster and with fewer errors than people with lower SpA. They have recommended that personnel with higher spatial ability should be selected to operate UVs. Baber *et al.* [49] have shown that participants with low spatial ability also exhibit greater deterioration in secondary task performance when monitoring 5 UVs; this could relate to the challenge of dividing attention across several UVs.

Chen *et al.* [48] found that people with higher SpA performed better on a situational awareness task than people with lower SpA. Chen and Terrence [65] also found a significant correlation between SpA and performance when performing concurrent gunnery, robotics control and communication tasks but when aided target recognition was available for their gunnery task, the participants with low SpA performed as well as those with high SpA. However, the test used to measure the relationship between spatial ability and performance is also important. For example, Chen and Joyner [66] used two different tests to measure spatial ability: the Cube Comparison Test (CCT) [82] and the Spatial Orientation Test (SOT) based on Gugerty and Brooks' [83] cardinal direction test. They found that the SOT was an accurate predictor of performance but that the CCT was not. One possibility is that these two tests measured the two sub-components of spatial ability and that the task to be completed by the participants correlated with one sub-component but not the other (see [74, 84] for a discussion of tests measuring each of the sub-components).

Research has shown spatial ability to be a good predictor of performance on navigation tasks. Moreover, individuals with higher spatial ability have been shown to perform target detection tasks better than people with lower spatial capability [46]. Chen [85] also found that participants with higher SpA performed better in a navigation task and that these participants also reported a lower perceived workload compared to participants with lower SpA. Similarly, Neumann [86] also showed that higher SpA was associated with a lower perceived workload in a teleoperating task.

Chen and Terrence [65, 71] observed that the type of AiTR display preferred by operators is correlated with their spatial ability, such that individuals with low SpA prefer visual cueing over tactile cueing. However, in environments that draw heavily on visual processing, tactile displays would enhance performance and be more appropriate as operators would be able to allocate visual attention to the tasks required and not to the automated alerts.

5.8.3   Sense of Direction

Another individual difference that has been studied is sense of direction (SoD). Contrary to expectations, SoD and spatial ability are only moderately, if at all, correlated [87]. Self-reported SoD is measured with a questionnaire and is defined by Kozlowski and Bryant [88, p. 590] as 'people's estimation of their own spatial orientation ability'. A study on route-learning strategies by Baldwin and Reagan [89] compared people with good and bad SoD. SoD was measured with a questionnaire developed by Takeuchi [90] and later refined by Kato and Takeuchi [91]. To do so, they used participants who scored at least one standard deviation above or below the mean of the larger sample including 234 respondents. This resulted in a sample of 42 participants, 20 participants with good SoD and 22 participants with poor SoD. The experiment involved a route-learning task and a concurrent interference task (verbal or spatial). Individuals classified as having good SoD traversed the routes faster and with fewer errors when learned under conditions of verbal interference relative to under conditions of visuospatial interference. Conversely, individuals with poor SoD were faster to go through the routes when they were learned under conditions of visuospatial interference relative to verbal interference. This interaction between SoD and the type of interference can be explained by the fact that good navigators tend to make greater use of survey strategies (cardinal directions, Euclidean distances and mental maps) which rely on visuospatial working memory whilst poor navigators tend to use ego-centred references to landmarks which rely on verbal working memory. When the interference task taps onto the same type of working memory used to perform navigation the performance drops, whereas the performance is less influenced by the other interference tasks. Also, individuals with better SoD learned the routes in fewer trials than individuals with poorer SoD. Finally, as with spatial ability, Baldwin and Reagan found that individuals with poor SoD reported a higher perceived workload than individuals with good SoD and the two SoD groups didn't differ in spatial ability as measured by the Mental Rotation Task [92].

Chen [85] also looked at SoD in a navigation task and found that it was correlated with map-marking scores, with participants having a higher SoD performing better. SoD was also negatively correlated with target search time, with participants having a better SoD taking less time than poorer participants. There was also an interaction between participants' SoD and the lighting conditions. In the night condition, those with poor SoD took significantly more time than those with good SoD to find targets. However, in the day condition, no difference was found between participants with poor and good SoD. These results are in line with those reported in Baldwin and Reagan [89] showing that participants with poor SoD relied on landmarks and verbal working memory to perform a navigation task. As landmarks are a lot more difficult to use at night, the difference between participants using visuospatial and verbal working memory became apparent in this condition whereas during the day, the participants using verbal working memory (poor SoD) managed to perform as well as the participants using visuospatial working memory (good SoD).

5.8.4   Video Games Experience

Video games experience can have an effect on an operator's performance when remotely operating a UAV. A laboratory-based study by De Lisi and Cammarano [93] looked at the effect of playing video games on spatial ability. Participants who took part in this study were administered the Vanderberg Test of Mental Rotation (VTMR), a paper-and-pencil assessment of mental rotation, before and after two 30-minute experimental sessions. During these two sessions, half of the participants played the computer game *Blockout* (which requires mental rotation of geometric figures) whilst the other half played *Solitaire* (which is a game that does not require mental rotation). The interesting result was that the average score on the VTMR increased significantly in the space of a week and with only two sessions of 30 minutes when participants were playing *Blockout*. Moreover, the score on the pretest was positively correlated with the participant's reported computer usage. Similarly, Neumann [86] found that video games experience correlated with the number of targets detected and the number of collisions in a teleoperating task but did not correlate with the time to complete the task.

Finally, a study by Chen [85] found a significant difference in map-marking between men and women, with the men performing better than the women. When looking only at the men's performance, those who played video games more frequently performed better in map-marking accuracy than those who played video games less frequently. Moreover, 61% of the male participants and only 20% of the female participants played video games frequently, which could point to an effect of video games experience rather than a pure gender effect.

5.9   Conclusions

Many of the human factors challenges of interacting with UAVs have been comprehensively reviewed by Chen *et al.* [94]. These issues include the bandwidth, frame rates and motion artefacts of video imagery presented to the operator; time lags between control actions and system response; lack of proprioception and tactile feedback to operators controlling UVs; frame of reference and two-dimensional views for operators monitoring UVs; switching attention between different displays. In this chapter, we offered additional considerations of the potential uses of multimodal interaction with UVs to support attention to several sources of information, the impact of managing multiple UAVs on operators' task-switching abilities, and the role of spatial ability and sense of direction. Empirical evidence related to the design challenges associated with the level and type of automation for information-gathering and decision-making functions was also considered. The chapter concluded with a discussion of the potential role of video games experience on UAV future operations.

As the range of applications of UVs increases in parallel with their increases in sophis- tication and autonomy, there is a likelihood that the operator will be required to perform fewer tasks related to the direct control of the UAV but a far wider range of tasks that are consequent upon the activity of the UAV. That is, rather than directly manipulating the UAV, the operator will control payload and respond to the data coming from the UAV (e.g. image interpretation and analysis). This is likely to lead to an increase in cognitive load for the operators, with a reduction in physical tasks. As the number of tasks increases, particularly when many of these might draw upon visual processing, then the capability to divide attention between tasks or UVs becomes important. Research shows that automating every feature of a UV is not the best strategy, especially when the human operator has to interact with the automation. Rather, a more efficient strategy is to design a flexible system that responds to context, operator requirements and the demands of the situation. This flexibility is provided by adaptive or adaptable automation, which may help the human operator more than inflexible static automation. However, designers of automation for UAS need to find a balance between how involved the user is in system modification in comparison to the system. Unpredictability increases if operators have little involvement with system modification and there is an increase in workload if an operator is too highly involved. Automation support, if used appropriately, has been shown to lead to better performance in multitasking environments involving the supervision and management of multiple UVs.

The fact that individual differences in attentional control impact on the effectiveness of automation needs to be considered in system designs. Operators with low attentional control find performing multiple tasks concurrently more difficult than those with higher attentional control, resulting in their overreliance on automation when available. The design of system displays also influences the ability of operators to divide attention between UVs. For example, tactile displays enhance performance in highly visual environments, allowing operators to allocate visual attention to the tasks and not to the automated cues.

A central theme of this chapter has been the challenge relating to 'human-in-the-loop' and automation. This is a significant issue for human factors and will continue to play a key role in the advancement of UAV design. Given the range of interventions in UAV operations that will involve human operators from imagery interpretation to maintenance, from mission planning to decisions on engaging with targets, the question is not whether the human can be removed from the 'loop', but rather: How will the operator(s) and UAV(s) cooperate as a team to complete the mission given some tasks can be automated? Appreciating some of the ways in which human performance can degrade, and the ways in which the adaptability, ingenuity and creativity of the human-in-the-loop can be best supported, will lead to a new generation of UAVs.

References

1. Cooper, J.L. and Goodrich, M.A., 'Towards combining UAV and sensor operator roles in UAV-enabled visual search', in *Proceedings of ACM/IEEE International Conference on Human–Robot Interaction*, ACM, New York, pp. 351–358, 2008.

2. Finn, A. and Scheding, S., *Developments and Challenges for Autonomous Unmanned Vehicles: A Compendium*, Springer-Verlag, Berlin, 2010.

3. Schulte, A., Meitinger, C. and Onken, R., 'Human factors in the guidance of uninhabited vehicles: oxymoron or tautology? The potential of cognitive and co-operative automation', *Cognition, Technology and Work*, 11, 71–86, 2009.

4. Zuboff, S., *In the Age of the Smart Machine: The Future of Work and Power*, Basic Books, New York, 1988.

5. Alexander, R.D., Herbert, N.J. and Kelly, T.P., 'The role of the human in an autonomous system', 4th IET International Conference on System Safety incorporating the SaRS Annual Conference (CP555), London, October 26–28, 2009.

6. Johnson, C.W., 'Military risk assessment in counter insurgency operations: a case study in the retrieval of a UAV Nr Sangin, Helmand Province, Afghanistan, 11th June', Third IET Systems Safety Conference, Birmingham, 2008.

7. Bainbridge, L., 'Ironies of automation', in J. Rasmussen, K. Duncan and J. Leplat (eds), *New Technology and Human Error*, John Wiley & Sons, New York, 1987.

8. Ballard, J.C., 'Computerized assessment of sustained attention: a review of factors affecting vigilance', *Journal of Clinical and Experimental Psychology*, 18, 843–863, 1996.

9. Defense Science Study Board, *Unmanned Aerial Vehicles and Uninhabited Combat Aerial Vehicles*, Office of the Under Secretary of Defense for Acquisition, Technology and Logistics, Washington, DC, 20301-3140, 2004.

10. Wickens, C.D., *Engineering Psychology and Human Performance*, Harper-Collins, New York, 1992.
11. Wiener, E.L. and Curry, E.R., 'Flight deck automation: problems and promises', *Ergonomics*, 23, 995–1012, 1980.
12. Duggan, G.B., Banbury, S., Howes, A., Patrick, J. and Waldron, S.M., 'Too much, too little or just right: designing data fusion for situation awareness', in *Proceedings of the 48th Annual Meeting of the Human Factors and Ergonomics Society, Santa Monica, CA*, HFES, 528–532, 2004.
13. Sheridan, T.B., 'Forty-five years of man–machine systems: history and trends', in G. Mancini, G. Johannsen and L. Martensson (eds), *Analysis, design and evaluation of man–machine systems. Proceedings of the 2nd IFAC/IFIP/FORS/ IEA Conference*, Pergamon Press, Oxford, pp. 1–9, 1985.
14. Baber, C., Grandt, M. and Houghton, R.J., 'Human factors of mini unmanned aerial systems in network-enabled capability', in P.D. Bust (ed.), *Contemporary Ergonomics*, Taylor and Francis, London, pp. 282–290, 2009.
15. Dekker, S. and Wright, P.C., 'Function allocation: a question of task transformation not allocation', in *ALLFN'97: Revisiting the Allocation of Functions Issue*, IEA Press, pp. 215–225, 1997.
16. Andes, R.C., 'Adaptive aiding automation for system control: challenges to realization', in *Proceedings of the Topical Meeting on Advances in Human Factors Research on Man/Computer Interactions: Nuclear and Beyond*, American Nuclear Society, LaGrange Park, IL, pp. 304–310, 1990.
17. Ruff, H.A., Narayanan, S. and Draper, M.H., 'Human interaction with levels of automation and decision-aid fidelity in the supervisory control of multiple simulated unmanned air vehicles', *Presence*, 11, 335–351, 2002.
18. Ruff, H.A., Calhoun, G.L., Draper, M.H., Fontejon, J.V. and Guilfoos, B.J., 'Exploring automation issues in supervisory control of multiple UAVs', paper presented at 2nd Human Performance, Situation Awareness, and Automation Conference (HPSAA II), Daytona Beach, FL, 2004.
19. Cummings, M.L. and Guerlain, S., 'Developing operator capacity estimates for supervisory control of autonomous vehicles', *Human Factors*, 49, 1–15, 2007.
20. Miller, C., 'Modeling human workload limitations on multiple UAV control', Proceedings of the Human Factors and Ergonomics Society 48th Annual Meeting, New Orleans, LA, pp. 526–527, September 20–24, 2004.
21. Galster, S.M., Knott, B.A. and Brown, R.D., 'Managing multiple UAVs: are we asking the right questions?', Proceedings of the Human Factors and Ergonomics Society 50th Annual Meeting, San Francisco, CA, pp. 545–549, October 16–20, 2006.
22. Taylor, R.M., 'Human automation integration for supervisory control of UAVs', Virtual Media for Military Applications, Meeting Proceedings Rto-Mp-Hfm-136, Paper 12, Neuilly-Sur-Seine, France, 2006.
23. Cummings, M.L., Bruni, S., Mercier, S. and Mitchell, P.F., 'Automation architectures for single operator, multiple UAV command and control', *The International C2 Journal*, 1, 1, 1–24, 2007.
24. Liu, D., Wasson, R. and Vincenzi, D.A., 'Effects of system automation management strategies and multi-mission operator-to-vehicle ratio on operator performance in UAV systems', *Journal of Intelligent Robotics Systems*, 54, 795–810, 2009.
25. Mitchell, D.K. and Chen, J.Y.C., 'Impacting system design with human performance modelling and experiment: another success story', Proceedings of the Human Factors and Ergonomics Society 50th Annual Meeting, San Francisco, CA, pp. 2477–2481, 2006.
26. Pashler, H., 'Attentional limitations in doing two tasks at the same time', *Current Directions in Psychological Science*, 1, 44–48, 1992.
27. Pashler, H., Carrier, M. and Hoffman, J., 1993, 'Saccadic eye-movements and dual-task interference', *Quarterly Journal of Experimental Psychology*, 46A, 51–82, 1993.

28. Monsell, S., 'Task switching', *Trends in Cognitive Sciences*, 7, 134–140, 2003.

29. Rubinstein, J.S., Meyer, D.E. and Evans, J.E., 'Executive control of cognitive processes in task switching', *Journal of Experimental Psychology: Human Perception and Performance*, 27, 763–797, 2001.

30. Schumacher, E.H., Seymour, T.L., Glass, J.M., Fencsik, D.E., Lauber, E.J., Kieras, D.E. and Meyer, D.E., 'Virtually perfect time sharing in dual-task performance: uncorking the central cognitive bottleneck', *Psychological Science*, 12, 101–108, 2001.

31. Squire, P., Trafton, G. and Parasuraman, R., 'Human control of multiple unmanned vehicles: effects of interface type on execution and task switching times', Proceedings of ACM Conference on Human–Robot Interaction, Salt Lake City, UT, pp. 26–32, March 2–4, 2006.

32. Chadwick, R.A., 'Operating multiple semi-autonomous robots: monitoring, responding, detecting', Proceedings of The Human Factors and Ergonomics Society 50th Annual Meeting, San Francisco, CA, pp. 329–333, 2006.

33. Cummings, M.L., 'The need for command and control instant message adaptive interfaces: lessons learned from tactical Tomahawk human-in-the-loop simulations', *CyberPsychology and Behavior*, 7, 653–661, 2004.

34. Dorneich, M.C., Ververs, P.M., Whitlow, S.D., Mathan, S., Carciofini, J. and Reusser, T., 'Neuro-physiologically-driven adaptive automation to improve decision making under stress', Proceedings of the Human Factors and Ergonomics Society, 50th Annual Meeting, San Francisco, CA, pp. 410–414, October 16–20, 2006.

35. Parasuraman, R., Cosenzo, K.A. and De Visser, E., 'Adaptive automation for human supervision of multiple uninhabited vehicles: effects on change detection, situation awareness, and mental workload', *Military Psychology*, 21, 270–297, 2009.

36. Chen, J.Y.C., Barnes, M.J. and Harper-Sciarini, M., Supervisory Control of Unmanned Vehicles, Technical Report ARL-TR-5136, US Army Research Laboratory, Aberdeen Proving Ground, MD, 2010.

37. Cummings, M.L., Brzezinski, A.S. and Lee, J.D., 'The impact of intelligent aiding for multiple unmanned aerial vehicle schedule management', *IEEE Intelligent Systems*, 22, 52–59, 2007.

38. Dixon, S.R. and Wickens, C.D., 'Control of multiple-UAVs: a workload analysis', 12th International Symposium on Aviation Psychology, Dayton, OH, 2003.

39. Maza, I., Caballero, F., Molina, R., Pena, N. and Ollero, A., 'Multimodal interface technologies for UAV ground control stations: a comparative analysis', *Journal of Intelligent and Robotic Systems*, 57, 371–391, 2009.

40. Trouvain, B. and Schlick, C.M., 'A comparative study of multimodal displays for multirobot supervisory control', in D. Harris (ed.), *Engineering Psychology and Cognitive Ergonomics*, Springer-Verlag, Berlin, pp. 184–193, 2007.

41. Wickens, C.D., Dixon, S. and Chang, D., 'Using interference models to predict performance in a multiple-task UAV environment', Technical Report AHFD-03-9/Maad-03-1, 2003.

42. Helleberg, J., Wickens, C.D. and Goh, J., 'Traffic and data link displays: Auditory? Visual? Or Redundant? A visual scanning analysis', 12th International Symposium on Aviation Psychology, Dayton, OH, 2003.

43. Donmez, B., Cummings, M.L. and Graham, H.D., 'Auditory decision aiding in supervisory control of multiple unmanned aerial vehicles', *Human Factors*, 51, 718–729, 2009.

44. Chen, J.Y.C., Drexler, J.M., Sciarini, L.W., Cosenzo, K.A., Barnes, M.J. and Nicholson, D., 'Operator workload and heart-rate variability during a simulated reconnaissance mission with an unmanned ground vehicle', Proceedings of the 2008 Army Science Conference, 2008.

45. Luck, J.P., McDermott, P.L., Allender, L. and Russell, D.C., 'An investigation of real world control of robotic assets under communication', Proceedings of 2006 ACM Conference on Human–Robot Interaction, pp. 202–209, 2006.

46. Chen, J.Y.C., 'Concurrent performance of military and robotics tasks and effects of cueing in a simulated multi-tasking environment', *Presence*, 18, 1–15, 2009.

47. Draper, M., Calhoun, G., Ruff, H., Williamson, D. and Barry, T., 'Manual versus speech input for unmanned aerial vehicle control station operations', Proceedings of the 47th Annual Meeting of the Human Factors and Ergonomics Society, Santa Monica, CA, pp. 109–113, 2003.

48. Chen, J.Y.C., Durlach, J.P., Sloan, J.A. and Bowens, L.D., 'Human robot interaction in the context of simulated route reconnaissance missions', *Military Psychology*, 20, 135–149, 2008.

49. Baber, C., Morin, C., Parekh, M., Cahillane, M. and Houghton, R.J., 'Multimodal human–computer interaction in the control of payload on multiple simulated unmanned vehicles', *Ergonomics*, 54, 792–805, 2011.

50. Parasuraman, R., Barnes, M. and Cosenzo, K., 'Decision support for network-centric command and control', *The International C2 Journal*, 1, 43–68, 2007.

51. Crocoll, W.M. and Coury, B.G., 'Status or recommendation: selecting the type of information for decision adding', Proceedings of the Human Factors Society, 34th Annual Meeting, Santa Monica, CA, pp. 1524–1528, 1990.

52. Rovira, E., McGarry, K. and Parasuraman, R., 'Effects of imperfect automation on decision making in a simulated command and control task', *Human Factors*, 49, 76–87, 2007.

53. Parasuraman, R. and Riley, V.A., 'Humans and automation: use, misuse, disuse, abuse', *Human Factors*, 39, 230–253, 1997.

54. Parasuraman, R., 'Designing automation for human use: empirical studies and quantitative models', *Ergonomics*, 43, 931–951, 2000.

55. Scerbo, M., 'Adaptive automation', in W. Karwowski (ed.), *International Encyclopedia of Ergonomics and Human Factors*, Taylor and Francis, London, pp. 1077–1079, 2001.

56. Parasuraman, R. and Miller, C., 'Delegation interfaces for human supervision of multiple unmanned vehicles: theory, experiments, and practical applications', in N. Cooke, H.L. Pringle, H.K. Pedersen and O. Conner (eds), *Human Factors of Remotely Operated Vehicles*, Advances in Human Performance and Cognitive Engineering, Vol. 7, Elsevier, Oxford, pp. 251–266, 2006.

57. Barnes, M., Parasuraman, R. and Cosenzo, K., 'Adaptive automation for military robotic systems', in *RTO-TR-HFM-078 Uninhabited Military Vehicles: Human factors issues in augmenting the force*, NATO Technical Report, NATO Research and Technology Organization, Brussels, pp. 420-440, 2006.

58. Parasuraman, R., Galster, S., Squire, P., Furukawa, H. and Miller, C.A., 'A flexible delegation interface enhances system performance in human supervision of multiple autonomous robots: empirical studies with RoboFlag', *IEEE Transactions on Systems, Man & Cybernetics – Part A: Systems and Humans*, 35, 481–493, 2005.

59. Wilson, G. and Russell, C.A., 'Performance enhancement in a UAV task using psycho-physiologically determined adaptive aiding', *Human Factors*, 49, 1005–1018, 2007.

60. Miller, C. and Parasuraman, R., 'Designing for flexible interaction between humans and automation: delegation interfaces for supervisory control', *Human Factors*, 49, 57–75, 2007.

61. Mitchell, D.K., 'Soldier workload analysis of the Mounted Combat System (MCS) platoon's use of unmanned assets', Technical Report ARL-TR-3476, US Army Research Laboratory, Aberdeen Proving Ground, MD, 2005.

62. Kaber, D.B., Endsley, M.R. and Onal, E., 'Design of automation for telerobots and the effect on performance, operator situation awareness and subjective workload', *Human Factors and Ergonomics in Manufacturing*, 10, 409–430, 2000.

63. Parasuraman, R., Sheridan, T.B. and Wickens, C.D., 'A model for types and levels of human interaction with automation', *IEEE Transactions on Systems, Man & Cybernetics – Part A: Systems and Humans*, 30, 286–297, 2000.

64. Sheridan, T. and Parasuraman, R., 'Human–automation interaction', *Reviews of Human Factors and Ergonomics*, 1, 89–129, 2006.

65. Chen, J.Y.C. and Terrence, P.I., 'Effects of tactile cueing on concurrent performance of military and robotics tasks in a simulated multitasking environment', *Ergonomics*, 51, 1137–1152, 2008.

66. Chen, J.Y.C. and Joyner, C.T., 'Concurrent performance of gunner's and robotic operator's tasks in a multitasking environment', *Military Psychology*, 21, 98–113, 2009.

67. Wickens, C.D., Dixon, S.R., Goh, J. and Hammer, B., 'Pilot dependence on imperfect diagnostic automation in simulated UAV flights: an attentional visual scanning analysis', Technical Report AHFD-05-02/MAAD-05-02, University of Illinois, Urbana-Champaign, IL, 2005.

68. Levinthal, B.R. and Wickens, C.D., 'Management of multiple UAVs with imperfect automation', Proceedings of the Human Factors and Ergonomics Society 50th Annual Meeting, San Francisco, CA, pp 1941–1944, 2006.

69. Dixon, S.R., Wickens, C.D. and McCarley, J., 'On the independence of compliance and reliance: are automation false alarms worse than misses?', *Human Factors*, 49, 564–572, 2007.

70. Derryberry, D. and Reed, M., 'Anxiety-related attentional biases and their regulation by attentional control', *Journal of Abnormal Psychology*, 111, 225–236, 2002.

71. Chen, J.Y.C. and Terrence, P.I., 'Effects of imperfect automation and individual differences on concurrent performance of military and robotics tasks in a simulated multitasking environment', *Ergonomics*, 52, 907–920, 2009.

72. Ekstrom, R.B., French, J.W., Harman, H. and Dermen, D., 'Kit of factor-referenced cognitive tests', Educational Testing Service, Princeton, NJ, 1976.

73. Pak, R., Rogers, W.A. and Fisk, A.D., 'Spatial ability subfactors and their influences on a computer-based information search task', *Human Factors*, 48, 154–165, 2006.

74. Kozhevnikov, M. and Hegarty, M., 'A dissociation between object-manipulation and perspective-taking spatial abilities', Memory & Cognition, 29, 745–756, 2001.

75. Pellegrino, J.W., Alderton, D.L. and Shute, V.J., 'Understanding spatial ability', *Educational Psychologist*, 19, 239–253, 1984.

76. Alderton, D.L., Wolfe, J.H. and Larson, G.E., 'The ECAT battery', *Military Psychology*, 9, 5–37, 1997.

77. Stanney, K.M. and Salvendy, G., 'Information visualization: assisting low spatial individuals with information access tasks through the use of visual mediators', *Ergonomics*, 38, 1184–1198, 1995.

78. Chen, C., Czerwinski, M. and Macredie, R., 'Individual differences in virtual environments – introduction and overview', *Journal of American Society for Information Science*, 51, 499–507, 2000.

79. Eyal, R. and Tendick, F., 'Spatial ability and learning the use of an angled laparoscope in a virtual environment', in *Proceedings of Medicine Meets Virtual Reality (MMVR)*, IOS Press, pp. 146–153, 2001.

80. Menchaca-Brandan, M.A., Liu, A.M., Oman, C.M. and Natapoff, A., 'Influence of perspective-taking and mental rotation abilities in space teleoperation', Proceedings of the 2007 ACM Conference on Human–Robot Interaction, Washington, DC, pp. 271–278, March 8–11, 2007.

81. Lathan, C.E. and Tracey, M., 'The effects of operator spatial perception and sensory feedback on human–robot teleoperation performance', *Presence*, 11, 368–377, 2002.

82. Educational Testing Service, Cube Comparison Test, 2005.

83. Gugerty, L. and Brooks, J., 'Reference-frame misalignment and cardinal direction judgments: group differences and strategies', *Journal of Experimental Psychology: Applied*, 10, 75–88, 2004.

84. Hegarty, M. and Waller, D., 'A dissociation between mental rotation and perspective-taking spatial abilities', *Intelligence*, 32, 175–191, 2004.

85. Chen, J.Y.C., 'UAV-guided navigation for ground robot teleoperation in a military reconnaissance environment', *Ergonomics*, 53, 940–950, 2010.

86. Neumann, J., 'Effect of operator control configuration on unmanned aerial system trainability', Doctoral dissertation, University of Central Florida, 2006.

87. Hegarty, M., Richardson, A.E., Montello, D.R., Lovelace, K. and Subbiah, I., 'Development of a self-report measure of environmental spatial ability', *Intelligence*, 30, 425–447, 2002.

88. Kozlowski, L.T. and Bryant, K.J., 'Sense of direction, spatial orientation, and cognitive maps', *Journal of Experimental Psychology: Human Perception and Performance*, 3, 590–598, 1977.

89. Baldwin, C.L. and Reagan, I., 'Individual differences in route-learning strategy and associated working memory resources', *Human Factors*, 51, 368–377, 2009.

90. Takeuchi, Y., 'Sense of direction and its relationship with geographical orientation, personality traits and mental ability', *Japanese Journal of Educational Psychology*, 40, 47–53, 1992.

91. Kato, Y. and Takeuchi, Y., 'Individual differences in wayfinding strategies', *Journal of Environmental Psychology*, 23, 171–188, 2003.

92. Cooper, L.A. and Shepard, R.N., 'Chronometric studies of the rotation of mental images', in W.G. Chase (ed.), *Visual Information Processing*, Academic Press, New York, pp. 75–176, 1973.

93. De Lisi, R. and Cammarano, D.M., 'Computer experience and gender differences in undergraduate mental rotation performance', *Computers in Human Behavior*, 12, 351–361, 1996.

94. Chen, J.Y.C., Haas, E.C. and Barnes, M.J., 'Human performance issues and user interface design for teleoperated robots', *IEEE Transactions on Systems, Man & Cybernetics – Part C: Applications and Reviews*, 37, 1231–1245, 2007.

Part III

SAA METHODOLOGIES

6

Sense and Avoid Concepts: Vehicle-Based SAA Systems (Vehicle-to-Vehicle)

Štěpán Kopřiva, David Šišlák and Michal Pěchouček
Czech Technical University, Prague, Czech Republic

6.1 Introduction

The various scenarios of unmanned aerial vehicles (UAVs) deployment require the ability to navigate UAVs in unknown terrain. The UAV, while fulfilling its mission objectives, has to avoid static obstacles as well as moving obstacles like other UAVs, airplanes, balloons or areas with bad weather forecast or bad weather conditions. Furthermore, if the UAV enters commercially controlled airspace, it needs to be able to sense and avoid the potential conflicts considering the air-traffic control regulations.

The concepts for development of automated systems providing the sense and avoid capability (also referred to as collision detection and resolution systems, CDR) came mainly from two domains. The first one is the air-traffic management domain, where automated tools like Traffic Collision Avoidance System (TCAS) [1] and Precision Runway Monitor (PRM) [2] are used to increase safety and fluency of the air-traffic. The second one is the artificial intelligence research and particularly robotics, where scientists investigated the trajectory planning and obstacle avoidance algorithms for aerial, ground, and maritime systems.

Various approaches to CDR systems, differences and similarities and categorization of the systems have been introduced in the literature. Krozel [3] and Kuchar [4] presented surveys of collision detection and resolution methods. Zeghal [5] conducted a survey of force field

Sense and Avoid in UAS: Research and Applications, First Edition. Edited by Plamen Angelov.
© 2012 John Wiley & Sons, Ltd. Published 2012 by John Wiley & Sons, Ltd.

collision detection and resolution methods and finally Albaker [6] introduced the survey of CDR methods for UAVs.

6.2   Conflict Detection and Resolution Principles

To discuss the conflict detection and resolution principles, the conflict has to be defined first. The *conflict* is an event in which the horizontal or vertical Euclidian distance between two aircrafts breaks the minimal defined separation criterion. The criterion varies based on the airspace the UAV operates in and may also be different for different UAVs. For illustration, currently in civilian air-traffic the en-route horizontal separation criterion is 5 NM and the vertical one is 1,000 ft for the airspace between the ground and the flight level 290. The horizontal and vertical separation criteria form a cylindrical airspace volume around the UAV called the *safety zone* (SZ). The safety zone may not under any circumstances be violated by any other UAV. For different CDR systems different horizontal and vertical criteria may be applied.

The function of the collision detection and resolution system is to *detect* the collision and provide the *resolution* in the form of an evasion maneuver which is executed by the UAV's autopilot. The general block diagram of the CDR unit is presented in Figure 6.1. The CDR system has five basic functions: sensing, trajectory prediction, conflict detection, conflict resolution, and evasion maneuver generation.

6.2.1   Sensing

The CDR system monitors the surrounding environment for both static and dynamic obstacles using onboard sensors represented by the sensors block in Figure 6.1. There are two types of sensor: cooperative and non-cooperative ones.

The cooperative sensors provide the ability to sense the environment and to communicate with aircrafts equipped with the same type of sensors by establishing a communication link. One example of the cooperative sensor is the Automatic Dependent Surveillance Broadcast (ADS-B) [7]. This device transfers the longitude, latitude, altitude, speed, and UAV identification. Some other cooperative sensors even allow exchanging the whole flight plans.

The non-cooperative sensors sense the environment in order to get information about the obstacles and airplanes. There are no communication links among the airplanes and the sensor

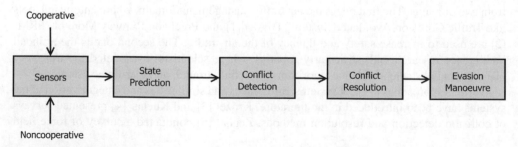

Figure 6.1   The general architecture of the collision detection and resolution system

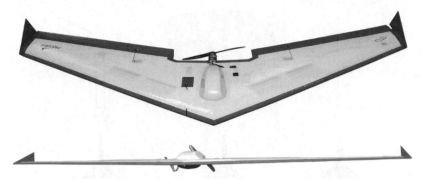

Figure 6.2 The Procerus UAV. Note the passive sensor – gimbal moving camera in the bottom part of the airframe

information needs to be processed in order to get the correct environment state knowledge. Currently used non-cooperative sensor technologies include inertial measurement unit, laser range finders, stereo camera systems, single moving cameras, and radars. Using the active radar is limited to the large UAV systems, whereas small cameras together with picture recognition software may be used even on small systems like the Procerus UAV (Figure 6.2). The non-cooperative sensors for UAVs are described in [8].

6.2.2 Trajectory Prediction

To detect and resolve a conflict, it is necessary to compare the trajectory of the UAV and the trajectory of the sensed object. The trajectory is produced by the trajectory computation unit from the raw sensor information gathered on the input. In [4], there are three basic models of trajectory prediction methods and we add one more.

In the nominal method (Figure 6.3(a)), the trajectory is predicted directly from the sensor data without considering the possibility of any uncertainty or change. The output of the nominal trajectory predictor is a single trajectory computed from several last sensor scans. The trajectory may be computed using different methods, i.e. linear prediction, Taylor series prediction, or prediction using the Kalman filter. The nominal method prediction is suitable for short-term predictions, where the probability of change is relatively low.

The worst-case prediction (Figure 6.3(b)), is another modeling approach. This method covers the whole range of maneuvers the aircraft may perform and using a lookahead time parameter computes the area where the aircraft may occur. This area is then considered as the predicted trajectory.

In the probabilistic prediction approach, the uncertainties are used to model the potential variations in the trajectory. To construct this model, all possible trajectories are generated (like for the worst-case prediction) and each trajectory is evaluated by the probability function. The probabilistic prediction is a trade-off between the nominal method and the worst-case prediction. The probabilistic prediction method is the most general one, the decisions may be made based on the likelihood of the conflict.

The flight plan exchange approach is bound to cooperative sensors only. The aircrafts exchange parts of their flight plans, the precise trajectory is known and no prediction is needed.

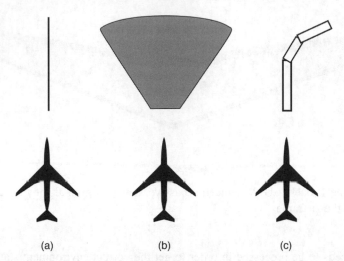

(a)   (b)   (c)

Figure 6.3 Trajectory prediction based on the samples from sensors: (a) the nominal method – the trajectory is predicted directly from the sensor data; (b) the worst-case prediction – range of possible maneuvers performed by the opponent aircraft; (c) the flight plan exchange – the flight plan representation using the safety zone and waypoints

The plans are exchanged as a set of way-points together with the safety zone parameters (Figure 6.3(c)). The advantage of this approach is the exact knowledge of the future trajectory. The disadvantage is the higher bandwidth required for the data transmission.

6.2.3   Conflict Detection

Conflict is detected based on the flight plan representation obtained from the trajectory prediction unit. The unit checks the flight plans of both airplanes and checks whether the safety zone of any airplane has been violated. If so, the parameters of the conflict (position of the conflicting airplanes and times of possible conflicts) are passed to the conflict resolution unit. In Figure 6.4, there is a conflict with its start and end time.

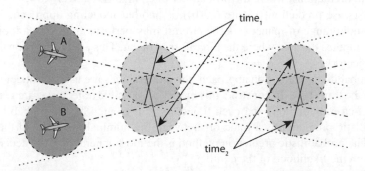

Figure 6.4   Conflict detection on the exchanged flight plans of two aircrafts. The conflict is detected when one (or both) aircraft violates the safety zone of the other one

6.2.4   Conflict Resolution

The conflict resolution block resolves the collision using one of the collision avoidance methods. The methods covered in this survey are: rule-based methods (RB), game theory methods (GT), field methods (F), geometric methods (G), numerical optimization methods (NO), combined methods (C), multi-agent methods (MA), and other methods (O).

The rule-based methods use a set of prescribed rules to avoid conflict. The sets of rules are fixed during the system design phase and originally were inspired by the visual flight rules (VFR) known from the civilian air-traffic domain. Even though the rule-based methods optimize the solution in the phase where the rules are designed, the set of pre-described rules is then fixed and has to be the same for all airplanes in the shared airspace. It is not possible to integrate further intentions and needs of a particular airplane into the conflict resolution process. The major advantage of the rule-based methods is the simplicity and fast implementation during the flight. There is no or very limited communication flow required among airplanes.

In the game theory methods, the authors model the conflict as a two-player differential game. These algorithms are useful mainly for non-cooperative conflict resolution. The methods are used for short-term conflict resolution, because they work with airplane states only. The cooperative conflict resolution methods are typically mid-term and long-term ones.

The field methods treat each airplane as a charged particle and are very close to a reactive control mechanism. Based on the current configuration (e.g. position of other airplanes, weather condition, considered uncertainty), the field is computed. Then, the UAV based on its own position in this field applies the control action depending on the current state of the airplane with respect to the field. The evasion maneuvers are then generated based on the repulsive forces between the fields. The advantage of the method is its relatively easy implementation. The major disadvantage of these methods is the computationally intensive phase where the field is computed. The field has to be updated when the current configuration is updated. Another major disadvantage appears if the UAV physical parameters are considered – field methods may come with solutions that produce maneuvers which may not be executed by the UAVs and have to be filtered. Similarly to the previous game approaches, these methods are not suitable for long-term airplane operation optimization.

Geometrical approaches consider in many cases the whole trajectory of the airplane while it avoids observed solution. Various approaches optimize given objective functions while they are searching for the proper evasion maneuver. In many cases, these methods are considering only collision of two airplanes and for multi-collision situations they come with sub-optimal approaches or sequential application of collision of two airplanes where safety is not guaranteed. Geometrical optimization is a very complex problem, especially if all available actions like change of heading, velocity, and altitude are considered.

The numerical optimization methods use a kinematic model of the vehicle together with a set of constraints and use cost metrics for the maneuver generation. The optimal evasion maneuver is then computed based on the most desired constraints. The major benefit of these methods is that the optimization criterion is clearly formalized and the final control is optimized with respect to the given problem definition. With increasing number of airplanes the problem becomes analytically unsolvable. The limitation of considered time horizon simplifies the problem. However, it is very hard to integrate airspace limitations like excluded airspaces (special use airspaces) and ground surface. Similarly, there cannot be integrated weather conditions.

The multi-agent methods use the multi-agent framework for solution generation. Each aircraft is controlled by one agent. The agents are able to communicate together and to negotiate about the solution using various utility functions.

6.2.5   Evasion Maneuvers

The evasion maneuver block in Figure 6.1 is responsible for the realization of the evasion maneuver proposed by the conflict resolution block. The basic evasion maneuvers are speed-up, slow-down, keep the same speed, turn-left, turn-right, climb, and descend. It is also possible to combine these maneuvers together, i.e. change speed while performing the turn maneuver.

6.3   Categorization of Conflict Detection and Resolution Approaches

In this section, the comprehensive categorization of recently published CDR methods is provided.

6.3.1   Taxonomy

The taxonomy tree taking the most important CDR design attributes has been designed, see Figure 6.5. In the top scheme level there are five basic attributes: sensing technology, trajectory prediction, conflict detection, conflict resolution, and maneuver realization. These attributes correspond to the five basic blocks in the conflict detection and resolution general architecture as presented in Figure 6.1. We have used these basic attributes with minor changes to classify CDR methods. For the CDR methods, it is often the case that the concepts have not been validated on the real hardware yet. We can hardly classify which sensors are utilized by the given method, so we have omitted the sensor attribute. We have classified the methods based on the conflict resolution method, conflict detection, and supported evasion maneuvers. Please check Table 6.1 for the values of the attributes.

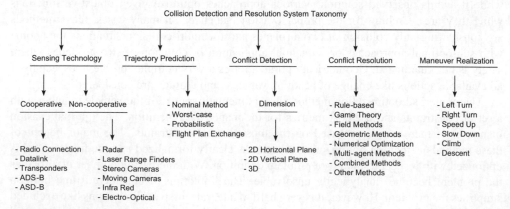

Figure 6.5   The conflict resolution and detection taxonomy. The top five attributes are used for model classification

Table 6.1 Attributes and abbreviations used for the CDR classification

Conflict resolution methods	Rule-based (RB) Game theoretical (GT) Field (F) Geometric (G) Numerical optimization (NO) Multi-agent (MA) Combined (C) Others (O)
Conflict detection dimensions	2D-horizontal plane (HP) 2D-vertical plane (VP) 3D (3D)
Trajectory prediction	Nominal (N) Worst-case (WC) Probabilistic (P) Flight plan exchange (FP)
Evasion maneuvers	Speed change (SC) Turn (T) Altitude change (AC)

The following part describes the evaluated models. The models are presented in sets based on the used conflict resolution methods. For each method, there is also a table that categorizes the models. The models have been divided into the following categories, where each category is described in its own section. Section 6.3.2 contains rule-based approaches where a conflict resolution maneuver is pre-described for a certain type of observed potential conflict. Section 6.3.3 provides a review of algorithms based on the game theoretic approach where all potential maneuvers of opposing airplanes are considered while searching for its own maneuver. Section 6.3.4 presents force, potential, and vortex field approaches which are used for establishing the current airplane control inputs. Section 6.3.5 summarizes algorithms considering geometrical representation of a collision situation while searching for an evasion maneuver. Section 6.3.6 contains conflict resolution algorithms formulated as an optimization problem minimizing an objective function. Section 6.3.7 presents collision avoidance methods combining various algorithms. Section 6.3.8 summarizes existing multi-agent approaches for air-traffic conflict resolution. Section 6.3.9 contains all other methods related to air-traffic control which do not explicitly belong to one of the previous categories.

6.3.2 Rule-Based Methods

Schild [9] designed a set of rules for autonomous separation for en-route air-traffic. His approach considers the following three evaluation objectives: (i) safe inter-aircraft separation, (ii) minimum number of maneuvers used, and (iii) maximum efficiency in additional time and fuel provided. The rules are derived from many optimization tasks considering optimization functions based on these three objectives. Thus, the result of this approach is a set of rules where for each rule there is defined an activation condition specified in the form of mutual

Table 6.2 Rule-based models

Model	Resolution method	Detection	Trajectory prediction	Evasion
Schild	RB	2D-HP	P	VC, T
Hwang	RB	2D-HP	P	VC, T

position and orientation of airplanes. These rules are defined for two airplanes. For multiple airplanes, these rules are applied sequentially and the stability of this rule-based system is validated in his work. The presented system may be used for the UAVs as well.

Hwang et al. [10, 11] described the protocol-based N-aircraft method for multiple-aircraft conflict avoidance. It is assumed that each airplane's position, heading, and velocity are available to all involved airplanes, nominal trajectories of airplanes are at constant altitude and heading, and all airplanes initiate conflict resolution maneuvers at the same time and at once. The velocity is then constant along the maneuver. The multi-airplane conflict resolution is presented as (i) an *exact conflict* and (ii) an *inexact conflict*. In the first case, the exact conflict, original trajectories of all airplanes collide at a point which helped to derive a closed-form analytic solution for the required heading change. This case is unrealistic for more than two airplanes but motivates the solution for the general case. In the second case, the inexact conflict, conflict points of multiple airplanes do not coincide. The velocity change is considered for the inexact conflict.

Hwang et al. construct a finite partition of the airspace around the conflict and derive a protocol for resolving the worst-case conflict within each partition. Thus the conflict resolution method is formulated as a rule which is easily understandable and implemented by all aircrafts involved in the conflict. It is shown that the solution is robust to uncertainties in the airplane's position, heading, velocity, and also with respect to asynchronous maneuvers where airplanes do not necessarily change their heading at the same time. The resulting method is not optimal in the sense of the deviation from the desired trajectory. However, it is implementable in real time, and always guarantees a safe conflict resolution. The method requires only few numerical computations.

Classification of rule-based methods according to the defined taxonomy is in Table 6.2. Even though these approaches optimize the solution in the phase where the rules are designed, the set of pre-described rules is then fixed and has to be the same for all airplanes in the shared airspace. It is not possible to integrate further intentions and needs of a particular airplane into the conflict resolution process. The major advantage of the rule-based approach is its simplicity and fast implementation during the flight. It requires no communication or very limited communication flow among UAVs.

6.3.3 Game Theory Methods

Lachner [12] investigated a worst-case approach based on pursuit-evasion differential games [13]. An evader (correct UAV) tries to avoid a collision against all possible pursuer's maneuvers. The solution of such a differential game provides a strategy which guarantees a best possible control against all unknown disturbances of an opponent's actions. He represents an optimal conflict resolution strategy along a lot of optimal paths which is used to synthesize a conflict resolution strategy globally by means of Taylor series expansions.

Table 6.3 Game theory models

Model	Resolution method	Detection	Trajectory prediction	Evasion
Lachner	GT	HP	WC	SC, T
Zhang	GT	HP	N	–
Bayen	GT	HP	P	SC, T

Zhang *et al.* [14] formulated the conflict resolution problem as a differential game too. They apply the Lie–Poisson reduction on the dual of the Lie algebra of the special Euclidean group. The reduced Hamiltonian dynamics is then derived and can be integrated explicitly backwards in time. They use a hybrid automata to describe the solution to the reduced dynamics as well as the aircraft maneuvers in the game. The safe portion of the target frontier set is calculated and the conflict resolution solution is derived as the safe portion of this boundary.

Later, Bayen *et al.* [15] applied a differential game formulated for a two-vehicle collision avoidance problem to the derivation of an alerting logic for conflict in high-altitude air-traffic. Using computational methods based on level sets, the three-dimensional solution of the game is calculated. The solution is employed to define unsafe regions for each pair of aircraft in the relevant airspace, and these regions are used as a metric to indicate if loss of separation could occur. Their approach considers the worst case based on the airplane kinematic configuration. Classification of these methods is presented in Table 6.3.

6.3.4 Field Methods

Duong *et al.* [16, 17] presented a technique in which a force field generated by an intruding airplane produces a conflict resolution action and a force from the flight plan generates an attracting action. The work further extends [18], where a distributed conflict is proposed using a resolution algorithm based on a symmetrical force field method. The resolution maneuver is generated using a relatively simple equation. However, the resolution maneuver may have several discontinuities which aircraft cannot follow and the safety cannot be proven for multi-airplane maneuvers. A review of different approaches based on force fields for airborne conflict resolution is described in [5].

Hu *et al.* [19] calculated the probability of conflict between two airplanes by modeling the airplane motion as a deterministic trajectory with addition in a scaled Brownian motion perturbation. The authors have considered only the two-dimensional case, where both airplanes fly at the same altitude. The probability of a conflict becomes the probability that a Brownian motion escapes from a time-varying safe region. Brownian motion integrates a measure of the probability of each path, where paths of large deviation are less likely than paths of small deviation. Their approach provides a closed-form formalism with approximations considering both finite and infinite time-horizon cases, which makes its implementation computationally inexpensive and enables fast derivation of a resolution algorithm.

Eby *et al.* [20–22] proposed the distributed conflict resolution method derived from the potential field model. Their self-organizational approach utilizes the concept of the force field. Airplanes are represented by positive charges and airplanes' destinations by negative charges.

Table 6.4 Field models

Model	Resolution method	Detection	Trajectory prediction	Evasion
Duong	F	3D	N	SC, T, AC
Hu	F	HP	P	T
Eby	F	3D	N	SC, T, AC
Prandini	F	3D	P	SC, T, AC

Positive charges tend to be drawn toward the fixed negative charge because of the mutual attraction of their opposite charges. At the same time, the positive particles tend to maintain distance between each other because of the mutual repulsion of their like charges. The authors demonstrate that their approach is robust in a complex multi-airplane conflict. The algorithm is more complex when compensating distance dependency between an airplane and its destination. Also, there is a modification which guarantees a minimum separation among airplanes. They also studied communication failures and restricted maneuverability constraints.

Prandini *et al*. [23–25] studied the safety in the three-dimensional airplane flight. In their study, the airplane future position during the flight is predicted based on a stochastic model that incorporates the information on the airplane flight plan and takes into account the presence of wind as the main source of uncertainty on the airplane motion. In [26], this approach is extended with Markov chain approximation of the stochastic processes modeling the motion of the aircrafts for the mid-term model. For the predicted aircraft positions along some finite time horizon, the authors integrate additive wind perturbations to the airplane's velocity. The spatial correlation structure for the wind perturbations is driven by distance of airplanes – for two closer airplanes, they use stronger correlation between the perturbations to airplanes' velocities. The probability of conflict is computed using the Markov chain state space which is obtained by discretizing the region around the position where the encounter occurs. In [27], they introduced a switching diffusion model to predict the future positions of an airplane following a given flight plan adopting the modeling framework of stochastic hybrid systems [28]. The weak approximation of the switching diffusion through a Markov chain is used to develop a numerical algorithm for computing an estimate of the probability that the airplane enters an unsafe region of the airspace or comes too close to another airplane.

In [29], Prandini *et al*. proposed conflict resolution where the design of an optimal conflict resolution maneuver is based on the airplane intent information. The intent information is made robust against the uncertainty affecting the airplane future positions by a randomized stochastic optimization method. In such a way, they account for a probabilistic description of the uncertainty affecting the aircraft motion and avoid the excessive computational load of a pure Monte Carlo stochastic optimization method. Table 6.4 summarizes field methods and presents their classification according to the taxonomy.

6.3.5   Geometric Methods

Ota *et al*. [30] proposed a method which produces an avoidance trajectory in both the horizontal and vertical planes of motion based on the geometric relationship between the airplane and the threats. The authors have introduced the new concept 'threat map,' which is used to describe moving threats as static threats. Thus, motion planning for dynamic threats is reduced to path planning for stationary threat avoidance. The threat map is regularly updated and the velocity direction of the airplane is calculated to avoid the threats. They use the 'risk space' to model a threat. Airplanes avoiding the risk space keep a safe separation from threats.

Chiang et al. [31] proposed an approach based on computational geometry, where the flight is represented as a Delaunay diagram [32]. Their resolution algorithm is computationally intensive in construction of non-intersecting tubes in that diagram representation.

Bilimoria [33] proposed a geometric optimization approach to the aircraft conflict resolution, utilizing information on current positions and velocity vectors. The resolutions are optimal in the sense that they minimize the velocity vector changes required for conflict resolution, which results in minimum deviations from the nominal trajectory. His approach utilizes the geometric characteristics of airplane trajectories along with intuitive reasoning to provide conflict resolution maneuvers from a closed-form analytic solution. The algorithm provides an optimal combination of heading and speed changes for conflict resolution in the horizontal plane.

For two airplanes, it is shown that the analytical solution is optimal. The solution has been validated by comparison with numerical solutions from a compute-intensive optimization process utilizing a semi-definite programming approach. Multi-airplane conflicts are resolved sequentially – each aircraft resolving its most immediate conflict at each update cycle. The successive application of pairwise conflict resolution cannot guarantee safety for multiple-aircraft conflicts, as shown in [10].

Hu et al. [34, 35] designed optimal coordinated maneuvers for the multiple airplane conflict resolution. They proposed an energy function to select, among all the conflict-free maneuvers, the optimal one. Their cost function incorporates a priority mechanism that favors those maneuvers where the airplane with lower priority assumes more responsibility in resolving the predicted conflicts. The resolution maneuvers involve changes in heading, speed, and altitude as well. For the conflict of two airplanes, they provide a geometric construction and a numerical algorithm for computing the optimal resolution maneuvers. For the multi-airplane case, they use an approximation scheme to compute a sub-optimal solution.

Christodoulou et al. [36] formalized the three-dimensional air-traffic collision problem as a mixed-integer nonlinear programming problem. The optimization function is defined as the total flight time of all airplanes avoiding all possible conflicts. They considered only maneuvers considering velocity changes. In [37], they used neural networks trained with examples prepared through the nonlinear programming to avoid three-dimensional collisions. However, this extension works with velocity maneuvers which optimize the velocity changes. Their neural network predicts the optimal velocity change of many airplanes in order to avoid imminent conflict. The authors train the neural network by randomly generated conflict situations along with the generated optimal solution computed as a nonlinear optimization.

Luongo et al. [38] proposed an optimal three-dimensional geometrical solution for aircraft non-cooperative collision avoidance. The authors derived the analytical solution as a proper kinematic optimization problem based on a geometric approach. Their approach combines all control variables (velocity, heading, and vertical changes). In a pairwise collision avoidance, they continuously update the velocity of the airplane in order to keep safety surrounding the other airplane which is considered as an intruder. The algorithm doesn't consider fixed obstacles and the limited airspace, e.g. terrain surface, special use airspaces. Up to now, their solution handles collisions of two airplanes only.

Table 6.5 Geometric methods

Model	Resolution method	Detection	Trajectory prediction	Evasion
Ota	G	3D	N	SC, T, CD
Chiang	G	3D	N	SC, T, CD
Bilimoria	G	HP	N	SC, T
Hu II	G	3D	N	SC, T, CD
Christodoulou	G	3D	–	SC
Luongo	G	3D	N	SC, T, CD
Pappas	G	3D	N	SC, T, CD

Pappas *et al.* [39] presented the method for crossing the metering fix by a set of aircrafts based on the mixed integer geometric programming. For ordering of the aircrafts above the fixed point they consider obtaining the utility function from the airlines.

In [40] the authors describe the method of CDR that uses a simple geometric approach. Two UAVs are treated as point masses with constant velocity. The authors discuss en-route aircrafts that build the information database based on the ADS-B updates. The algorithm calculates the PCA (point of closest approach) and evaluates the earliest collision between two UAVs. The paper proposes one resolution maneuvering logic called the *vector sharing resolution*. Using the miss distance vector in PCA, the algorithm proposes directions for two UAVs to share the conflict region. With these directions, UAVs are going to maneuver cooperatively. Taxonomy of presented geometric methods is provided in Table 6.5.

6.3.6 Numerical Optimization Approaches

Durand *et al.* [41] proposed predefined maneuvers which are used to construct a multi-airplane conflict resolution maneuver to solve en-route conflicts. The authors formally define conflict as a constrained optimization problem minimizing the delays due to the conflict resolution. Their method uses an optimal problem solver which is based on a stochastic optimization technique using genetic algorithms to generate each resolution maneuver successively. They demonstrate the method on many experiments where a genetic algorithm is very efficient and solves conflicts in a real-time situation. Later in [42], they use the genetic algorithm for training a neural network which is then used to solve conflicts between two airplanes.

Menon *et al.* [43] proposed a conflict resolution algorithm based on the quasi-linearization method. They used the nonlinear point-mass airplane models with integrated operational constraints. They compute a conflict resolution trajectory with two different costs: (i) deviation from the original trajectory as a square of the perturbation from the nominal trajectories and (ii) a linear combination of total flight time and fuel consumption. The optimization process gives the three-dimensional optimal multi-airplane conflict resolution but the optimization process is in general computationally intensive.

Frazzoli *et al.* [44] used randomized searches to choose one from all possible cross patterns and uses a convex optimization on the chosen cross pattern to obtain a resolution maneuver minimizing energy. They show that the planar multi-airplane conflict resolution problem accounting for all possible crossing patterns among airplanes can be recast as a non-convex quadratically constrained program. They show that there exist efficient numerical relaxations for this type of problem. These relaxations lead to a random search technique to compute feasible and locally optimal conflict-free strategies.

Bicchi *et al.* [45] proposed the planning optimal conflict resolution method for kinematic models of airplanes flying in the horizontal plane with constant cruise speed and curvature bounds. The conflict resolution is formulated as an optimal control problem to minimize the flight time. The solution was used as the decentralized hybrid system for mobile agents evolving on the plane [46, 47].

Pallotino *et al.* [48, 49] described two different formulations of the multi-airplane conflict resolution as a mixed-integer linear program. In the first case, only velocity changes are considered. In the second case, only heading maneuver changes are planned. The linear formalization of the problem is quickly solved with existing solvers in contrast to the nonlinear model presented in [45]. The primary benefit of this approach is its real-time implementation. The authors prove that a decentralized adaptation of the algorithm is possible with a given look-ahead distance considering the worst-case maneuvering requirements during state transitions as an other airplane becomes visible.

Raghunathan *et al.* [50] described the problem of the optimal cooperative three-dimensional conflict resolution with multiple airplanes as the rigorous numerical trajectory optimization – minimize a certain objective function while the safe separation between each airplane pair is maintained. They model an airplane as nonlinear point-mass dynamics. The optimal control problem is converted to a finite-dimensional nonlinear program by the use of collocation on finite elements. The nonlinear problem is solved by the use of an interior point algorithm that incorporates a line search method. They propose a reliable initialization strategy that yields a feasible solution.

Šišlák *et al.* [51] present the non-cooperative CDR used in the cases when the communication between planes is not possible. Such a situation can happen, for example, when the communication device on board the UAV is broken or if the other aircraft intentionally refuses to communicate (an enemy). Classical non-cooperative collision avoidance utilizing optimization algorithms like [52, 53] can optimally solve a collision with only one non-cooperative object. Such methods can fail for more non-cooperative objects. Therefore, the authors have designed the method based on the dynamic no-flight zones. The designed method is based on the path planning using A* algorithm [54], which is capable of planning a flight path which doesn't intersect defined no-flight zones. The algorithm is responsible for the coordination of all operations needed to avoid potential future collisions of the UAV and an object representing a non-cooperative one. The event that triggers the conflict resolution process is information obtained from the sensors (radar) providing the position of an unknown object in the local area. The observation is used for the update of the solver knowledge base. The collision point is defined by an intersection of the current UAV's flight plan and the predicted flight trajectory of the non-cooperative object. The algorithm uses the linear prediction estimating the future object trajectory including current velocity, which requires two last positions with time information. The prediction provides both predicted collision point position and time information. The detected collision point is wrapped by a dynamic no-flight zone. All no-flight zones are implemented as binary octant trees [55].

Šišlák *et al.* [56, 57] present the decentralized collision avoidance algorithm utilizing a solution of the defined optimization problem where efficiency criteria collision penalties and airplanes' missions are integrated in an objective function. The optimal control for UAVs is a set of actions which minimize the objective function and thus solve collisions as well.

Table 6.6 Numerical optimization methods

Model	Resolution method	Detection	Trajectory prediction	Evasion
Durand	Nu	3D	N	T
Menon	Nu	3D	N	SC, T, CD
Frazzoli	Nu	HP	N	SC, T
Bicchi	Nu	HP	N	T
Pallotino	Nu	HP	N	T(SC)
Raghunathan	Nu	3D	–	–
Šišlák I	Nu	3D	N	SC, T, AC
Šišlák II	Nu	2D	N	T

They use the *probability collectives* (PC) framework [58, 59] as an optimization solver. The PC is a stochastic optimizer using probabilistic operators optimizing over a variable space. The major benefit of the PC optimizer is that the whole optimization process can be distributed among several agents controlling airplanes – several parts can be performed simultaneously. Two different implementation approaches are then used for PC deployment: (i) *parallelized optimization* and (ii) *centralized optimization*. In the first approach, the PC optimization process is done cooperatively by a group of UAVs. Each optimized variable from the PC is mapped to one UAV. This approach can fully profit from a parallelized execution of the PC optimization but on the other hand it requires a complex negotiation protocol. The second approach requires collection of optimization inputs, selection of a host where the optimization will be performed, and distribution of the solution to all involved UAVs. The process integrated mechanism (PIM) [60] was used, which automatically takes care of synchronization and communication issues utilizing migration of the coordination process among UAVs. On the other hand, such an approach cannot utilize the parallelization potential of the stochastic optimization. Taxonomy of these methods is included in Table 6.6.

6.3.7   Combined Methods

Pappas *et al.* [61, 62] proposed an automated decentralized conflict resolution scheme based on a hybrid system including both (i) non-cooperative dynamic game and (ii) coordinated resolution based on predefined control laws. The basic component of their collision avoidance architecture is the long-range conflict prediction component. This component identifies the set of airplanes involved in potential conflicts. Once this set of airplanes is identified, their approach assumes that no new airplane will enter the set until all conflicts are resolved. If a new airplane is likely to enter that set, an extended problem including this airplane is formulated.

The first attempt to resolve the conflict is to perform the non-cooperative collision avoidance with no coordination among airplanes. In this case, airplanes are considered as players in a non-cooperative, zero-sum dynamic game [63]. Each airplane is aware of the possible actions of others. Actions of other airplanes are modeled as disturbances. Assuming a saddle solution to the game exists, the airplane chooses an optimal policy assuming the worst possible disturbance. The resulting solution involves switching between different modes of operation and can be represented as a hybrid automaton. The performance requirements for each airplane are encoded in various cost functions and the game is won whenever the cost function exceeds a certain threshold. The performance requirement is encoded by the distance between two agents which should never fall below a minimum threshold – known as a separation distance. If the saddle solution of the game is safe (the cost function exceeds a certain threshold), the airplane follows the policy dictated by the saddle solution and no coordination is used.

If the saddle solution of the game is not safe, partial coordination between airplanes is used in order to reduce the disturbance set. During the partial coordination, homogenous airplanes exchange their disturbance sets for which a safe solution exists for their game. If the intersection of these reduced disturbance sets is nonempty then collision can be avoided by simply reducing the possible actions of each airplane. With heterogenous airplanes, where each airplane is ranked with unique priority, airplanes may choose their own policy as long as it doesn't interfere with the policies of the higher ranked agents. Where the reduction of disturbance sets still doesn't lead to the safe solution, the full cooperative collision avoidance is applied. In this case, airplanes follow predefined maneuvers which are proven to be safe – for example, the right-hand rule. The coordination among airplanes is in the form of predefined communication protocols and control laws which have been proven to satisfy performance requirements.

In [61, 62], only the heading control for airplanes is considered. In the subsequent extensions of this work [64, 65], there is addressed the non-cooperative part of the hybrid architecture described above where the game is modeled by a finite automata with differential equations associated with each state and results in the hybrid system which is safe by design. The automata implementing a constant altitude conflict resolution is provided, considering both (i) resolution by angular velocity (heading changes) and (ii) resolution by linear velocity (speed changes). In [66, 67], verification is provided that conflict resolution maneuvers resolve the initial conflict and are safe. Košecká et al. [68, 69] used potential and vector field techniques for multiple airplane motion planning. The hybrid system technique introduced above utilizes the coordination between airplanes using a series of horizontal and vertical planar avoidance maneuvers resulting in two-and-a-half-dimensional solutions.

Krozel et al. [70] proposed an optimal control theory [71] to maximize the miss distance for short-term conflicts. Their solution considers only short-term conflict for two airplanes in a deterministic setting. They introduced a tactical alert zone around an airplane which is used by the optimal control strategy to provide a resolution maneuver based on an economical conflict resolution strategy with safety integrated as a constraint. The resolution strategy is the result of the optimization function where the closest approach is maximized. They apply Euler–Lagrange equations for the optimal control. Moreover, initially only the two-dimensional case is considered which provides heading or velocity maneuvers only. Later in [72], they extended the approach to provide the fully three-dimensional solution considering also altitude maneuvers for the conflict resolution.

In [3, 73], they extended the tactical approach with the strategic level which provides the conflict resolution for mid-term collisions. They model the strategic conflict detection with a non-deterministic analysis by introducing a conflict probability map. The strategic strategy optimizes economics while it maintains safety as the constraint. The strategic conflict resolution strategy analyzes the geometry of heading, speed, and altitude maneuvers and estimates the direct operating constant for these maneuvers.

Gross et al. [74] used a mixed geometric and collision cone approach known from ground robotics [75] for collision avoidance of two airplanes in the three-dimensional environment. For the most general cases, they derived analytical results using numerical optimization

Table 6.7 Combined methods

Model	Resolution method	Detection	Trajectory prediction	Evasion
Pappas, Tomlin	C	HP	N	T
Kosecka	C	HP	N	SC, T
Krozel	C	3D	N	SC, T, AC
Gross	C	3D	N	SC, T, AC
Šišlák III	C	3D	FP	SC, T
Angelov	C	2D	WC	T

techniques. The results provided by their algorithm are optimal as they tend to minimize the velocity vector changes and thus result in minimum deviations from the nominal trajectory and avoid the conflict. Similarly to [38], the mixed approach combines changes in velocity, heading, and vertical changes together. They integrate implicit bounds on the airspeed and turning rates to enforce realistic scenarios.

Šišlák *et al.* [76] present a rule-based approach to conflict detection and resolution using the multi-agent framework to model the system. The rule-based collision solver is a domain-dependent algorithm. It is based on the visual flight rules defined by the FAA. The airspace around the aircraft is divided into four sectors. First, the type of collision between airplanes is identified and the specific conflict type is detected based on the angle between the direction vectors of the concerned airplanes projected to the ground plane. For head-on conflict the airplanes avoid each other by both of them turning to the right. For a rear collision, there are two subcases: (i) the front aircraft is faster and the airplanes do not change their current flight plans; (ii) the rear airplane is faster and the airplane has to change its intended flight plan so that it turns to the right and passes the front airplane without endangering it. For a side collision, one of the aircrafts (say airplane A) needs to slow down its speed so that it reaches the collision point later than the other airplane (say airplane B). If this is not possible due to the minimal flight speed defined for each airplane type, then airplane A slows down as much as possible and shifts its intended flight plan point to the right so there is no collision between the two flight plans. For a side collision, airplane B has lower traffic priority. Aircraft A changes its flight plan by increasing its flight speed so that it passes the collision point before airplane B. Airplane A only accelerates as much as needed.

Angelov *et al.* [77] present a passive approach to conflict detection and resolution. A collision is detected based only on the bearings of the aircraft. The method estimates the risk based on the current and passed bearings. The risk estimator uses a Gaussian. Once a collision is detected, the maneuver based on the worst-case scenario is initiated. After avoiding the collision, an optimal return to the pre-planned route is executed. All combined methods are summarized in Table 6.7.

6.3.8 Multi-agent Methods

Wangermann *et al.* [78–82] used principled negotiation [83] among agents with different interests for air-traffic management by time slot negotiations. In their approach, agents generate options and assess proposals that are best-suited to their problem. So, each agent optimizes its own actions. The actions of all agents are incorporated in a declarative, procedural, and reflexive model of behavior. Principled negotiation allows agents to search options that would not be available otherwise, improving the utility function of all agents.

Rong et al. [84] described a cooperative agent-based solution based on constraint satisfaction problems. Conflicting airplanes negotiate pairwise until a mutually acceptable resolution is found. A pairwise and argument-based negotiation approach is established for the airplane to search for a multilaterally acceptable conflict resolution. The proposed algorithm can fail and no solution is provided. The ground-based air-traffic controller is always included in their system and acts as a high-level supervisor and coordinator. He has the authority to approve or override any proposal from any airplane. If negotiation fails, the centralized controller forces its own resolution.

Wollkind et al. [85, 86] defined various protocols for a fully distributed solution based on agent-to-agent negotiation. They propose the solution for two-airplane conflict using the monotonic concession protocol [87] with aircraft-to-aircraft data links. The airplane looks 20 minutes into the future for prediction of conflicts. If a new conflict is going to occur, the airplane initiates negotiation with the other airplane. Airplanes exchange alternative trajectories along with utility scores integrating preferences of the airplane. After this exchange, airplanes initiate the monotonic concession protocol to select one of the deals from the negotiation set.

Resmerita et al. [88–91] used the airspace partitioning into static cells occupied by only one airplane at a given time. The conflict resolution is carried out by finding a conflict-free path through the cells. Two trajectories of airplanes are in conflict if they occupy the same cell at the same time. These cells become the vertices of an undirected graph whose edges are paths between cells. Agent (airplane) trajectories are directed, timed graphs that overlay the airspace graph. Before an aircraft enters the system, it registers itself with a central controller that maintains a list of all airplanes and their trajectories. The controller then distributes resources as planes request them, eliminating any communication between agents.

Conflict resolution becomes necessary when an agent desires a resource that has already been allocated. First, the agent attempts to alternate the paths. If no acceptable path can be constructed, the controller will request that agents holding the resources free them by choosing alternate paths of their own. Such a cascade of resource shuffling can free necessary resources or fail, and the airplane cannot enter the system at all. Each airplane follows one of its optimal paths; if an optimal path cannot be found, it does not enter. This algorithm is computationally intense and depends on a centralized controller with full knowledge.

Approaches provided by Wangermann et al. [78], Resmerita et al. [88], and Jonker et al. [92] are related to conflict resolution but do not adopt the free flight concept which has many benefits in comparison to algorithms working with the current ATM structures. There are pairwise negotiations where in each negotiation a collision between two airplanes is resolved.

Šišlák et al. [51, 93] presents the utility-based iterative peer-to-peer algorithm for cooperative CDR. It is a domain-independent algorithm, where the maneuver implementation is domain-dependent. A utility-based avoidance mechanism provides the solution for a pair of airplanes. First, the participating airplanes generate a set with various modified flight trajectories and each variation is evaluated with the utility value. The utility function is used to include the aircraft's intention in the proposed solutions of the conflict. Variations of their flight trajectories are generated using seven parameterized conflict resolution maneuvers: straight maneuver (no change in the flight plan), turn right, turn left, climb, descend, speed up, and slow down. The best possible conflict resolution maneuver is identified by a variation of the monotonic concession protocol [87].

Hill *et al.* [94–97] used an approach based on satisficing game theory [98, 99]. Satisficing game theory is the concept based on dual social utility: selectability and rejectability. Selectability characterizes the effectiveness in reaching the goal regardless of the cost and rejectability describes the amount of resources consumed. Unlike conventional game theory models maximizing self-interest metrics, they propose a satisficing extension where the airplanes take into consideration the preferences of others. Their algorithm provides heading changes only. Each airplane divides all others into subgroups with specified priorities. Based on these priorities, each airplane computes the complete selectability of higher ranked airplanes using its own incomplete knowledge. This approach is very complex.

Thus, they introduce a simplified model, where the airplanes are divided into five groups according to possible heading changes. The number of airplanes in each group is taken as a weight for the group. The final decision is made according to whether the airplane is risk averse or risk seeking. Risk-averse airplanes select the option with the lowest rejectability utility and risk-seeking airplanes select the option with the highest selectability utility. The mode of the airplane is selected according to the difference between selectability and rejectability utilities for each airplane.

In [93, 100], Šišlák *et al.* propose the CDR method based on the creation of groups of airplanes which together solve one or more conflicts. In the denser airspace, this method provides better utilization of the airspace. Let's imagine a situation where two airplanes have a conflict but for them it is difficult to avoid the conflict as other airplanes are close to them. The situation can be so difficult that they have only two options, either dramatically deviate from their course or deviate only slightly but make their flight plans collide with another airplane's flight plans. However, they can create a group with the other airplanes and solve the collision together. Basically, we can say that the two colliding airplanes will ask other airplanes to make space for their evasion maneuvers. The basic idea behind the proposed multiparty algorithm is to search the state space of possible applications of sequences of evasion maneuvers in the flight plans of airplanes. The goal of the search is to solve a multi-conflict with respect to given criteria evaluating the fitness of solution. Again the algorithm utilizes the A* algorithm [54]. Categorization of listed multi-agent methods is in Table 6.8.

Table 6.8 Multi-agent methods

Model	Resolution method	Detection	Trajectory prediction	Evasion
Wangermann	MA	3D	P	SC, T, CD
Rong	MA	3D	N	SC, T, CD
Wollkind	MA	3D	N	3D
Resmerita	MA	3D	N	SC, T, CD
Šišlák IV	MA	3D	FP	SC, T, AC
Hill	MA	HP	N	T
Šišlák V	MA	3D	FP	SC, T, AC

6.3.9 Other Methods

Erzberger *et al.* [101] presented the approach combining deterministic trajectory prediction and stochastic conflict analysis to achieve reliable conflict detection. The conflict probability algorithm is a three-dimensional extension of their previous algorithm [102]. They formulate error models for trajectory prediction and estimate conflict probability as a function of an encounter geometry. Then, they use motion equations to generate four-dimensional trajectories for automated conflict resolution with constraints on the conflict probability.

Viebahn *et al.* [103] developed the method for detecting and avoiding flight hazards that combines all potential external threats for an airplane into a single system. Their method is based on an airplane-surrounding airspace model consisting of discrete volume elements. For each element of the volume the threat probability is derived from the sensor output. The position of the own airplane is predicted by utilizing a probability distribution. Such an approach ensures that all potential positions of the airplane within the near future are considered during weighting the most likely flight path. The conflict resolution maneuver is generated taking into account all potential hazards in its surrounding.

Alliot *et al.* [104–106] solved multi-airplane conflicts by generating resolution maneuvers for each aircraft sequentially with a token allocation strategy. Each collision pair is solved so that there is no collision for the next 5 minutes. They used the A* algorithm to select the optimal maneuver from a set of predefined maneuvers. Problems may occur when one attempts to apply successively the pairwise resolution to multiple aircraft conflicts. The performance depends highly on the quality of the ordering. It is extremely difficult to find the optimal ordering [41, 107].

Burdun *et al.* [108] designed an artificial intelligence knowledge model which supports automated conflict management for the free flight operation. A group of potentially conflicting airplanes is represented as an autonomous process. This process incorporates the principles of collective behavior inspired by nature, like bird flocking, fish schooling, insect swarming, etc. These principles are used to manage potential conflicts based on kinematic and geometric constraints. Beside these principles, it integrates comprehensive knowledge of system dynamics. They combine self-organization and physics-based prediction because a flocking model alone is not sufficient to avoid collisions. Based on certain conditions they employ one of the methods.

Hu *et al.* [109] used the braid theory to categorize the types of multi-airplane resolutions. They classify the type of conflict resolution for airplane encounters according to their state in the joint space-time coordinates. The probabilistic resolution algorithm is integrated as a random type chooser so that the combinatorial optimization problem is solved with a randomized solution. The result of this classification assigns correspondence to that of the pure braid groups. They construct the resolution maneuvers by solving a convex optimization problem to minimize the total length of trajectories included in the cost function for a given type of maneuver.

For two airplanes, they use analytic expressions for construction of optimal resolution maneuvers. For the multiple airplane case, they use the convex optimization technique to find the optimal two-legged resolution maneuver within each categorized collision type. When increasing the number of airplanes, their solution becomes computationally complex. Due to the usage of randomized algorithms, their approach could produce different solutions to the same conflict problems.

Krozel *et al.* [110] described the distributed algorithm using constant speed heading changes only. They resolve the future collisions in a pairwise manner using finite time horizon. The solution of the previous conflict can cause further conflict, which is solved within the next iteration. The colliding airplane is passed in front or behind the conflicting airplane using two different strategies. The myopic strategy determines the most efficient resolution for the conflict by choosing the front-side or back-side maneuver that requires the minimum heading change. The look-ahead strategy first determines the most efficient maneuver (front- or back-side) and then checks if this maneuver would create a new conflict earlier than the current one. If no such conflict is found, it executes the selected maneuver (similar to myopic strategy). However, if such a conflict is found, it checks the opposite solution to see whether it is conflict-free. If so, it executes that solution. Otherwise, it searches for the maneuver heading in 2-degree increments starting from the original variant until it finds a conflict-free path. This results in a minimum separation distance in a given time horizon.

Mao *et al.* [111–115] addressed the conflict resolution problem for two perpendicular airplane flows intersecting at a fixed point. They proved that the resolution scheme based on one minimum heading change formulated as the heading change maneuver upon entering the airspace provides a stable solution for this conflict as an extension of their previous work [116]. Further work was extended to analyze the multiple intersecting flows where simple decentralized conflict avoidance rules may not handle all traffic scenarios [117]. The conflict resolution problem is decomposed into a sequence of subproblems. Each involves only two intersecting flows of airplanes. The strategy for achieving the decomposition is to laterally displace the airplane flows so that they intersect in pairs and the resulting conflict zones are non-overlapping. They define the conflict zone as a circular area that is centered at the intersection of a pair of flows. They formulate an optimization problem to minimize the lateral displacements of the airplane flows. The non-convex problem is then converted to a closed-form solution for three intersecting flows.

Bayen *et al.* [118] used a theoretical model of sector-based air-traffic flow which is derived using the hybrid automata theory. Their model is based on trajectory-dependent aggregate quantities such as the average number of aircrafts in a portion of airspace. Using the finite set of simple instructions for airplanes provided by the air-traffic control system, their model tries to reach thee maximum authorized number of aircraft in sectors. The system is very close to the current air-traffic control based on predefined airways where the system primarily controls the distribution of airplanes along these airways. The defined hybrid automaton used for airplane control also integrates airways changes represented by shortcut and detour actions.

Visintini *et al.* [119, 120] defined the conflict resolution problem as an optimization of expected value criterion. Their model includes the levels of uncertainty using a stochastic simulator [121]. The iterative procedure based on Markov chain Monte Carlo is used for the conflict resolution in a stochastic setting. Their approach is motivated by Bayesian statistics [122]. They consider an expected value resolution criterion that takes into account separation. The approach selects the most efficient maneuver which is safe with high enough probability. Such a constrained optimization problem they approximated by an optimization problem with an expected value criterion so that the optimal maneuver ensures a high probability of constraint satisfaction. The method is validated for terminal and approach sectors.

Grabbe *et al.* [123] set the domain in terms of a job shop scheduling problem over user-preferred routes to reduce trajectory crossing points. For the specified job shop scheduling problem, they use a 0–1 integer programming model to calculate the optimal departure and en-route control. The model is enhanced with a ration-by-schedule-based heuristic to transform the original model into a subset of problems.

Paielli [124] provided an algorithm for computing vertical resolution maneuvers [125] to resolve air-traffic conflicts within the next two minutes. He uses several procedures including rejection of altitude amendments, temporary altitudes, step altitudes, level-off confirmation.

Grabbe et al. [126, 127] proposed a sequential optimization method to manage the air-traffic flow under uncertainty in airspace capacity. They developed a decision support system by integrating a deterministic integer programming model for assigning delays to airplanes under en-route capacity constraints to reactively account for system uncertainties. They address the departure control model which assigns the pre-departure delays to individual flights based on deterministic estimates of the flight schedule, airport capacities, and airspace capacities.

Kim et al. [128] present a real-time algorithm for CDR which uses 3D point – mass aircraft models. The conflict probability is calculated by using Monte Carlo simulation. The time of the collision and the distance between the aircrafts in the given time are compared to given threshold values to determine the collision probability. For the conflict resolution, the predefined maneuvers are used. The best maneuver for resolution is selected and the acceleration command is transformed into thrust, load, and bank angle of the aircraft.

Gariel and Feron [129] presented a 3D conflict resolution algorithm in the presence of uncertainty in the case of failure in communication, navigation, or surveillance systems. The authors propose using a minimal set of maneuvers–heading change, speed change, and flight level change. The mixed integer program is used to determine the set of maneuvers to be used in conflict resolution. Uncertainties are simply modeled as an increase in the requirements for the safety zone size.

Table 6.9 Other methods

Model	Resolution method	Detection	Trajectory prediction	Evasion
Erzberger	O	3D	N	VC, T, AC
Viebahn	O	3D	P	VC, T, AC
Alliot	O	3D	N	VC, T, AC
Burdun	O	3D	P	VC, T, AC
Hu II	O	3D	P	VC, T, AC
Krozel	O	2D-HP	N	T
Mao	O	2D-HP	N	T
Bayen	O	2D	N	–
Visintini	O	3D	P	VC, T, AC
Grabbe I	O	3D	P	VC, T, AC
Paielli	O	2D-HP	N	AC
Grabbe II	O	3D	P	VC, T, AC
Kim	O	3D	P	VC, T, AC
Gariel & Feron	O	3D	N	VC, T, AC
Kouzegharani	O	–	P	–
Van Daalen	O	3D	P	VC, T, AC

Kouzegharani in his dissertation [130] models the CDR process as a hybrid system – interaction of the continuous dynamics of the aircraft used for collision prediction together with the discrete conflict detection logic. A hybrid hidden Markov model is used to enable the prediction of the stochastic aircraft states by combining elements of the probabilistic timed input–output automaton and partially observable Markov decision process frameworks.

In the dissertation of van Daalen [131] the author introduces probabilistic collision detection using the flow of probability through the boundary of the conflict region. The actual conflict detection is computed using adaptive numerical integration. The conflict resolution method uses the kino-dynamic motion planning algorithms with probabilistic road-maps. Taxonomy of all other methods is presented in Table 6.9.

Acknowledgments

The research in this chapter has been sponsored by the Czech Ministry of Education grant number 6840770038, the Czech Ministry of Defence grant number OVCVUT2010001, and by the Air Force Office of Scientific Research, Air Force Material Command, USAF, under grant number FA8655-06-1-3073. The views and conclusions contained herein are those of the authors and should not be interpreted as representing the official policies or endorsements, either expressed or implied, of the Air Force Office of Scientific Research or the US Government.

References

1. W. H. Harman, 'TCAS: A system for preventing midair collisions,' *Lincoln Laboratory Journal*, vol. 2, no. 3, pp. 437–457, 1989.

2. FAA, Document DOT/FAA/RD-91/5, *Precision Runway Monitor Demonstration Report*, February 1991.

3. J. Krozel, M. Peters, and G. Hunter, *Conflict detection and resolution for future air transportation management*. Technical Report NASA CR-97-205944, April 1997.

4. J. Kuchar and L. Yang, 'A review of conflict detection and resolution modeling methods,' *IEEE Transactions on Intelligent Transportation Systems*, vol. 1, pp. 179–189, December 2000.

5. K. Zeghal, 'A review of different approaches based on force fields for airborne conflict resolution,' in *Proceedings of the AIAA Guidance, Navigation, and Control Conference* (Boston, MA), pp. 818–827, August 1998.

6. B. Albaker and N. Rahim, 'A survey of collision avoidance approaches for unmanned aerial vehicles,' in *Technical Postgraduates (TECHPOS), 2009 International Conference for*, 2009.

7. R. Holdsworth, *Autonomous In-Flight Path Planning to replace pure Collision Avoidance for Free Flight Aircraft using Automatic Dependent Surveillance Broadcast*. PhD thesis, Swinburne University, Melbourne, Australia, November 2003.

8. J. Kim, S. Sukkarieh, and S. Wishart, 'Real-time navigation, guidance, and control of a UAV using low-cost sensors,' in *Springer Tracts in Advanced Robotics*, 2006.

9. R. Schild, *Rule optimization for airborne aircraft separation*. PhD thesis, Technical University Vienna, Vienna, Austria, November 1992.

10. I. Hwang, J. Kim, and C. Tomlin, 'Protocol-based conflict resolution for air traffic control,' *Air Traffic Control Quarterly*, vol. 15, no. 1, pp. 1–34, 2007.

11. I. Hwang and C. Tomlin, 'Protocol-based conflict resolution for finite information horizon,' in *Proceedings of the American Control Conference*, pp. 748–753, 2002.

12. R. Lachner, 'Collision avoidance as a differential game: Real-time approximation of optimal strategies using higher derivatives of the value function,' in *Proceedings of the IEEE International Conference on Systems, Man, and Cybernetics*, vol. 3, pp. 2308–2313, October 1997.

13. R. Isaacs, *Differential Games*. New York: R. E. Krieger, 1965.

14. J. Zhang and S. Sastry, 'Aircraft conflict resolution: Lie–Poisson reduction for game on SE(2),' in *Proceedings of the 40th IEEE Conference on Decision and Control*, vol. 2, pp. 1663–1668, 2001.

15. A. Bayen, S. Santhanam, I. Mitchell, and C. Tomlin, 'A differential game formulation of alert levels in ETMS data for high-altitude traffic,' in *Proceedings of the AIAA Guidance, Navigation, and Control Conference* (Austin, TX), August 2003.

16. V. Duong and K. Zeghal, 'Conflict resolution advisory for autonomous airborne separation in low-density airspace,' in *Proceedings of the 36th IEEE Conference on Decision and Control*, vol. 3, pp. 2429–2434, December 1997.

17. V. Duong and E. Hoffman, 'Conflict resolution advisory service in autonomous aircraft operations,' in *Proceedings of the 16th Digital Avionics System Conference* (Irvine, CA), pp. 9.3.10–9.3.17, October 1997.

18. K. Zeghal, 'Toward the logic of an airborne collision avoidance system which ensures coordination with multiple cooperative intruders,' in *Proceedings of the International Council of the Aeronautical Sciences*, vol. 3 (Anaheim, CA), pp. 2208–2218, September 1994.

19. J. Hu, J. Lygeros, M. Prandini, and S. Sastry, 'Aircraft conflict prediction and resolution using Brownian motion,' in *Proceedings of the 38th IEEE Conference on Decision and Control*, vol. 3, pp. 2438–2443, 1999.

20. M. Eby and W. Kelly, 'Free flight separation assurance using distributed algorithms,' in *Proceedings of the IEEE Aerospace Conference* (Snowmass, CO), pp. 429–441, March 1999.

21. W. Kelly and M. Eby, 'Advances in force field conflict resolution algorithms,' in *Proceedings of the AIAA Guidance, Navigation, and Controls Conference* (Denver, CO), August 2000.

22. M. S. Eby, 'A self-organizational approach for resolving air traffic conflicts,' *Lincoln Laboratory Journal*, vol. 7, no. 2, pp. 239–254, 1994.

23. M. Prandini, J. Hu, J. Lygeros, and S. Sastry, 'A probabilistic approach to aircraft conflict detection,' *IEEE Transactions on Intelligent Transportation Systems*, vol. 1, pp. 199–220, December 2000.

24. M. Prandini, J. Lygeros, A. Nilim, and S. Sastry, 'A probabilistic framework for aircraft conflict detection,' in *Proceedings of the AIAA Guidance, Navigation, and Control Conference* (Portland, OR), August 1999.

25. M. Prandini, J. Lygeros, A. Nilim, and S. Sastry, 'Randomized algorithms for probabilistic aircraft conflict detection,' in *Proceedings of the 38th IEEE Conference on Decision and Control*, vol. 3, pp. 2444–2449, 1999.

26. J. Hu, M. Prandini, and S. Sastry, 'Aircraft conflict prediction in the presence of a spatially correlated wind field,' *IEEE Transactions on Intelligent Transportation Systems*, vol. 6, pp. 326–340, September 2005.

27. M. Prandini and J. Hu, 'Application of reachability analysis for stochastic hybrid systems to aircraft conflict prediction,' in *Proceedings of the 47th IEEE Conference on Decision and Control*, pp. 4036–4041, December 2008.

28. H. A. P. Blom and J. Lygeros, *Stochastic hybrid systems: Theory and safety applications, volume 337 of Lecture Notes in Control and Informations Sciences*. Berlin: Springer, 2006.

29. M. Prandini, L. Piroddi, and J. Lygeros, 'A two-step approach to aircraft conflict resolution combining optimal deterministic design with Monte Carlo stochastic optimization,' in *Proceedings of the European Control Conference* (Budapest, Hungary), August 2009.

30. T. Ota, M. Nagati, and D. Lee, 'Aircraft collision avoidance trajectory generation,' in *Proceedings of the AIAA Guidance, Navigation, and Control Conference* (Boston, MA), pp. 828–837, August 1998.

31. Y. J. Chiang, J. T. Klosowski, C. Lee, and J. S. B. Mitchell, 'Geometric algorithms for conflict detection/resolution in air traffic management,' in *Proceedings of the IEEE Conference on Decision and Control*, pp. 1835–1840, December 1997.

32. S. Fortune, *Handbook of Discrete and Computational Geometry*, ch. Voronoi diagrams and Delaunay triangulations, pp. 377–388. Boca Raton, FL: CRC Press LLC, 1997.

33. K. Bilimoria, 'A geometric optimization approach to aircraft conflict resolution,' in *Proceedings of the AIAA Guidance, Navigation, and Control Conference* (Denver, CO), August 2000.

34. J. Hu, M. Prandini, A. Nilim, and S. Sastry, 'Optimal coordinated maneuvers for three dimensional aircraft conflict resolution,' *Proceedings of the AIAA Journal of Guidance, Control and Dynamics*, vol. 25, pp. 888–900, 2002.

35. J. Hu, M. Prandini, A. Nilim, and S. Sastry, 'Optimal coordinated maneuvers for three dimensional aircraft conflict resolution,' *Proceedings of the AIAA Guidance, Navigation and Control Conference*, August 2001.

36. M. A. Christodoulou and S. G. Kodaxakis, 'Automatic commercial aircraft-collision avoidance in free flight: The three-dimensional problem,' *IEEE Transactions on Intelligent Transportation Systems*, vol. 7, pp. 242–249, June 2006.

37. M. A. Christodoulou and C. Kontogeorgou, 'Collision avoidance in commercial aircraft free flight via neural networks and non-linear programming,' *International Journal of Neural Systems*, vol. 18, no. 5, pp. 371–387, 2008.

38. S. Luongo, C. Carbone, F. Corraro, and U. Ciniglio, 'An optimal 3D analytical solution for collision avoidance between aricraft,' in *Proceedings of the IEEE International Conference on Mechatronics*, 2009.

39. J. Ny and G. J. Pappas, 'Geometric programming and mechanism design for air traffic conflict resolution,' in *American Control Conference*, 2010.

40. J. Park, H. Oh, and M. Tahk, 'UAV collision avoidance based on geometric approach,' in *SICE Annual Conference*, 2008.

41. N. Durand, J.-M. Alliot, and O. Chansou, 'An optimizing conflict solver for ATC,' *Journal of Air Traffic Control*, vol. 3, 1995.

42. N. Durand, J.-M. Alliot, and F. Medioni, 'Neural nets trained by genetic algorithms for collision avoidance,' *Applied Artificial Intelligence*, vol. 13, no. 3, 2000.

43. P. K. Menon, G. D. Sweriduk, and B. Sridhar, 'Optimal strategies for free flight air traffic conflict resolution,' *AIAA Journal of Guidance, Control, and Dynamics*, vol. 22, no. 2, pp. 202–211, 1997.

44. E. Frazzoli, Z. Mao, J.-H. Oh, and E. Feron, 'Resolution of conflicts involving many aircraft via semi-definite programming,' *Journal of Guidance, Control, and Dynamics*, vol. 24, pp. 79–86, February 1999.

45. A. Bicchi and L. Pallottino, 'On optimal cooperative conflict resolution for air traffic management systems,' *IEEE Transactions on Intelligent Transportation Systems*, vol. 1, pp. 221–232, December 2000.

46. E. Frazzoli, L. Pallottino, V. Scordio, and A. Bicchi, 'Decentralized cooperative conflict resolution for multiple nonholonomic vehicles,' in *Proceedings of the AIAA Guidance, Navigation and Control Conference*, August 2005.

47. L. Pallottino, V. Scordio, E. Frazzoli, and A. Bicchi, 'Probabilistic verification of a decentralized policy for conflict resolution in multi-agent systems,' in *Proceedings of the International Conference on Robotics and Automation* (Orlando, FL), pp. 2448–2453, 2006.

48. L. Pallottino, E. Feron, and A. Bicchi, 'Conflict resolution problems for air traffic management systems solved with mixed integer programming,' *IEEE Transactions on Intelligent Transportation Systems*, vol. 3, pp. 3–11, March 2002.

49. L. Pallottino, A. Bicchi, and E. Feron, 'Mixed integer programming for aircraft conflict resolution,' in *Proceedings of the AIAA Guidance, Navigation, Control Conference* (Montreal, Canada), August 2001.

50. A. Raghunathan, V. Gopal, D. Subramanian, L. Biegler, and T. Samad, 'Dynamic optimization strategies for three-dimensional conflict resolution of multiple aircraft,' *AIAA Journal of Guidance, Control, and Dynamics*, vol. 27, no. 4, pp. 586–594, 2004.

51. D. Šišlák, P. Volf, A. Komenda, J. Samek, and M. Pěchouček, 'Agent-based multi-layer collision avoidance to Unmanned Aerial Vehicles,' in *Proceedings of International Conference on Integration of Knowledge Intensive Multi-Agent Systems (KIMAS)* (Piscataway, NJ), pp. 365–370, IEEE, 2007.

52. C. Tomlin, G. J. Pappas, and S. Sastry, 'Conflict resolution for air traffic management: A study in multi-agent hybrid systems,' *IEEE Transactions on Automatic Control*, vol. 43, pp. 509–521, 1998.

53. S.-C. Han and H. Bang, 'Proportional navigation-based optimal collision avoidance for UAVs,' in *Second International Conference on Autonomous Robots and Agents* (S. C. Mukhopadhyay and G. S. Gupta, eds), pp. 76–81, Massey University, New Zealand, 2004.

54. P. Hart, N. Nilsson, and B. Raphael, 'A formal basis for the heuristic determination of minimum cost paths,' *IEEE Transactions on Systems Science and Cybernetics*, no. 2, pp. 100–107, 1968.

55. S. Frisken and R. Perry, 'Simple and efficient traversal methods for quadtrees and octrees,' 2002.

56. D. Sislak, P. Volf, M. Pechoucek, and N. Suri, 'Automated conflict resolution utilizing probability collectives optimizer,' *IEEE Transactions on Systems, Man, and Cybernetics – Part C: Applications and Reviews*, vol. 41, pp. 365–375, May 2011.

57. D. Šišlák, P. Volf, M. Pěchouček, N. Suri, D. Nicholson, and D. Woodhouse, 'Optimization based collision avoidance for cooperating airplanes,' in *Proceedings of the IEEE/WIC/ACM International Conference on Intelligent Agent Technology (IAT)* (Los Alamitos, CA), IEEE Computer Society, 2009.

58. C. F. Lee and D. H. Wolpert, 'Product distribution theory for control of multi-agent systems,' in *AAMAS '04: Proceedings of the Third International Joint Conference on Autonomous Agents and Multiagent Systems* (Washington, DC), pp. 522–529, IEEE Computer Society, 2004.

59. D. H. Wolpert, 'Information theory – the bridge connecting bounded rational game theory and statistical physics,' in *Complex Engineered Systems* (D. Braha, A. A. Minai, and Y. Bar-Yam, eds), (Berlin), pp. 262–290, Springer, 2006.

60. K. M. Ford, N. Suri, K. Kosnar, P. Jisl, P. Benda, M. Pechoucek, and L. Preucil, 'A game-based approach to comparing different coordination mechanisms,' in *Proceedings of the IEEE International Conference on Distributed Human-Machine Systems (DHMS)*, IEEE, 2008.

61. G. J. Pappas, C. Tomlin, and S. Sastry, 'Conflict resolution in multi-agent hybrid systems,' in *Proceedings of the IEEE Conference on Decision and Control*, vol. 2, pp. 1184–1189, December 1996.

62. C. Tomlin, G. Pappas, J. Lygeros, D. Godbole, and S. Sastry, 'Hybrid control models of next generation air traffic management,' in *Hybrid Systems IV, Lecture Notes in Computer Science*, pp. 378–404, Springer-Verlag, 1997.

63. J. Lygeros, D. N. Godbole, and S. Sastry, 'A game theoretic approach to hybrid system design,' in *Lecture Notes in Computer Science 1066*, pp. 1–12, Springer-Verlag, 1995.

64. C. Tomlin, G. J. Pappas, and S. Sastry, 'Noncooperative conflict resolution,' in *Proceedings of the IEEE Conference on Decision and Control* (San Diego, CA), pp. 1816–1821, December 1997.

65. C. Tomlin, Y. Ma, and S. Sastry, 'Free flight in 2000: Games on Lie groups,' in *Proceedings of the 37th IEEE Conference on Decision and Control*, vol. 2, pp. 2234–2239, December 1998.

66. C. Tomlin, I. Mitchell, and R. Ghosh, 'Safety verification of conflict resolution manoeuvres,' *IEEE Transactions on Intelligent Transportation Systems*, vol. 2, pp. 110–120, June 2001.

67. C. Tomlin, G. Pappas, J. Košecká, J. Lygeros, and S. Sastry, 'Advanced air traffic automation: A case study in distributed decentralized control,' in *Control Problems in Robotics and Automation*, pp. 261–295, Springer-Verlag, 1998.

68. J. Košecká, C. Tomlin, G. Pappas, and S. Sastry, 'Generation of conflict resolution manoeuvres for air traffic management,' in *Proceedings of the Intelligent Robots and Systems Conference*, vol. 3, pp. 1598–1603, September 1997.

69. J. Košecká, C. Tomlin, G. Pappas, and S. Sastry, '2-1/2 D conflict resolution maneuvers for ATMS,' in *Proceedings of the 37th IEEE Conference on Decision and Control* (Tampa, FL), pp. 2650–2655, 1998.

70. J. Krozel, T. Mueller, and G. Hunter, 'Free flight conflict detection and resolution analysis,' in *Proceedings of the AIAA Guidance and Control Conference* (San Diego, CA), July 1996.

71. A. Bryson and Y. Ho, *Applied Optimal Control*. New York: Hemisphere, 1975.

72. J. Krozel and M. Peters, 'Conflict detection and resolution for free flight,' *Air Trafic Control Quarterly Journal*, 1997.

73. J. Krozel and M. Peters, 'Strategic conflict detection and resolution for free flight,' in *Proceedings of the IEEE Conference on Decision and Control* (San Diego, CA), pp. 1822–1828, December 1997.

74. J. Gross, R. Rajvanshi, and K. Subbarao, 'Aircraft conflict detection and resolution using mixed geometric and collision cone approaches,' in *Proceedings of the AIAA Guidance, Navigation, and Control Conference* (Rhode Island), 2004.

75. A. Chakravarthy and D. Ghose, 'Obstacle avoidance in a dynamic environment: A collision cone approach,' *IEEE Transactions on Systems, Man and Cybernetics, Part A: Systems and Humans*, vol. 28, pp. 562–574, September 1998.

76. M. Pěchouček, D. Šišlák, D. Pavlíček, and M. Uller, 'Autonomous agents for air-traffic deconfliction,' in *Proceedings of the 5th International Joint Conference on Autonomous Agents and Multiagent Systems (AAMAS)* (New York), pp. 1498–1505, ACM, 2006.

77. P. Angelov, C. D. Bocaniala, C. Xydeas, C. Pattchett, D. Ansell, M. Everett, and G. Leng, 'A passive approach to autonomous collision detection and avoidance in uninhabited aerial systems,' in *Tenth International Conference on Computer Modeling and Simulation, 2008. UKSIM 2008*, 2008.

78. J. P. Wangermann and R. F. Stengel, 'Optimization and coordination of multiagent systems using principled negotiation,' *Journal of Guidance, Control, and Dynamics*, vol. 22, no. 1, pp. 43–50, 1999.

79. J. P. Wangermann and R. F. Stengel, 'Principled negotiation between intelligent agents: A model for air traffic management,' *Artificial Intelligence in Engineering*, vol. 12, no. 3, pp. 177–187, 1998.

80. J. P. Wangermann and R. F. Stengel, 'Optimization and coordination of multi-agent systems using principled negotiation,' in *Proceedings of the AIAA Guidance, Navigation, and Control Conference* (San Diego, CA), pp. 43–50, July 1996.

81. J. P. Wangermann and R. F. Stengel, 'Principled negotiation between intelligent agents: A model for air traffic management,' in *Proceedings of the ICAS*, vol. 3 (Anaheim, CA), pp. 2197–2207, September 1994.

82. K. Harper, S. Mulgund, S. Guarino, A. Mehta, and G. Zacharias, 'Air traffic controller agent model for free flight,' in *Proceedings of the AIAA Guidance, Navigation, and Control Conference* (Portland, OR), pp. 288–301, August 1999.

83. R. Fisher and W. Ury, *Negotiating Agreement Without Giving In*. New York: Penguin, 1981.

84. J. Rong, S. Geng, J. Valasek, and T. R. Ioerger, 'Air traffic control negotiation and resolution using an onboard multi-agent system,' in *Proceedings of the Digital Avionics Systems Conference*, vol. 2, pp. 7B2–1–7B2–12, 2002.

85. S. Wollkind, J. Valasek, and T. R. Ioerger, 'Automated conflict resolution for air traffic management using cooperative multiagent negotiation,' in *Proceedings of the American Institute of Aeronautics and Astronautics Conference on Guidance, Navigation, and Control* (Providence, RI), 2004.

86. S. Shandy and J. Valasek, 'Intelligent agent for aircraft collision avoidance,' in *Proceedings of the AIAA Guidance, Navigation, and Control Conference* (Montreal, Canada), August 2001.

87. G. Zlotkin and J. S. Rosenschein, 'Negotiation and task sharing among autonomous agents in cooperative domains,' in *Proceedings of the 11th International Joint Conference on Artificial Intelligence* (San Mateo, CA), pp. 912–917, Morgan Kaufmann, 1989.

88. S. Resmerita, M. Heymann, and G. Meyer, 'Towards a flexible air traffic management: Dealing with conflicts,' in *Proceedings of the 11th World Conference on Tranport Research* (UC Berkeley), June 2007.

89. S. Resmerita and M. Heymann, 'Conflict resolution in multi-agent systems,' in *Proceedings of the IEEE Conference on Decision and Control*, vol. 2, pp. 2537–2545, 2003.

90. S. Resmerita, M. Heymann, and G. Meyer, 'A framework for conflict resolution in air traffic management,' in *Proceedings of the 42nd IEEE Conference on Decision and Control*, vol. 2, pp. 2035–2040, December 2003.

91. S. Resmerita, *A multi-agent approach to control of multi-robotic systems*. PhD thesis, Department of Computer Science, Technion - Israel Instite of Technology, Israel, 2003.

92. F. Jonker and J. Meyer, 'Achieving cooperation among selfish agents in the air traffic management domain using signed money,' in *Proceedings of the Sixth International Joint Conference on Autonomous Agents and Multi-Agent Systems*, May 2007.

93. D. Šišlák, P. Volf, and M. Pěchouček, 'Agent-based cooperative decentralized airplane collision avoidance,' *IEEE Transactions on Intelligent Transportation Systems*, vol. 12, pp. 36–46, March 2011.

94. J. K. Archibald, J. C. Hill, N. A. Jepsen, W. C. Stirling, and R. L. Frost, 'A satisficing approach to aircraft conflict resolution,' *IEEE Transactions on Systems, Man, and Cybernetics, Part C: Applications and Reviews*, vol. 38, no. 4, pp. 510–521, 2008.

95. J. C. Hill, F. R. Johnson, J. K. Archibald, R. L. Frost, and W. C. Stirling, 'A cooperative multi-agent approach to free flight,' in *Proceedings of the 4th International Joint Conference on Autonomous agents and Multiagent systems (AAMAS)* (New York), pp. 1083–1090, ACM Press, 2005.

96. F. R. Johnson, J. C. Hill, J. K. Archibald, R. L. Frost, and W. C. Stirling, 'A satisficing approach to free flight,' in *Proceedings of the IEEE Networking, Sensing and Control*, pp. 123–128, March 2005.

97. J. C. Hill, J. K. Archibald, W. C. Stirling, and R. L. Frost, 'A multi-agent architecture for air traffic control,' in *Proceedings of the 2005 AIAA Guidance, Navigation, and Control Conference* (San Francisco, CA), 2005.

98. J. K. Archibald, J. C. Hill, F. R. Johnson, and W. C. Stirling, 'Satisficing negotiations,' *IEEE Transactions on Systems, Man and Cybernetics, Part C: Applications and Reviews*, vol. 36, no. 1, pp. 4–18, 2006.

99. W. C. Stirling, *Satisficing Games and Decision Making: With Applications to Engineering and Computer Science*. Cambridge: Cambridge University Press, 2003.

100. D. Šišlák, J. Samek, and M. Pěchouček, 'Decentralized algorithms for collision avoidance in airspace,' in *Proceedings of the 7th International Converence on Autonomous Agents and Multi-Agent Systems (AAMAS)* (New York), pp. 543–550, ACM Press, 2008.

101. H. Erzberger, R. A. Paielli, D. R. Isaacson, and M. M. Eshow, 'Conflict detection and resolution in the presence of prediction error,' in *Proceedings of the 1st USA/Euroupe Air Traffic Management Research Development Seminar* (Saclay, France), June 1997.

102. R. A. Paielli and H. Erzberger, 'Conflict probability estimation for free flight,' *AIAA Journal of Guidance, Control, and Dynamics*, vol. 20, pp. 588–596, 1997.

103. H. von Viebahn and J. Schiefele, 'A method for detecting and avoiding flight hazards,' in *Proceedings of the SPIE Meeting on Enhanced Synthetic Vision* (Bellingham, WA), pp. 50–56, April 1997.

104. J.-M. Alliot, N. Durand, and G. Granger, 'FACES: a Free flight Autonomous and Coordinated Embarked Solver,' in *Proceedings of the 2nd USA/EUROPE Air Traffic Management R&D Seminar*, December 1998.

105. N. Durand and J.-M. Alliot, 'Optimal resolution of en route conflicts,' in *Proceedings of the Seminaire Europe/USA* (Saclay, France), 1997.

106. G. Granger, N. Durand, and J.-M. Alliot, 'Optimal resolution of en route conflicts,' in *Proceedings of Air Traffic Management*, 2001.

107. G. Granger, N. Durand, and J.-M. Alliot, 'Token allocation strategy for free-flight conflict solving,' in *Proceedings of the Thirteenth Conference on Innovative Applications of Artificial Intelligence Conference*, pp. 59–64, AAAI Press, 2001.

108. I. Burdun and O. Parfentyev, 'AI knowledge model for self-organizing conflict prevention/resolution in close free-flight air space,' in *Proceedings of the Aerospace Conference*, vol. 2, pp. 409–428, 1999.

109. J. Hu, M. Prandini, and S. Sastry, 'Optimal maneuver for multiple aircraft conflict resolution: A braid point of view,' in *Proceedings of the 39th IEEE Conference on Decision and Control*, vol. 4, pp. 4164–4169, 2000.

110. J. Krozel, M. Peters, K. D. Bilimoria, C. Lee, and J. S. Mitchel, 'System performance characteristics of centralized and decentralized air traffic separation strategies,' in *Proceedings of the 4th USA/Europe Air Traffic Management R&D Seminar* (Stanta Fe, NM), December 2001.

111. Z. H. Mao, D. Dugail, E. Feron, and K. Bilimoria, 'Stability of intersecting aircraft flows using heading-change maneuvers for conflict avoidance,' *IEEE Transactions on Intelligent Transportation Systems*, vol. 6, pp. 357–369, December 2005.

112. D. Dugail, E. Feron, and K. Bilimoria, 'Stability of intersecting aircraft flows using heading change maneuvers for conflict avoidance,' in *Proceedings of the American Control Conference*, pp. 760–766, 2002.

113. D. Dugail, Z. Mao, and E. Feron, 'Stability of intersecting aircraft flows under centralized and decentralized conflict avoidance rules,' in *Proceedings of the AIAA Guidance, Navigation, and Control Conference*, August 2001.

114. Z. Mao and E. Feron, 'Stability and performance of intersecting aircraft flows under sequential conflict resolution,' in *Proceedings of the 2001 American Control Conference*, pp. 722–729, June 2001.

115. Z. Mao and E. Feron, 'Stability of intersecting aircraft flows under decentralized conflict avoidance rules,' in *Proceedings of the AIAA Guidance, Navigation and Control Conference*, August 2000.

116. Z. Mao, E. Feron, and K. Bilimoria, 'Stability and performance of intersecting aircraft flows under decentralized conflict avoidance rules,' *IEEE Transactions on Intelligent Transportation Systems*, vol. 2, pp. 101–109, June 2001.

117. K. Treleaven and Z.-H. Mao, 'Conflict resolution and traffic complexity of multiple intersecting flows of aircraft,' *IEEE Transactions on Intelligent Transportation Systems*, vol. 9, pp. 633–643, December 2008.

118. A. Bayen, P. Grieder, G. Meyer, and C. Tomlin, 'Lagrangian delay predictive model for sector based air traffic flow,' in *AIAA Journal of Guidance, Control, and Dynamics*, vol. 28, pp. 1015–1026, 2005.

119. A. L. Visintini, W. Glover, J. Lygeros, and J. Maciejowski, 'Monte Carlo optimization for conflict resolution in air traffic control,' *IEEE Transactions on Intelligent Transportation Systems*, vol. 7, pp. 470–482, December 2006.

120. A. Lecchini, W. Glover, J. Lygeros, and J. M. Maciejowski, 'Air-traffic control in approach sectors: Simulation examples and optimisation,' in *Proceedings of the 8th International Workshop on Hybrid Systems: Computation and Control* (Zurich, Switzerland), pp. 433–448, March 2005.

121. W. Glover and J. Lygeros, 'A stochastic hybrid model for air traffic control simulation,' in *Proceedings of the 7th International Workshop on Hybrid Systems: Computation and Control* (Philadelphia, PA), pp. 372–386, March 2004.

122. P. Muller, 'Simulation based optimal design,' *Bayesian Statistics*, vol. 6, 1998.

123. S. Grabbe and B. Sridhar, 'Central East Pacific flight scheduling,' in *Proceedings of the AIAA Guidance, Navigation and Control Conference and Exhibit*, August 2007.

124. R. A. Paielli, 'Tactical conflict resolution using vertical maneuvers in enroute airspace,' *AIAA Journal of Aircraft*, vol. 45, no. 6, 2008.

125. R. A. Paielli, 'Modeling maneuver dynamics in air traffic conflict resolution,' *Journal of Guidance, Control, and Dynamics*, vol. 26, no. 3, pp. 407–415, 2003.

126. S. Grabbe, B. Sridhar, and A. Mujkerjee, 'Sequential traffic flow optimization with tactical flight control heuristics,' in *Proceedings of the AIAA Guidance, Navigation, Control Conference* (Honolulu, HI), August 2008.

127. S. Grabbe, B. Sridhar, and A. Mujkerjee, 'Integrated traffic flow decision making,' in *Proceedings of the AIAA Guidance, Navigation, and Control Conference* (Chicago, IL), August 2009.

128. K. Kim, J. Park, and M. Tahk, 'A probabilistic algorithm for multi-aircraft collision detection and resolution in 3-d,' in *KSAS International Journal*, 2008.

129. M. Gariel and E. Feron, '3d conflict avoidance under uncertainties,' in *Digital Avionics System Conference*, 2009.

130. A. N. Kouzehgarani, *Mode Identification Using Stochastic Hybrid Models with Applications to Conflict Detection and Resolution*. PhD thesis, University of Illinois at Urbana-Champaign, 2010.

131. C. E. van Daalen, *Conflict Detection and Resolution for Autonomous Vehicles*. PhD thesis, Stellenbosch University, 2010.

7

UAS Conflict Detection and Resolution Using Differential Geometry Concepts

Hyo-Sang Shin, Antonios Tsourdos and Brian White
Cranfield University, UK

7.1 Introduction

The large-scale military application of unmanned aircraft systems (UAS) is generating operating experience and technologies that will enable the next phases of utilisation. An important driver for future growth of UAS is in the civil commercial sector, which could emerge as the largest user in due course. The employment of UAS for military operations and civil commercial use remains restricted to operations mostly in segregated airspace. Integrating UAS into non-segregated airspace is the key enabler for dramatic development of many new and innovative applications. Therefore, it is likely that UAS integration into non-segregated airspace will be enabled as demand from operators materialises. Collision detection and resolution (CD&R) – i.e., sense and avoid – would be a major step towards allowing UAS integration into non-segregated airspace.

The CD&R algorithm has been considered an important problem ever since aircraft were developed. Moreover, there have been numerous studies and implementations of it. Thus far most CD&R algorithms have been performed by air traffic control (ATC) from the ground station [1]. However, the ground station-based air traffic control would have limited coverage capability to cope with dramatic increases in air traffic [2]. Free flight including an autonomous CD&R algorithm, operated by an onboard computer, can reduce the burden on the ground

---

Sense and Avoid in UAS: Research and Applications, First Edition. Edited by Plamen Angelov.
© 2012 John Wiley & Sons, Ltd. Published 2012 by John Wiley & Sons, Ltd.

station-based air traffic control [3]. This has become a possible option to consider due to the development of onboard computer and sensor technology.

A number of different approaches have been applied to the CD&R problem [4]. Collision avoidance using a potential function has been investigated since the first study by Khatib [5]. In essence, this method uses an artificial potential field which governs UAS kinematics, but cannot always guarantee that the relative distance is greater than a minimum safe separation distance due to the difficulty of predicting the minimum relative distance. Intelligent control has also been introduced as the new control theory [6] and implemented in the CD&R algorithm [7, 8]. A CD&R algorithm based on the hybrid system has been studied at Stanford University and UC Berkeley [2, 9]. In this research, cooperating vehicles resolve the conflict by specific manoeuvres such as level flight, coordinate turn, or both under the assumption that all information is fully communicated. There have also been studies of a hybrid CD&R algorithm using the Hamilton–Jacobi–Bellan equation and hybrid collision avoidance with ground obstacles using radar [10, 11].

Since traffic collision alerting system (TCAS) equipment with an onboard pilot has been used in civil aviation, a number of studies and flight tests on TCAS-like CD&R for UAS have been investigated over the past decades [12–14]. TCAS provides resolution advisories with vertical manoeuvre command, and these advisories are based on various experiences and a great database, not analytical verification [2]. However, it is possible to guarantee the high reliability of TCAS by statistical analysis. The accident rate is lower than 1% per year with respect to 50 possible accidents in the case of aircraft not equipped with TCAS [13]. UAS Battlelab has studied to develop TCAS on high-performance UAS such as Global Hawk in non-military airspace. Cho *et al.* [13] proposed the TCAS algorithm on UAS by converting TCAS vertical commands into UAS autopilot input and analysed its performance by numerical examples. Lincoln Laboratory, MIT published the result in 2004 [15]. Furthermore, the Environmental Research Aircraft and Sensor (ERAST) has investigated collision avoidance systems with TCAS-II and radar by experimentation. As a new concept of collision avoidance, air vehicles communicate their information using ADS-B. The Swedish Civil Aviation Administration (SCAA) performed flight tests for a medium-altitude long-endurance (MALE) UAS – the Eagle – produced by European Aeronautic Defense and Space Company (EADS) and Israel Aircraft Industries (IAI) to make a landing from one civil airport to another outside Kiruna with the concept of IFR (instrument flight rules) using a remote pilot in 2002 [16]. The Eagle UAS was equipped with a standard Garmin ATC transponder as well as a ADS-B VDLm transponder [4]. However, it is difficult to analytically verify these approaches and consider limitations caused by physical and operational constraints such as a relatively low decent/climb or turning rate.

The main issue with the CD&R is whether the algorithm can guarantee collision avoidance by strict verification, because the CD&R algorithm is directly related to the safety of the aerial vehicle. In this study, the cases for both single and multiple conflicts are considered for a single UAS. Two CD&R algorithms are proposed using the differential geometry concepts: one controls the heading angle alone and the other controls it with the ground speed. The proposed algorithms also use the principles of airborne collision avoidance systems [17, chapter 14] conforming to TCAS. Moreover, their stability and feasibility are examined using rigorous mathematical analysis, not using the statistical analysis as in the TCAS algorithm. In order to design the algorithms, we first introduce the definitions of the conflict, conflict detection and conflict resolution by using the same concepts as in [18, 19], such as the closest approach distance (CAD) and the time to closest point of approach (TCPA). Then, conflict resolution guidance will be proposed after deriving the geometric requirements to detect and resolve the conflict. The proposed algorithms are modified from previous research undertaken by the authors and presented in [20, 21].

The study limits the analysis to non-cooperating UAS and intruders, which is denoted as aircraft in this chapter, as this is more challenging in the CD&R problem. It is also assumed that:

- Vehicle dynamics are represented by point mass in Cartesian coordinates on R^2.

- Aircraft are non-manoeuvring for collision avoidance.

- UAS can obtain the deterministic position and velocity vectors of aircraft by using sensors, a communication system or estimator.

This implies that UAS predict the trajectories and future state information using the current position and velocity vector and their linear projections. Note that these assumptions are for ease of analysis and are not a restriction of the approach.

7.2 Differential Geometry Kinematics

Consider the scenario shown in Figure 7.1, where a UAS is following a prescribed path and an aerial vehicle is crossing the path with the possibility of interception.

The UAS senses the aircraft and establishes a sightline between it and the aircraft. Assuming the velocity of the aircraft is known (usually using a motion estimator), the motion geometry can be defined. Several axis sets can be defined. A tangent and normal basis vector set $(\mathbf{t}_s \ \mathbf{n}_s)$ defines the sightline, with $(\mathbf{t}_u \ \mathbf{n}_u)$ defining the UAS and $(\mathbf{t}_a \ \mathbf{n}_a)$ defining the aircraft. From Figure 7.1, the sightline range vector is given by

$$\mathbf{r} = \mathbf{r}_a - \mathbf{r}_u. \tag{7.1}$$

If the assumption is made that both the UAS and the aircraft velocity is constant, then the differential of equation (7.1) yields

$$\dot{r}\mathbf{t}_s + r\dot{\theta}_s\mathbf{n}_s = v_a\mathbf{t}_a - v_u\mathbf{t}_u. \tag{7.2}$$

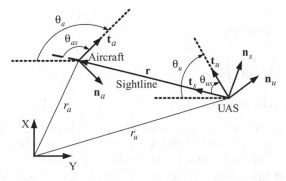

Figure 7.1 Guidance geometry (modified from [21])

This equation represents the components of the aircraft velocity relative to the UAS. Components of the relative velocity along and normal to the sightline are given by projection onto the basis vectors $\mathbf{t}_s$ and $\mathbf{n}_s$. Hence:

$$\dot{r} = v_a \mathbf{t}_s \cdot \mathbf{t}_a - v_u \mathbf{t}_s \cdot \mathbf{t}_u,$$
$$r\dot{\theta}_s = v_a \mathbf{n}_s \cdot \mathbf{t}_a - v_u \mathbf{n}_s \cdot \mathbf{t}_u. \tag{7.3}$$

The UAS-to-aircraft relative acceleration is given by differentiating equation (7.2) and noting

$$\dot{\mathbf{t}}_s = \dot{\theta}_s \mathbf{n}_s,$$
$$\dot{\mathbf{n}}_s = -\dot{\theta}_s \mathbf{t}_s, \tag{7.4}$$

to give

$$\left(\ddot{r}\mathbf{t}_s + \dot{r}\dot{\theta}_s \mathbf{n}_s\right) + \left(\dot{r}\dot{\theta}_s \mathbf{n}_s + r\ddot{\theta}_s \mathbf{n}_s - r\dot{\theta}_s^2 \mathbf{t}_s\right) = (v_a \dot{\mathbf{t}}_a - v_u \dot{\mathbf{t}}_u). \tag{7.5}$$

The Serret–Frenet equations for the UAS and the aircraft can be rewritten in terms of a constant velocity trajectory in the form

$$\dot{\mathbf{t}}_i = \kappa_i v_i \mathbf{n}_i = \dot{\theta}_i \mathbf{n}_i$$
$$\dot{\mathbf{n}}_i = -\kappa_i v_i \mathbf{t}_i = -\dot{\theta}_i \mathbf{t}_i \quad , i = a, u, \tag{7.6}$$

where κ_i is the curvature of the trajectory and $\dot{\theta}_i$ is the instantaneous rotation rate of the Serret–Frenet frame about the bi-normal vector $\mathbf{b}_i$. The normal vector $\mathbf{n}_i$ is a unit vector that defines the direction of the curvature of the trajectory (cf. Figure 7.1) and the bi-normal vector $\mathbf{b}_i$ is orthonormal to $\mathbf{t}_i$ and $\mathbf{n}_i$, forming a right-handed triplet $(\mathbf{t}_i, \mathbf{n}_i, \mathbf{b}_i)$. Hence:

$$\left(\ddot{r} - r\dot{\theta}_s^2\right)\mathbf{t}_s + (r\ddot{\theta}_s + 2\dot{r}\dot{\theta}_s)\mathbf{n}_s = v_a^2 \kappa_a \mathbf{n}_a - v_u^2 \kappa_u \mathbf{n}_u. \tag{7.7}$$

Components along and normal to the sightline can be determined by projection onto the basis vectors $\mathbf{t}_s$ and $\mathbf{n}_s$, to give

$$\left(\ddot{r} - r\dot{\theta}_s^2\right) = v_a^2 \kappa_a \mathbf{t}_s \cdot \mathbf{n}_a - v_u^2 \kappa_u \mathbf{t}_s \cdot \mathbf{n}_u,$$
$$(r\ddot{\theta}_s + 2\dot{r}\dot{\theta}_s) = v_a^2 \kappa_a \mathbf{n}_s \cdot \mathbf{n}_a - v_u^2 \kappa_u \mathbf{n}_s \cdot \mathbf{n}_u. \tag{7.8}$$

Equation (7.8) describes the acceleration kinematics of the engagement and equation (7.3) describes the velocity kinematics.

7.3   Conflict Detection

7.3.1   Collision Kinematics

In order to develop conflict detection and resolution algorithms, the collision conditions are first investigated. The geometry and matching conditions of a non-manoeuvring aircraft with a direct, straight line UAS collision trajectory are shown in Figure 7.2(a, b).

Note that the intercept triangle **AIU** does not change shape, but shrinks as the UAS and aircraft move along their respective straight-line trajectories. The UAS-to-sightline angle θ_{us}

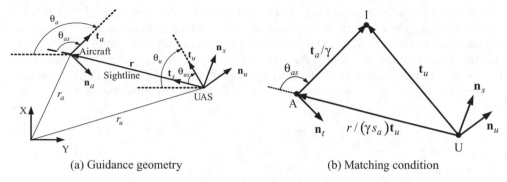

(a) Guidance geometry   (b) Matching condition

Figure 7.2 Collision geometry (modified from [21])

and the aircraft-to-sightline angle θ_{as} remain constant over the whole engagement. As shown in Figure 7.3, a useful interpretation of this condition can be visualised if the relative velocity of the UAS with respect to the aircraft is considered.

In the figure, the relative velocity of the UAS with respect to the aircraft, $\mathbf{v}_r$, is denoted

$$\mathbf{v}_r = \mathbf{v}_u - \mathbf{v}_a. \tag{7.9}$$

The collision condition is shown to be such that the relative velocity vector should lie along the sightline. This ensures that the sightline will not change direction, but only change in length and so the geometry remains the same over time.

From the intercept triangle in Figure 7.2(b), we have

$$s_u \mathbf{t}_u = r \mathbf{t}_s + s_a \mathbf{t}_a. \tag{7.10}$$

Noting that

$$\frac{s_u}{s_a} = \frac{v_u}{v_a} = \gamma, \tag{7.11}$$

gives

$$\mathbf{t}_u = \frac{1}{\gamma}\left[\frac{r}{s_a}\mathbf{t}_s + \mathbf{t}_a\right]. \tag{7.12}$$

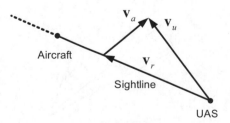

Figure 7.3 Relative velocity for collision (from [21])

Equation (7.12) can be visualised as a vector addition, and is shown in Figure 7.2(b). It is in a non-dimensional form and will thus represent the solution for all ranges between the UAS and the aircraft. The ratio r/s_a is fixed for the whole solution and thus as the range r decreases, so will the aircraft arc length s_a. Given the geometry of the aircraft basis vector t_a and the range basis vector t_s, the direction of the UAS basis vector t_u is fixed. In equation (7.12), the ratio r/s_a can be obtained by applying the cosine rule in Figure 7.2(b). From the figure, we have

$$\left(\frac{r}{s_a}\right)^2 + 2\cos(\theta_{as})\left(\frac{r}{s_a}\right) - (\gamma^2 - 1) = 0. \tag{7.13}$$

This quadratic in r/s_a can be solved explicitly, to give

$$\frac{r}{s_a} = -\cos(\theta_{as}) \pm \sqrt{\gamma^2 - \sin^2(\theta_{as})}. \tag{7.14}$$

Given that both $r > 0$ and $s_a > 0$, the solution will exist for any γ such that

$$\gamma > |\sin(\theta_{as})|. \tag{7.15}$$

The collision direction can easily be determined from Figure 7.3. This is because, when the collision geometry meets the kinematic condition in equation (7.13) and hence the geometric condition in equation (7.12), the geometry is invariant and the relative velocity v_r defined in equation (7.9) and shown in Figure 7.3 defines the approach direction.

7.3.2   Collision Detection

Based on the concept of collision geometry, an algorithm detecting the danger of collision is established. If the distance between the UAS and the aircraft is or will be smaller than the minimum separation of d_m within a specific time, it is said that there is a conflict in which the UAS and the aircraft experience a loss of minimum separation. Although this does not in itself mean that there exists a danger of collision, it represents the level of danger. In this study, the minimum separation, the CAD and TCPA are used to detect the conflict between the UAS and the aircraft.

The sightline geometry is shown in Figure 7.4.

For a non-manoeuvring aircraft, the CAD d_c can be derived by projecting the relative position vector along the sightline:

$$d_c = r\sin(\theta), \tag{7.16}$$

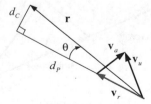

Figure 7.4   Sightline geometry for single UAS and aircraft

where θ denotes the angle from the sightline to the relative velocity vector. TCPA τ is, thus, determined as

$$\tau = \frac{d_p}{v_r}, \tag{7.17}$$

where d_p denotes the relative distance travelled to the CPA in the form

$$d_p = r\cos(\theta). \tag{7.18}$$

From the CAD and TCPA, the conflict is defined as: the UAS and aircraft are said to be in conflict if the CAD is strictly smaller than the minimum separation of d_m and the TCPA is in the future but before the look-ahead time T, i.e.

$$d_c < d_m \quad \text{and} \quad \tau \in [0, T). \tag{7.19}$$

Figure 7.5 shows a scenario in which a conflict exists between a UAS and an aircraft.

In the figure, the protected zone of the aircraft is a virtual region defined as the set P_a of points $\mathbf{x} \in R^2$ such that

$$P_a = \{\mathbf{x} | \, \|\mathbf{r}_a - \mathbf{x}\| < d_m\}, \tag{7.20}$$

where $\mathbf{r}_a$ is the position vector of the aircraft and $\|.\|$ represents the Euclidean norm. The boundary of the protected zone is defined as the minimum separation circle. Note that the conflict of a UAS and an aircraft can also be defined using the protected zone; it is said that a UAS and an aircraft are in conflict when the position of the UAS is or will be an element of the protected zone P_a.

Now, let us consider the scenario in which a UAS and multiple aircraft are in conflict at the same time. In this study, a UAS and multiple aircraft are defined to be in multiple conflicts if the UAS encounters losses of separation with more than one aircraft at the same time.

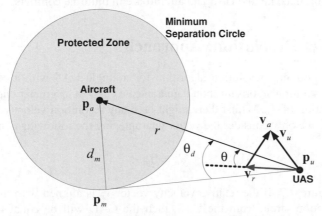

Figure 7.5 CD&R geometry (modified from [20])

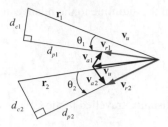

Figure 7.6 An example scenario of multiple conflicts

Figure 7.6 shows a simple example in which a UAS and two aircraft are on a course that will bring them to prescribed distances simultaneously.

In the figure, the relative motion information with subscript i represents the information with respect to the ith aircraft. The relative velocity vectors of the UAS with respect to the first and second aircraft are obtained as

$$
\begin{aligned}
\mathbf{v}_{r1} &= \mathbf{v}_u - \mathbf{v}_{a1}, \\
\mathbf{v}_{r2} &= \mathbf{v}_u - \mathbf{v}_{a2}.
\end{aligned}
\tag{7.21}
$$

The distances to the close points of approach for each aircraft are given by

$$
\begin{aligned}
d_{c1} &= r\sin(\theta_1), \\
d_{c2} &= r\sin(\theta_2).
\end{aligned}
\tag{7.22}
$$

When the distances are smaller than the minimum separation within a specific time T at the same time:

$$
\begin{cases}
d_{c1} < d_m \quad \text{and} \quad \tau_1 \in [0, T), \\
d_{c2} < d_m \quad \text{and} \quad \tau_2 \in [0, T),
\end{cases}
\tag{7.23}
$$

there exist two conflicts, i.e. the UAS and aircraft are in multiple conflicts.

7.4   Conflict Resolution: Approach I

In this section, a conflict resolution algorithm for a single UAS with a constant speed is proposed. The resolution algorithm should guarantee a CAD of d_c greater than or equal to the minimum separation of d_m. Whilst there might be many resolution velocity vectors meeting this requirement, we only consider the vector guaranteeing the following condition:

$$
d_c = d_m.
\tag{7.24}
$$

As shown in Figure 7.7, if the relative velocity vector $\mathbf{v}_r$ is aligned with the tangent to the minimum separation circle from the UAS, then the CAD will be equal to the minimum separation of d_m.

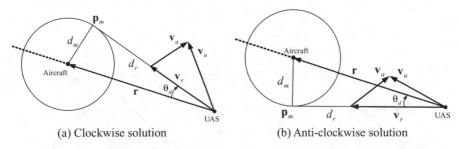

(a) Clockwise solution   (b) Anti-clockwise solution

Figure 7.7   Relative velocity for minimum separation (modified from [21])

If there exist uncertainties, an aircraft is manoeuvring, or both, this condition might be inappropriate for conflict resolution. Scaling up d_m will resolve this problem because it results in a CAD bigger than the minimum separation. However, we leave this issue for future study.

7.4.1   Collision Kinematics

For the conflict resolution, the direction of the relative velocity vector, θ_r, should become

$$\theta_r = \theta_m \equiv: \begin{cases} \theta_s + \theta_d & \text{for the clockwise solution,} \\ \theta_s - \theta_d & \text{for the anti-clockwise solution.} \end{cases} \tag{7.25}$$

The resolution geometry for a clockwise solution is shown in Figure 7.8.
    The matching condition is thus derived as

$$\mathbf{v}_u = \mathbf{v}_r + \mathbf{v}_a,$$
$$\mathbf{t}_u = \alpha \mathbf{t}_m + \frac{1}{\gamma} \mathbf{t}_a, \tag{7.26}$$

where the sightline basis set $(\mathbf{t}_s\ \mathbf{n}_s)$ is replaced by the separation basis set $(\mathbf{t}_m\ \mathbf{n}_m)$ and

$$\alpha \equiv v_r/v_u. \tag{7.27}$$

Calculating the ratio α will determine the matching condition of the resolution geometry.

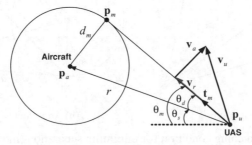

Figure 7.8   Conflict resolution geometry for clockwise rotation

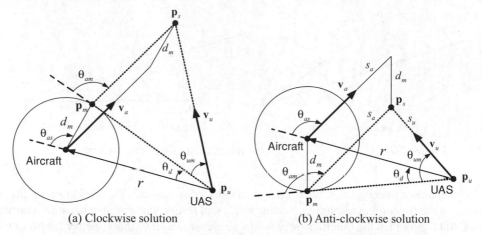

(a) Clockwise solution   (b) Anti-clockwise solution

Figure 7.9 Geometry for minimum separation (modified from [21])

In order to derive the velocity ratio α, the resolution geometry shown in Figure 7.9 is obtained by modifying the collision geometry.

The figure shows that the original collision triangle has been modified to the resolution triangle given by $\{\mathbf{p}_u, \mathbf{p}_m, \mathbf{p}_s\}$, for both clockwise and anti-clockwise. This resolution triangle will maintain its orientation and shape in a similar manner to the impact triangle. The matching conditions for the clockwise and anti-clockwise solutions are shown in Figure 7.10.

The vector sums for the UAS calculated with respect to the resolution triangle **MIU** yield the matching condition of the form

$$\mathbf{t}_u = \frac{1}{\gamma}\left[\frac{d_r}{s_a}\mathbf{t}_m + \mathbf{t}_a\right]. \tag{7.28}$$

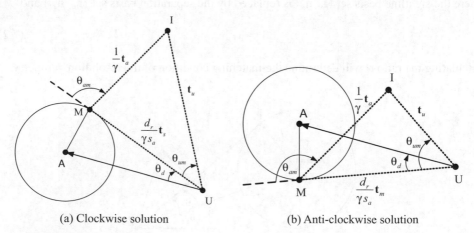

(a) Clockwise solution   (b) Anti-clockwise solution

Figure 7.10 Matching condition for minimum separation (modified from [21])

From equations (7.26) and (7.28), we have

$$\alpha = \frac{d_r}{\gamma s_a}. \tag{7.29}$$

Applying the cosine rule to the resolution geometry now gives

$$\left(\frac{d_r}{s_a}\right)^2 + 2\cos(\theta_{am})\left(\frac{d_r}{s_a}\right) - (\gamma^2 - 1) = 0, \tag{7.30}$$

where

$$d_r = \sqrt{r^2 - d_m^2}, \\ \theta_{am} = \theta_{as} \pm \theta_d, \tag{7.31}$$

and θ_d denotes the angle between the sightline and the tangent line on the minimum separation circle from the UAS position. Note that θ_d is either subtracted for a clockwise solution or added for an anti-clockwise solution. Now, we have

$$\cos(\theta_{am}) = \cos(\theta_{as} \pm \theta_d), \\ = \cos(\theta_{as})\cos(\theta_d) \mp \sin(\theta_{as})\sin(\theta_d) \tag{7.32}$$

where

$$\cos(\theta_d) = \frac{d_r}{r} = \frac{\sqrt{r^2 - d_m^2}}{r}, \\ \sin(\theta_d) = \frac{d_m}{r}. \tag{7.33}$$

Note that the geometry with respect to the sightline is now not fixed, but as the solution requires the relative velocity vector to lie along the tangent line from $\mathbf{p}_u$ to $\mathbf{p}_m$, this line will not rotate. For the UAS and other aircraft with constant speed, this implies that the triangle $\{\mathbf{p}_u, \mathbf{p}_m, \mathbf{p}_s\}$ is fixed in shape and orientation, and will shrink as the UAS and aircraft approach each other. Hence the ratio d_r/s_a will have a fixed solution, as will the angle subtended between the aircraft velocity vector and the tangent line θ_{am}. Hence, a solution to equation (7.30) is calculated as

$$\frac{d_r}{s_a} = -\cos(\theta_{am}) \pm \sqrt{\gamma^2 - \sin^2(\theta_{am})}. \tag{7.34}$$

Given γ and d_r/s_a in equation (7.34), the velocity ratio α is obtained as

$$\alpha = \frac{-\cos(\theta_{am}) \pm \sqrt{\gamma^2 - \sin^2(\theta_{am})}}{\gamma}. \tag{7.35}$$

7.4.2 Resolution Guidance

Since the UAS heading angle is different from the desired one satisfying the matching condition in equation (7.28), it is essential to design an algorithm to regulate the heading angle at the desired angle. Let us define the desired UAS tangent vector as $\hat{\mathbf{t}}_u$, then

$$\hat{\mathbf{t}}_u = \frac{1}{\gamma}\left[\left(\frac{d_r}{s_a}\right)\mathbf{t}_s + \mathbf{t}_a\right]. \tag{7.36}$$

A geometric interpretation of equation (7.36) is reproduced in Figure 7.11.

Figure 7.11 shows that as the geometry of the engagement changes due to mismatched UAS and aircraft tangent vectors, the solution $\hat{\mathbf{t}}_u$ will change and rotate around the circle. The figure also shows that the rotation of the solution vector $\hat{\mathbf{t}}_u$ and the rotation of the sightline vector $\hat{\mathbf{t}}_r (= \mathbf{t}_m)$ are related. In moving from solution **A** to solution **B**, the solution angle $\hat{\theta}_{um}$ increases.

Minimum separation can be met by regulating the heading error, θ_e:

$$\theta_e \equiv \hat{\theta}_{um} - \theta_{um} = \hat{\theta}_u - \theta_u, \tag{7.37}$$

where $\hat{\theta}_u$ and θ_u are the direction angles of the desired UAS tangent vector $\hat{\mathbf{t}}_u$ and the UAS tangent vector $\mathbf{t}_u$. The regulating algorithm of the heading angle can be determined by use of a simple Lyapunov function V of the form

$$V = \frac{1}{2}\theta_e^2. \tag{7.38}$$

The time derivative of the Lyapunov candidate function V is given by

$$\frac{dV}{dt} = \dot{\theta}_e\theta_e. \tag{7.39}$$

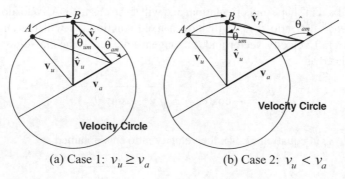

(a) Case 1: $v_u \geq v_a$ (b) Case 2: $v_u < v_a$

Figure 7.11 Geometry interpretation of conflict resolution for clockwise rotation (modified from [20])

For stability, it is required that

$$\dot{\theta}_e \theta_e < 0. \tag{7.40}$$

From the definition of θ_e, we have

$$\dot{\theta}_e = \dot{\hat{\theta}}_{um} - \dot{\theta}_{um}. \tag{7.41}$$

The first time derivative of the matching condition for minimum separation is obtained as

$$\dot{\hat{\theta}}_{um} = \frac{\cos(\theta_{am})}{\gamma \cos(\hat{\theta}_{um})} \dot{\theta}_{am} = \frac{\cos(\theta_{am})}{\sqrt{\gamma^2 - \sin^2(\theta_{am})}} \dot{\theta}_{am}. \tag{7.42}$$

This means

$$-\frac{1}{\gamma}|\dot{\theta}_{am}| \leq \dot{\hat{\theta}}_{um} \leq \frac{1}{\gamma}|\dot{\theta}_{am}|. \tag{7.43}$$

From the definition of θ_{am}, we have

$$\dot{\theta}_{am} = \dot{\theta}_a - \dot{\theta}_m, \tag{7.44}$$

where

$$\dot{\theta}_m = \dot{\theta}_s \pm \dot{\theta}_d. \tag{7.45}$$

Here θ_s is the sightline angle. Since $\dot{\theta}_a = 0$ for a non-manoeuvring aircraft, equation (7.51) becomes

$$\dot{\theta}_{am} = -\dot{\theta}_m. \tag{7.46}$$

Substituting equation (7.46) into equation (7.43) yields

$$-\frac{1}{\gamma}|\dot{\theta}_m| \leq \dot{\hat{\theta}}_{um} \leq \frac{1}{\gamma}|\dot{\theta}_m|. \tag{7.47}$$

From the resolution geometry, $\dot{\theta}_m$ is obtained as

$$\dot{\theta}_m = \frac{v_r}{\sqrt{r^2 - d_m^2}} \sin(\theta_d \mp \theta). \tag{7.48}$$

Equation (7.47) can be rewritten as

$$-\frac{1}{\gamma} \frac{v_r}{\sqrt{r^2 - d_m^2}} \leq \dot{\hat{\theta}}_{um} \leq \frac{1}{\gamma} \frac{v_r}{\sqrt{r^2 - d_m^2}}. \tag{7.49}$$

Hence, a resolution guidance algorithm is proposed as

$$\dot{\theta}_u = \left(1 + \frac{1}{\gamma}\right) \frac{v_r}{\sqrt{r^2 - d_m^2}} \text{sign}(\theta_e) + K\theta_e, \tag{7.50}$$

$$K > 0, \tag{7.51}$$

$$\text{sign}(\theta_e) = \frac{|\theta_e|}{\theta_e}, \tag{7.52}$$

to give

$$\frac{dV}{dt} = \theta_e \left[\dot{\hat{\theta}}_{um} - \frac{1}{\gamma} \frac{v_r}{\sqrt{r^2 - d_m^2}} \text{sign}(\theta_e) + \dot{\theta}_m - \frac{v_r}{\sqrt{r^2 - d_m^2}} \text{sign}(\theta_e)\right] - K\theta_e^2 \leq 0, \tag{7.53}$$

which is negative semi-definite. The curvature of UAS, κ_u can be obtained from equation (7.50) and

$$\kappa_u = \frac{\dot{\theta}_u}{V_u}. \tag{7.54}$$

7.4.3   Analysis and Extension

Now, let us analyse the proposed conflict resolution algorithm. In order to investigate the feasibility of conflict resolution, we should examine the velocity vectors. If the desired relative velocity is realisable from the combination of UAS and aircraft velocity vectors, the resolution algorithm is feasible. Otherwise, it is infeasible. From equation (7.34), it is possible to investigate the feasibility of the avoidance solutions for the UAS with a constant speed.

Lemma 7.1   For constant ground speed of the UAS, v_u, if it is greater than or equal to that of the aircraft, v_a, then the conflict resolution algorithm is feasible and the desired relative speed $\hat{v}_r$ is given by

$$\hat{v}_r = -v_a \cos(\theta_{am}) + \sqrt{\lambda(\hat{\theta}_{am})}, \tag{7.55}$$

where

$$\lambda(\hat{\theta}_{am}) = v_u^2 - v_a^2 \sin^2(\hat{\theta}_{am}). \tag{7.56}$$

Proof. Multiplying both sides of equation (7.34) by v_a yields

$$\hat{v}_r = -v_a \cos(\theta_{am}) \pm \sqrt{\lambda(\hat{\theta}_{am})}. \tag{7.57}$$

From the assumption that v_u is greater than or equal to v_a, we have

$$\lambda(\hat{\theta}_{am}) \geq v_a^2 \cos^2(\hat{\theta}_{am}),$$

$$\sqrt{\lambda(\hat{\theta}_{am})} \geq v_a| \cos(\hat{\theta}_{am})|,$$

(7.58)

for any values of the angle $\hat{\theta}_{am}$. Since v_r should be a positive value, equation (7.55) is true for $v_u \geq v_a$. Hence, the resolution algorithm is feasible.

Lemma 7.2   For constant ground speed v_u of the UAS less than the aircraft speed v_a:

- if $\lambda(\hat{\theta}_{ra})$ is less than zero, there is no feasible solution, i.e. the desired relative velocity which cannot be aligned along with the two tangent vectors;

- if $\lambda(\hat{\theta}_{ra})$ is equal to zero, there is only one feasible solution, i.e. one desired relative velocity vector can be aligned with a corresponding tangent vector;

- if $\lambda(\hat{\theta}_{ra})$ is greater than zero, there exist two feasible solutions, i.e. two relative velocity vectors can be aligned along the two tangent vectors.

Figure 7.12 shows an example in which only one avoidance solution is feasible.

As illustrated in Figure 7.12, only the anti-clockwise solution is feasible. Note that the velocity circle in this figure represents a feasible UAS heading for a constant speed: its radius is the UAS speed and its centre is in a velocity vector of the aircraft from the UAS position.

The desired heading angle of the UAS can be calculated simply from equation (7.36). Since there could be two possible solutions, the clockwise and anti-clockwise, we need to determine the turning direction of the UAS. Figure 7.13 represents two resolution velocity vectors for that, $v_u \geq v_a$.

In the figure, $\hat{\mathbf{t}}_{uc}$ and $\hat{\mathbf{t}}_{ua}$ denote the desired UAS tangent vectors for the clockwise and anti-clockwise rotation, and δ represents the collision sector which lies between the two tangent vectors. Note that if the tangent vector is located inside the collision sector and TCPA is less than a specific time, then the UAS and aircraft are in conflict. On the other hand, the UAS and

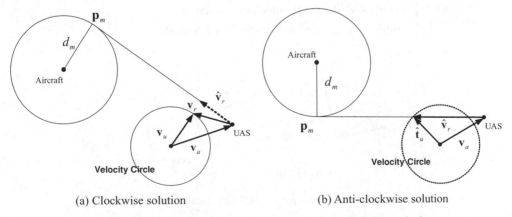

(a) Clockwise solution   (b) Anti-clockwise solution

Figure 7.12   An example of the existence of one feasible solution

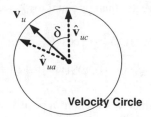

Figure 7.13 Geometry of UAS velocity for conflict resolution (from [20])

aircraft are not in conflict when the tangent vector is outside the sector. As shown in Figure 7.13, turning towards the vector $\hat{\mathbf{t}}_{uc}$ will make a clockwise turn and turning towards the vector $\hat{\mathbf{t}}_{ua}$ will produce an anti-clockwise turn. If the UAS turns towards the closest vector, then it will produce a monotonically increasing CAD. To determine the turn direction, it is possible to consider several methods fulfilling the requirements – such as obeying the rules of the air, or allowing more efficient path following after the resolution manoeuvre. In this study, turning towards the closest vector is among the possible solutions considered.

Now, let us consider a simple scenario where the UAS and only two aircraft are in multiple conflicts. Figure 7.14(a, b) shows the velocity circles of the UAS with respect to the first and second aircraft, respectively.

A conflict sector for multiple conflicts resolution δ is the union of two conflicts sectors δ_1 and δ_2:

$$\delta = \delta_1 \cup \delta_2. \tag{7.59}$$

In this scenario, the desired tangent vectors are u_{uc1} and u_{ua2}. Similarly, for n number of conflicts, the sector will be the union of all sectors:

$$\delta = \delta_1 \cup \delta_2 \cup \cdots \cup \delta_n. \tag{7.60}$$

The heading angle of the relative vector for the resolution can be simply obtained as:

$$\theta_r = \theta_m \equiv: \begin{cases} \max(\theta_{si} + \theta_{di}) & \text{for the clockwise solution} \\ \min(\theta_{si} - \theta_{di}) & \text{for the anti-clockwise solution} \end{cases}, \quad i = 1, \ldots, n, \tag{7.61}$$

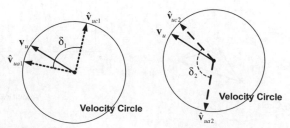

(a) With respect to first aircraft   (b) With respect to second aircraft

Figure 7.14 Velocity circles against two aircraft

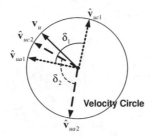

Figure 7.15 A turning direction issue for multiple conflicts resolution (modified from [20])

where n is the number of aircraft which are in conflict with the UAS, and the angles θ_{si} and θ_{di} are the angles θ_s and θ_d with respect to the ith aircraft. The desired relative velocity for resolution of multiple conflicts is then chosen as the one satisfying the condition in equation (7.42). However, the decision to change direction needs to be carefully verified, because turning to the closest resolution vector from the current UAS velocity vector could result in a problem. Figure 7.15 illustrates.

In the case shown in Figure 7.14, the UAS first tries to solve the second conflict and the velocity vector will be located outside δ_2, as shown in Figure 7.15. Since the proposed turning direction will rotate the velocity towards $\mathbf{v}_{ua1}$ in this case, the UAS and aircraft will be in multiple conflicts again. Therefore, this problem might cause chattering on the resolution command and unsafe trajectories. In order to resolve this problem, simple decision-making is proposed. If a resolution manoeuvre does not satisfy the condition

$$\mathbf{v}_r \cdot r < 0, \tag{7.62}$$

the UAS maintains the turning direction.

7.5 Conflict Resolution: Approach II

In the previous section, it was assumed that the ground speed of the UAS is constant to design a conflict resolution algorithm. Owing to this assumption, the resolution guidance controls the heading angle only. It also limits the feasibility region of the avoidance solutions: when the ground speed of the aircraft is greater than that of the UAS, some solutions are infeasible. An example scenario for an infeasible solution is shown in Figure 7.16.

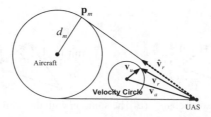

Figure 7.16   Example geometry in which the clockwise solution is infeasible

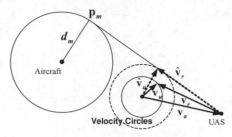

Figure 7.17   Concept of controlling heading and speed of UAS

Note that the velocity circle can represent the feasibility of the solution: as illustrated in the figure, the desired relative velocity $\hat{\mathbf{v}}_r$ is unrealisable regardless of the UAS heading. We thus propose a resolution algorithm which can resolve this problem. Since the proposed resolution algorithm with constant speed can be used when the conflict solution is feasible, we only consider the case where the aircraft speed is greater than the UAS speed in this section.

7.5.1   Resolution Kinematics and Analysis

As stated in Section 7.4, a relative velocity vector should be aligned with one of the two tangent vectors for conflict resolution. Controlling the speed with UAS heading could enlarge the feasibility region so as to generate a feasible solution as shown in Figure 7.17.

In the figure, the velocity circle with the solid line is the circle with the current UAS velocity and the dashed circle is a circle with a modified velocity vector. As depicted, the UAS is able to realise a desired relative velocity, $\hat{\mathbf{v}}_r$, not with the current speed, but with an increased speed.

There might be numerous avoidance solutions when the UAS speed can be controlled. Figure 7.18(a) shows a few possible solutions in a simple scenario.

In order to determine the avoidance solution, let us investigate the velocity relation. The velocity relation with an arbitrary relative speed $\hat{v}_r$ is given by

$$\begin{aligned}\hat{v}_u^2 &= \hat{v}_r^2 + v_a^2 - 2\hat{v}_r v_a \cos(\pi - \hat{\theta}_{am}) \\ &= \hat{v}_r^2 + v_a^2 + 2\hat{v}_r v_a \cos\hat{\theta}_{am}.\end{aligned} \qquad (7.63)$$

Since equation (7.63) can be rewritten as

$$\hat{v}_u^2 = (v_a \sin\hat{\theta}_{am})^2 + (\hat{v}_r + v_a \cos\hat{\theta}_{am})^2 \geq 0, \qquad (7.64)$$

(a) Hypothetical solutions   (b) The solution with the minimum UAS speed

Figure 7.18   Velocity relation for avoidance solutions

we have

$$\hat{v}_u = \sqrt{\hat{v}_r^2 + v_a^2 + 2\hat{v}_r v_a \cos \hat{\theta}_{am}}. \tag{7.65}$$

For aircraft with a constant speed, determining $\hat{v}_r$ defines the desired UAS speed and consequently the avoidance solution. The minimum UAS speed for the conflict resolution can be derived from the following condition:

$$\frac{d\hat{v}_u}{d\hat{v}_r} = 2\hat{v}_r + 2v_a \cos \hat{\theta}_{am} = 0. \tag{7.66}$$

If $\cos \hat{\theta}_{am} \geq 0$, the minimum UAS speed satisfying equation (7.66) is the aircraft speed. Otherwise, the minimum speed is obtained as

$$\min[\hat{v}_u] = \sqrt{v_a^2(1 - \cos^2 \hat{\theta}_{am})} = v_a \sin \hat{\theta}_{am}. \tag{7.67}$$

A geometric interpretation of equation (7.67) is shown in Figure 7.18(b): the relative speed yielding the minimum UAS speed is given by

$$\hat{v}_r = v_a \cos(\pi - \hat{\theta}_{am}). \tag{7.68}$$

As stated, the resolution algorithm controlling the speed and heading will be implemented only when the avoidance solution is infeasible. In this case, the UAS speed must be increased to make the solution feasible, as shown in Figure 7.17. In this study, therefore, the avoidance solution is derived from the minimum UAS ground speed to minimise deviation from the current speed of UAS resulting in the minimum fuel consumption. Note that the UAS velocity is likely to be chosen for a certain mission, thus the resolution algorithm obtained from the minimum UAS speed would also be desirable for mission accomplishment.

If the minimum solution among the desired UAS speeds is greater than the maximum bound of the UAS speed, there is no feasible solution for conflict resolution. In this case, the best resolution manoeuvre is to maximise the speed and align the UAS heading with the desired velocity vector as closely as possible.

7.5.2   Resolution Guidance

Owing to the speed and heading of UAS which are different from the desired ones for conflict resolution, it is necessary to develop a regulation algorithm – termed 'resolution guidance' in this study. In order to design an algorithm, the Lyapunov stability theory is used again. The stability of the conflict resolution algorithm can be determined by implementing a simple Lyapunov candidate function V, of the form

$$V = \frac{1}{2}\left(\theta_e^2 + v_e^2\right), \tag{7.69}$$

where

$$\begin{aligned}
\theta_e &= \hat{\theta}_{um} - \hat{\theta}_{um}, \\
v_e &= \hat{v}_u - v_u.
\end{aligned} \tag{7.70}$$

The time derivative of the function V is given by

$$\frac{dV}{dt} = \dot{\theta}_e \theta_e + \dot{v}_e v_e. \tag{7.71}$$

In order to guarantee the stability, the resolution guidance must satisfy the following equation:

$$\dot{\theta}_e \theta_e + \dot{v}_e v_e \leq 0. \tag{7.72}$$

From equation (7.67), the time derivative of $\hat{v}_u$ is derived as

$$\dot{\hat{v}}_u = v_a \dot{\theta}_{am} \cos \hat{\theta}_{am}. \tag{7.73}$$

Substituting equations (7.46) and (7.48) into equation (7.73) yields

$$\dot{\hat{v}}_u = \frac{v_a v_r}{\sqrt{r^2 - d_m^2}} \cos \hat{\theta}_{am} \sin(\theta_d \mp \theta). \tag{7.74}$$

Hence, we have

$$-\frac{v_a v_r}{\sqrt{r^2 - d_m^2}} \leq \dot{\hat{v}}_u \leq \frac{v_a v_r}{\sqrt{r^2 - d_m^2}}. \tag{7.75}$$

For the desired UAS speed equal to that of the aircraft, we have

$$\dot{\hat{v}}_u = \dot{v}_a = 0. \tag{7.76}$$

As shown in Figure 7.18(b), $\hat{\theta}_{um}$ is a right angle and its first time derivative is zero. Therefore, the resolution guidance algorithm is proposed as

$$\dot{\theta}_u = \frac{v_r}{\sqrt{r^2 - d_m^2}} \text{sign}(\theta_e) + K_1 \theta_e,$$

$$\dot{v}_u = \frac{v_a v_r}{\sqrt{r^2 - d_m^2}} \text{sign}(v_e) + K_2 v_e, \tag{7.77}$$

to give

$$\frac{dV}{dt} \leq \theta_e \left[\dot{\theta}_m - \frac{v_r}{\sqrt{r^2 - d_m^2}} \text{sign}(\theta_e) \right] + v_e \left[\dot{\hat{v}}_u - \frac{v_a v_r}{\sqrt{r^2 - d_m^2}} \text{sign}(v_e) \right] - K_1 \theta_e^2 - K_2 v_e^2 \leq 0 \tag{7.78}$$

for $K_1 > 0$ and $K_2 > 0$. The curvature of the UAS, κ_u, can be obtained from equation (7.54) and the tangential acceleration is given from

$$a_u = \dot{v}_u. \tag{7.79}$$

7.6   CD&R Simulation

In this section, the performance and reliability of the proposed CD&R algorithms are investigated using numerical examples. For the nonlinear simulation, it is assumed that UAS are able to obtain the following motion information of aircraft by any means:

- position vector;
- velocity vector.

Furthermore, the look-ahead time T is selected as 3 minutes and 3 km is considered as a minimum separation distance. Note that the FAA classified minimum vertical and horizontal separation distances considering several standards: minimum horizontal separation for aircraft safety is 5 nautical mile (nm) in the en-route environment and 3 nm in the terminal environment; minimum vertical separation is currently 1,000 ft at flight levels below 41,000 ft. The UAS heads to a waypoint of (20 km, 20 km) and its initial position and heading angle of UAS are (0 km, 0 km) and 0 degree.

7.6.1   Simulation Results: Approach I

For the numerical examples to examine the performance of the first approach, two scenarios are considered. In the scenarios, it is assumed that the initial ground speed of the UAS is 100 m/s and the physical constraints imposed on the UAS are given in Table 7.1.

The initial conditions of intruders (aircraft) are represented in Table 7.2.

Aircraft are non-manoeuvring in the first scenario, whereas in the second scenario they are manoeuvring with a constant turn rate give as

$$[\dot{\theta}_{a1}, \dot{\theta}_{a2}, \dot{\theta}_{a3}, \dot{\theta}_{a4}] = [-0.2, -0.2, -0.2, 0]. \tag{7.80}$$

Whilst the first scenario is to verify the performance of the conflict detection and the first resolution algorithms, the second scenario is to check their performance when the non-manoeuvring assumption is no longer valid. Note that the conflict detection and first resolution

Table 7.1   Physical constraints of UAS for the first resolution approach

Max. turn rate	Max. speed rate	Max. ground speed	Min. ground speed
5 deg/s	40 m/s^2	150 m/s	70 m/s

Table 7.2   Initial conditions of aircraft for the first resolution approach

Intruder	Position (km, km)	Heading angle (deg)	Ground speed (m/s)
Aircraft 1	(9.5, −1)	180	50
Aircraft 2	(10, 0.5)	180	50
Aircraft 3	(19, 8)	165	60
Aircraft 4	(15, −5.5)	90	55

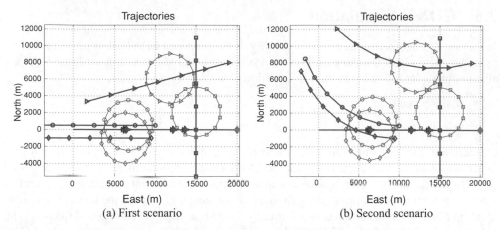

Figure 7.19   Original trajectories: the first resolution approach

algorithms are derived with a constant UAS speed. The original trajectories without conflict resolution are shown in Figure 7.19.

In this figure, the solid line represents the trajectory of the UAS, solid lines and circles with markers show the trajectories of aircraft and the minimum separation circles at the minimum distances from the UAS, and a diamond shape at (20 km, 0 km) illustrates the waypoint the UAS heads to. Note that the UAS shapes are depicted at the closest points from the aircraft, thus the UAS is in conflict with the aircraft if any UAS is located in the minimum separation circles. In the two scenarios, UAS is in conflicts with the first, second and fourth aircraft. As shown in Figure 7.19, one UAS and four aircraft are initially distributed over a rectangular air space of 20 km by 10 km. Since the simulated space is more stressed than that in a hypothetical conflict situation, it will allow the rigorous performance evaluation of the proposed CD&R algorithms.

In order to resolve the multiple conflicts, the conflict detection algorithm and the first resolution approach are implemented into UAS. The minimum distances between the UAS and aircraft are always greater than the minimum safe separation of 3 km as represented in Table 7.3, so the proposed algorithms effectively detect and resolve the conflicts.

Simulation results in the first scenario are shown in Figure 7.20.

As shown in Figure 7.20(a, b, d), not only can the first resolution approach avoid the collision, but also the heading change of the UAS is smooth. It is also shown in Figure 7.20(c, e) that the ground speed of the UAS is constant. Figure 7.21 illustrates the simulation results in the second scenario.

Table 7.3   Minimum relative distances of aircraft from the UAS: the first resolution approach

Intruder	First scenario	Second scenario
Aircraft 1	4.2040 km	4.0679 km
Aircraft 2	3.2297 km	3.1219 km
Aircraft 3	3.0752 km	3.0220 km
Aircraft 4	3.0680 km	3.0693 km

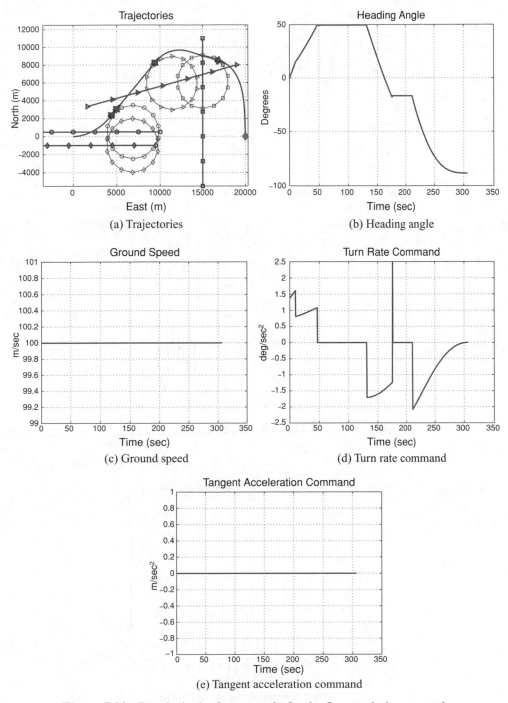

Figure 7.20 Results in the first scenario for the first resolution appoach

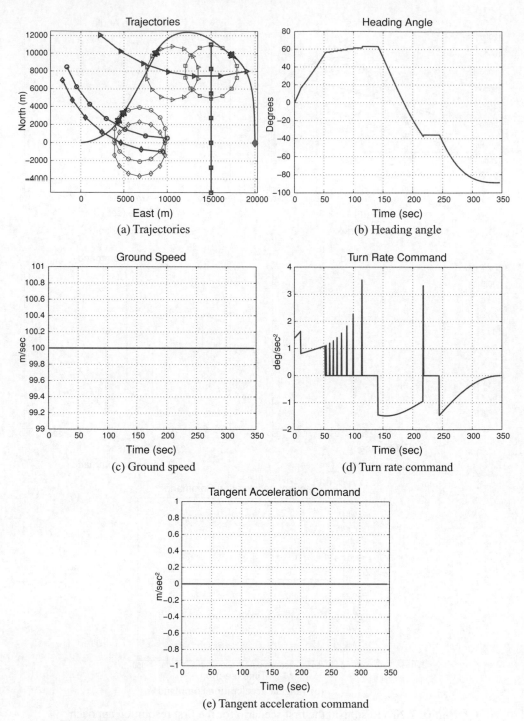

Figure 7.21   Results in the second scenario for the first resolution approach

Table 7.4 Physical constraints of UAS for the first resolution approach

Max. turn rate	Max. speed rate	Max. ground speed	Min. ground speed
5 deg/s	30 m/s^2	85 m/s	30 m/s

As depicted in these figures, the proposed algorithm resolves conflicts and maintains a constant ground speed. However, the turn rate command is spiky because the detection algorithm is derived against non-manoeuvring: aircraft manoeuvre will generate conflict conditions resulting from the altered aircraft heading.

7.6.2   Simulation Results: Approach II

In the numerical examples for the performance evaluation of the second approach, we assume that the initial ground speed of the UAS is 50 m/s and the physical constraints of the UAS are given in Table 7.4.

Table 7.5 represents the initial conditions of the aircraft.

Note that, unlike the numerical examples for the first resolution approach, the UAS speed is less than that of the aircraft. The turn rates of the aircraft for each scenario are the same as those for the first resolution approach. In these scenarios, without the resolution algorithm, conflicts exist between the UAS and the first, second and fourth aircraft as shown in Figure 7.22.

Table 7.6 represents the minimum distances between the UAS and the aircraft. From the distances which are bigger than the safe distance, it is shown that the conflicts detection and the second resolution approach work effectively.

Figure 7.23 shows the simulation results in the first scenario.

Since there was no feasible avoidance solution at the beginning of the simulation, the resolution algorithm increases the UAS speed. The second scenario is considered to check the effect when the assumption of non-manoeuvring aircraft is invalid and its results are illustrated in Figure 7.24.

The simulation results show that the second resolution approach successively resolves the conflicts. In the second scenario, the UAS speed is again increased to make an unrealisable solution feasible. Moreover, similar to the first resolution approach in the second scenario, the aircraft manoeuvre results in chattering on the turn rate command. In Figure 7.23(e) and Figure 7.24(e), the tangent acceleration commands seem to be chattering. Therefore, we check the command profiles in a smaller time window and they are depicted in Figure 7.25.

As shown in Figure 7.25, the tangent acceleration command is not in fact chattering.

Table 7.5 Initial conditions of aircraft for the first resolution approach

Intruder	Position (km, km)	Heading angle (deg)	Ground speed (m/s)
Aircraft 1	(6.5, −1)	180	100
Aircraft 2	(7.5, 0.5)	180	100
Aircraft 3	(18, 0)	165	95
Aircraft 4	(10, −15)	90	70

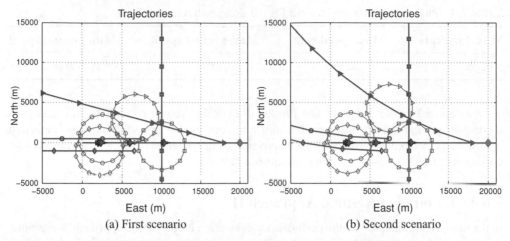

Figure 7.22   Original trajectories: the second resolution approach

7.7   Conclusions

In this chapter, a UAS conflict detection algorithm and two resolution algorithms have been introduced based on the differential geometry concepts. To develop all algorithms, it is assumed that aircraft are non-manoeuvring. The closest approach distance (CAD) and time to closest point of approach (TCPA) allow the detection algorithm to define the conflict. Whilst the first resolution algorithm controls the UAS heading only, the second algorithm controls both the ground speed and the heading. The feasibility and performance of two algorithms were also mathematically analysed. If the aircraft speed is greater than the UAS speed, the constant speed of the UAS may lead to infeasible solutions. The second algorithm resolves this problem and consequently expands the feasibility region by controlling the UAS speed. We also extended the proposed algorithms for the multiple conflicts in which a UAS and many aircraft are in danger of collision at the same time. The performance of the detection and resolution algorithms was also illustrated and validated through numerical simulations. The results of the nonlinear simulation have shown that the proposed algorithm performs effectively not only for non-manoeuvring aircraft, but also for manoeuvring aircraft. The analysis did not include the UAS dynamics and this, together with the extension to 3D, will be the subject of future study.

Table 7.6   Minimum relative distances of aircraft from the UAS: the second resolution approach

Intruder	First scenario	Second scenario
Aircraft 1	3.5952 km	3.5689 km
Aircraft 2	3.0096 km	3.0980 km
Aircraft 3	5.2482 km	4.1128 km
Aircraft 4	3.0177 km	3.0607 km

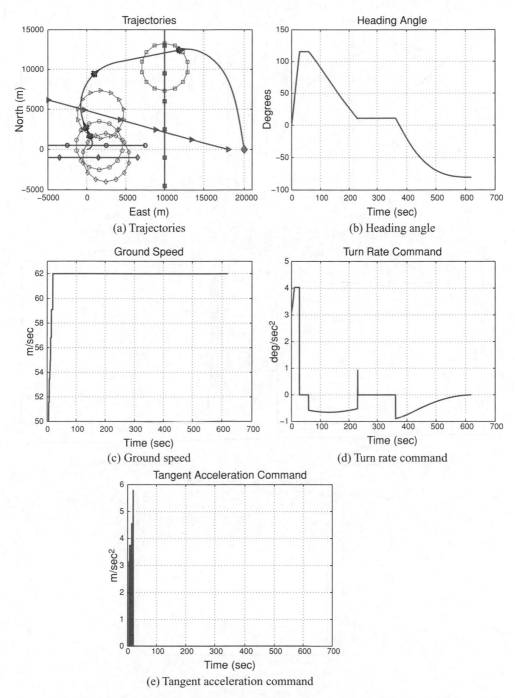

(a) Trajectories

(b) Heading angle

(c) Ground speed

(d) Turn rate command

(e) Tangent acceleration command

Figure 7.23   Results in the first scenario for the second resolution approach

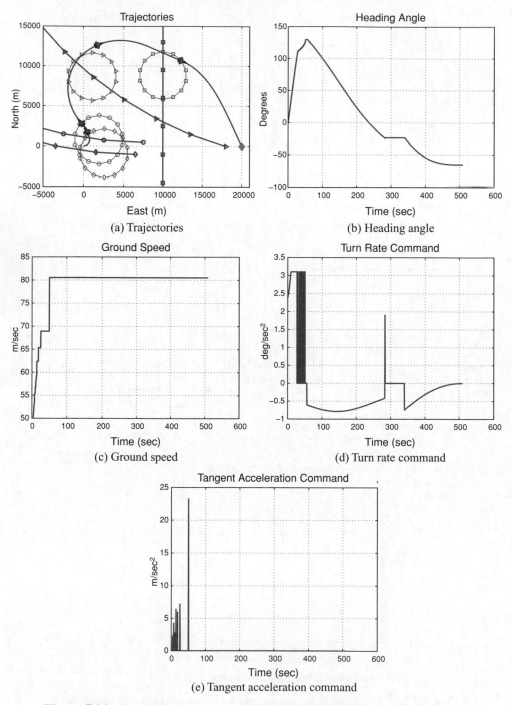

(a) Trajectories

(b) Heading angle

(c) Ground speed

(d) Turn rate command

(e) Tangent acceleration command

Figure 7.24   Results in the second scenario for the second resolution approach

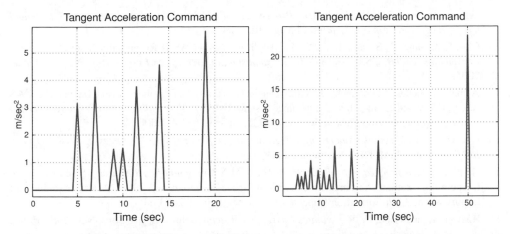

Figure 7.25   Tangent acceleration command in a time window

References

1. Han, S. C. and Bang, H. C., 'Proportional navigation-based optimal collision avoidance for UASs', *Proceedings of the 2nd International Conference on Autonomous Robots and Agents*, Palmerston North, New Zealand, 2004.

2. Tomlin, C., Pappas, G. J., and Sastry, S., 'Conflict resolution for air traffic management: a study in multi-agent hybrid systems', *IEEE Transactions on Automatic Control*, 43(4), 509–521, 1998.

3. RTCA TF 3, *Final report of RTCA Task Force 3: Free flight implementation*, RTCA Task Force 3, RTCA Inc., Washington, DC, 1995.

4. Kuchar, J. and Yang, L., 'Review of conflict detection and resolution modeling methods', *IEEE Transactions on Intelligent Transportation Systems*, 1(4), 179–189, 2000.

5. Khatib, O. and Burdick, A., 'Unified approach for motion and force of robot manipulators', *IEEE Journal of Robotics and Automation*, 3(1), 43–53, 1987.

6. Passino, K. M., 'Bridging the gap between conventional and intelligent control', *Control Systems Magazine, IEEE*, 13(3), 12–18, 1993.

7. Tang, P., Yang, Y., and Li, X., 'Dynamic obstacle avoidance based on fuzzy inference and transposition principle for soccer robots', *Proceedings of 10th International Conference on Fuzzy Systems*, Melbourne, Victoria, Australia, 2001.

8. Rathbun, D., Kragelund, S., Pongpunwattana, A., and Capozzi, B., 'An evolution based path planning algorithm for autonomous motion of a UAS through uncertain environments', *Proceedings of IEEE Digital Avionics Systems Conference*, 2002.

9. Ghosh, R. and Tomlin, C., 'Maneuver design for multiple aircraft conflict resolution', *Proceedings of American Control Conference*, Chicago, IL, 2000.

10. Kumar, B. A. and Ghose, D., 'Radar-assisted collision avoidance/guidance strategy for planar flight', *IEEE Transactions on Aerospace and Electronic System*, 37(1), 77–90, 2001.

11. Sunder, S. and Shiller, Z., 'Optimal obstacle avoidance based on the Hamilton–Jacobi–Bellman equation', *IEEE Transactions on Robotics and Automation*, 13(2), 305–310, 1997.

12. Kuchar, J., Andrews, J., Drumm, A., Hall, T., Heinz, V., Thompson, S., and Welch, J., 'A safety analysis process for the traffic alert and collision avoidance system (TCAS) and see-and-avoid

systems on remotely piloted vehicles', *Proceedings of AIAA 3rd 'Unmanned Unlimited' Technical Conference, Workshop and Exhibit*, Chicago, IL, 2004.

13. Cho, S. J., Jang, D. S., and Tahk, M. J., 'Application of TCAS-II for unmanned aerial vehicles', *Proceedings of 2005 JSASS-KSASS Joint Symposium on Aerospace Engineering*, Nagoya, Japan, 2005.

14. Asmat, J., Rhodes, B., Umansky, J., Villavicencio, C., Yunas, A., Donohue, G., and Lacher, A., 'UAS safety: unmanned aerial collision avoidance system (UCAS)', *Proceedings of the 2006 Systems and Information Engineering Design Symposium*, Charlottesville, VA, 2006.

15. DeGarmo, M. T., *Issues Concerning Integration of Unmanned Aerial Vehicles in Civil Airspace*, MP 04W0000323, MITRE, November, 2004.

16. Hedlin, S., *Demonstration of Eagle MALE UAS for Scientific Research*, Swedish Space Corporation Mattias Abrahamsson, Swedish Space Corporation, 2002. Available online at: http://www.neat.se/information /papers/NEAT_paper_Bristol_2003.pdf.

17. Kayton, M. and Fried, W. R., *Avionics Navigation Systems*, John Wiley & Sons, New York, 1996.

18. Dowek, G. and Muñoz, C., 'Conflict detection and resolution for 1, 2, . . . , N aircraft', *Proceedings of the 7th AIAA Aviation Technology, Integration and Operations Conference*, Belfast, Northern Ireland, 2007.

19. Galdino, A., Muñoz, C., and Ayala, M., 'Formal verification of an optimal air traffic conflict resolution and recovery algorithm', *Proceedings of the 14th Workshop on Logic, Language, Information and Computation*, 2007.

20. Shin, H. S., White, B. A., and Tsourdos, A., 'Conflict detection and resolution for static and dynamic obstacles', *Proceedings of AIAA GNC 2008*, August 2008, Honolulu, HI, AIAA 2008-6521 .

21. White, B. A., Shin, H. S., and Tsourdos, A., 'UAV obstacle avoidance using differential geometry concepts', *IFAC World Congress 2011*, Milan, Italy, 2011.

8

Aircraft Separation Management Using Common Information Network SAA

Richard Baumeister[1] and Graham Spence[2]

[1]Boeing, USA
[2]Aerosoft Ltd., UK

8.1 Introduction

Achieving continual safe separation distances between all aircraft, including unmanned aerial systems (UAS), is a critical requirement for integrating UAS and manned aircraft within controlled and uncontrolled airspace. Historically, this has been achieved for manned aircraft in controlled airspace by ground controllers mandating conservative safety volumes and strict flight plan adherence rules. Potential violations of these volumes can be determined if all aircraft in the controlled airspace are being tracked. If the safety volumes are in danger of being violated by an intruder, air traffic control (ATC) officers can request aircraft trajectory adjustments (usually issued by voice commands). This manual process may take from tens of seconds to minutes depending on: (i) the human controller workload; (ii) the ATC service available; (iii) the availability of decision support tools; (iv) the surveillance equipment such as the radar update rate; (v) the number of aircraft in conflict; and (vi) the time it takes for pilots to manually implement changes. This relatively large *airspace control latency* manifests itself in the application of correspondingly large en-route desired safe horizontal separation distances between aircraft in controlled airspace. A typical value for en-route safe separation is 5 nautical miles, although this value may vary from a few nautical miles near airports to

Sense and Avoid in UAS: Research and Applications, First Edition. Edited by Plamen Angelov.
© 2012 John Wiley & Sons, Ltd. Published 2012 by John Wiley & Sons, Ltd.

tens of nautical miles for trans-oceanic flights. More recently, the use of an air–ground data link implemented as the Controller–Pilot Data Link Communications (CPDLC) has started to be deployed to reduce the need for relatively slow and routine voice communication between pilots and ATC officers, replacing ATC clearances with data-linked messages. However, this improvement in communications and control latency does not affect the large latencies inherent in the surveillance, manual decision, and implementation process.

Controlled airspace for en-route aircraft is relatively *steady state*, with aircraft following static flight plans filed prior to departure, and with deviations from these being an exception. When tactical changes are required due to an unforeseen event such as weather, or the deconfliction of a potential loss of separation, the system can adjust in various ways, such as limiting new aircraft departures, re-routing aircraft, or limiting the number of flights within particular sectors.

In uncontrolled airspace there is, typically, no manned ground control authority to monitor safety volumes and, perhaps, no radar coverage. In such airspace, pilots conform to well-known safety rules and safety volumes are self-imposed by a pilot using *see and avoid*. For UAS operating in uncontrolled airspace, a pilot's see and avoid role would be fulfilled by *sense and avoid* systems. Standards for these systems are still being researched and developed, but there are no specific standards defining the expected performance (such as conformance to defined separation minima) and guidance behavior resulting from the sense and avoid function. In contrast with controlled airspace, uncontrolled airspace is more dynamic with aircraft crews not required to file and follow a prescribed flight plan. If a threat aircraft is observed or sensed, a pilot can determine and execute the control actions necessary to avoid breaching a loss of separation (LOS) threshold and avoid a potential aircraft collision.

There are two broad categories with regard to sense and avoid (SAA) systems. Here, we define *direct sensing* as those SAA methods that use onboard sensors to detect local aircraft. We also define *networked sensing* or *network sense and avoid*, in which a platform receives information about nearby aircraft via external communication data links. A non-exhaustive list of direct sensing methods includes: (a) pilot visual detection; (b) optical sensor detection; (c) long-wave/forward-looking infrared; and (d) onboard radar. Such sensors allow an aircraft to autonomously detect nearby aircraft. However, these methods suffer from issues including sensor field of view, effective range of detection, and, for certain UAS, there can be severe onboard sizing and weight limitations. Minimal onboard sensing has proven effective for aircraft collision and avoidance maneuvers when a threat aircraft is in very close proximity and emergency action is necessary to prevent a catastrophic collision [1].

Networked sense and avoid occurs when a platform (pilot, ground control operator, and/or onboard flight management system (FMS)) is continually provided with situational awareness and potential threat aircraft data via external communication channels. The networked SAA concept implies the existence of an automated system that maintains aircraft communication links and has access to aircraft tracking data within a region of interest. This tracking data can be monitored and, if a LOS breach is predicted or detected, flight path modifications determined and uplinked to ensure safe separation. In addition to flight plan modifications (and possible other control requests), situational awareness (SA) information can easily be provided to the aircraft pilot and, for potential use by an advanced FMS or a remote UAS operator to decide on suitable avoidance actions.

Networked SAA can help integrate UAS into controlled and uncontrolled airspace by providing separation management information directly to the UAS FMS and/or to the UAS controller. Such an SAA doesn't attempt to provide a vision system equivalent to a human, but rather provides an integrated situation awareness and separation management system applicable to manned and unmanned aircraft. The rationale behind testing networked SAA is supported in the '2009–2034 Unmanned Systems Integrated Roadmap' [2] (A.3.2.3, p. 99), which states that 'Since the purpose of this regulation [see and avoid] is to avoid mid-air collisions, this should be the focus of technological efforts to address the issue as it relates to UAS rather than trying to mimic and/or duplicate human vision avoiding mid-air collisions'.

One key advantage of networked sensing is that it is straightforward to network aircraft on a communications network using COTS hardware. Table 8.1 lists how networked sensing using persistent communication links to an external network can help overcome several challenges with direct sensing methods for SAA.

Table 8.1 Techniques in which networked sensing can overcome challenges with direct sensing

Direct sense and avoid challenges	Advantages of networked sensing
Aircraft sensor hardware implementation requirements including power, weight, and footprint.	Aircraft communications hardware can be lightweight, low power, and small footprint, including satellite and cellular transceivers.
Threat detection problems including threat range and direction of approach.	Tracked aircraft at all ranges and directions.
Each sensor is not typically designed with the ability for external communications with other aircraft.	Connectivity with other aircraft and Internet.

Networked SAA does not preclude the use of onboard target detection sensors on UAS that may augment safety margins. The data from these sensors can also be put on the network as an additional information source.

This chapter explores using networked SAA (in this chapter referred to simply as SAA unless explicitly indicated otherwise) to achieve safe separation thresholds for manned and unmanned aircraft, within a region of uncontrolled airspace using a common information network (CIN). The CIN is assumed to network cooperative airspace users with airspace information sources including an automated separation management process. The work presented here focuses on uncontrolled airspace for the following reasons:

- Controlled airspace has well-established aircraft separation procedures. This work is not a replacement for existing ATM systems and procedures for controlled airspace.

- Many future civil and commercial UAS operations could take place in uncontrolled airspace and at relatively low altitudes. Examples include surveillance, surveying, search and rescue.

- More accidents (specifically, mid-air collisions) with general aviation aircraft, near uncontrolled airports, and under visual meteorological conditions (VMC) [3]. Such airports could be candidates for usage by commercial UAS.

- To ensure safety of all airspace users in the event of future UAS operations in uncontrolled airspace.

Another reason for considering uncontrolled airspace is that civil airspace regulators have plans for implementing next generation ATM systems in controlled airspace which would include a specific implementation similar to networked SAA (NextGen [4] and SESAR [5]). Although developments in automatic dependent surveillance continue, there is at present no agreed way of integrating UAS into these future ATM systems, or of applying these systems to the general aviation community. It is hoped that the research presented in this chapter may help in these efforts.

This chapter will consider uncontrolled airspace in which cooperative aircraft (suitably equipped) share their state information in real time over a network which is connected to an automated separation manager (SM) process. Uncooperative aircraft (those lacking equipment or behaving uncooperatively) are assumed to have their positions tracked by other sensors (such as primary and secondary surveillance radar) and are also present on the network but do not connect with the CIN or the SM process. The SM ensures that safe separation thresholds are achieved for all cooperative aircraft by sending situational awareness data and suggested route deviations for all cooperative aircraft, while accounting for uncooperative aircraft in the suggested deviations.

The rest of this chapter is organized as follows. In Section 8.2 a generic decomposition of aircraft information flow on a CIN is illustrated. A specific implementation of this process for the Smart Skies Flight Test program is described in Section 8.3. Test results from the Smart Skies Flight Tests are given in Section 8.4. Finally, in Section 8.5, possible future uses of this approach are discussed.

8.2   CIN Sense and Avoid Requirements

In this section we present a generic process for information flow on a network-based SAA system on a CIN which continuously monitors aircraft tracks, issues SA information, and, if necessary, computes safe aircraft trajectory modification information. Throughout this chapter we limit our considerations of SAA using automated aircraft separation management performed by a centralized control center. In other words, the control center provides a hub for the airspace separation management system components, including the networked aircraft. This assumption is made so that the SAA process is easily visualized and reflects a system that has been flight tested by the authors. A decentralized and networked SAA system can also be constructed in a similar manner but will not be discussed.

It is advantageous to decompose this information processing and flow in a structured, hierarchical manner [6]. Figure 8.1 illustrates a high-level depiction of a closed-loop SAA information flow on a CIN. The flow contains seven segments. Segment 1, *User Data Source*, is the process by which current aircraft state data is estimated for an air vehicle and inserted into the CIN. For cooperative aircraft, Segment 1 could be implemented on an aircraft by merely deriving and transmitting GPS onboard navigational data. For uncooperative aircraft, Segment 1 could be implemented by deriving aircraft tracks from a ground-based radar system. In Segment 2, *Track Capture*, the transmitted data from Segment 1 is captured at a CIN communications node (such as a relay satellite, air-vehicle or ground station) for subsequent relay to a control center. Segments 3, 4, and 5 represent the process and flow at the centralized SAA control center. Segment 3, *Estimate True Track*, consolidates and extrapolates data received from all Segment 2 sources. Since one aircraft may have multiple sources of state

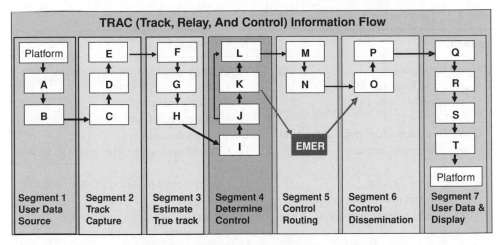

Figure 8.1   SSA information flow on a CIN

data, any redundant sources must be analyzed, compared, and resolved. The key output from Segment 3 is the best estimate of the present and future track of the tracked aircraft. This output is then used by Segment 4, *Determine Control Process*, which contains the actual SM algorithm. In Segment 4, present and predicted distances between all aircraft are estimated to determine if specified LOS safety thresholds between aircraft are violated. If any future violations occur, safe trajectory modifications are computed by the SM. In general, LOS violations will be anticipated far enough in the future so that there is ample time for the new trajectory deviations to be transmitted, received, and then executed by the aircraft. However, in cases where a LOS violation is imminent or LOS has already occurred, the SM can declare an emergency (EMER) condition and immediately send a safe (fall-back) maneuver to the aircraft. Segment 4 also includes SA information including aircraft states. Segment 5, *Control Routing*, within which the SM determines the optimum communication path for the situation awareness data and possible deviations to be sent to the user. These messages are disseminated to the user in Segment 6, *Control Dissemination*, and then displayed in Segment 7, *User Data & Display*.

The information flow shown in Figure 8.1 represents a real-time closed-loop control process. The track, relay, and control (TRAC) segments define the individual sub-processes of information flow at a high level. These TRAC segments, in turn, are broken down into 20 *segment functions* labeled A to T, plus the emergency action EMER function. The segments and associated segment functions are listed in Table 8.2. The purpose of this decomposition is to define and understand critical components in the flow of information on a CIN and to determine key requirements. One key requirement is understanding information latency of SAA on a CIN, which we define as the time necessary for information to flow from Segment 1 to Segment 7. Each segment function in Table 8.2 has an associated latency. If we make the conservative assumption that the flow is sequential, the total latency is simply the sum of the individual segment function latencies. It is interesting to compare the information flow latency for completely manual versus completely automated flows. This is accomplished in Figure 8.2, where a LOS scenario involving several aircraft must be monitored and safely separated. In

Table 8.2 Information flow segment functions

Segment	Key functions
1. *User Data Source*: Prepare and transmit AC state information	A. Prepare aircraft state information for transmission this cycle B. Transmit information
2. *Track Capture*: Communication of aircraft information to SAA control center	C. Capture aircraft information D. Process data stream E. Connect and relay data stream to ground center
3. *Estimate True Track*: Derive aircraft predicted states from potentially several sources	F. Aggregate information from all aircraft this cycle G. Integrate, filter, and route received information H. Predict aircraft states using current and historical information
4. *Determine Control*: Evaluate separation constraints and perform separation management	I. Evaluate separation constraints this cycle over appropriate time windows J. Determine controls needed to achieve safe separations, perform EMER action if necessary K. Generate situation awareness information for local display L. Prepare controls and SA information for transmission EMER. Prepare and route emergency information
5. *Control Routing*: Determine communications path to aircraft	M. Establish routing for nominal transmissions this cycle N. Route nominal transmissions O. Establish aircraft connection
6. *Control Dissemination*: Communications to aircraft	P. Capture control center information Q. Receive SA and control information
7. *User Data & Display*: Aircraft receives SA and control information and takes appropriate action	R. Process control information S. Integrate all information for decision T. Perform control this cycle

this figure the *x*-axis represents the segment functions A–T listed in Table 8.2. The *y*-axis represents the estimated latency associated with each segment function. Manual control using voice commands between a controller and a pilot is an intensive task, with some cases taking on the order of minutes. The top half of Figure 8.2 indicates the major contributions to the latency resulting from voice communications, the ability to predict and resolve the potential LOS, and the implementation of these new commands (segment functions B, H, I, J, Q, and S). The bottom half of the figure shows how this latency can be reduced to a few seconds by employing data links over a CIN and with computer algorithms performing the critical tasks

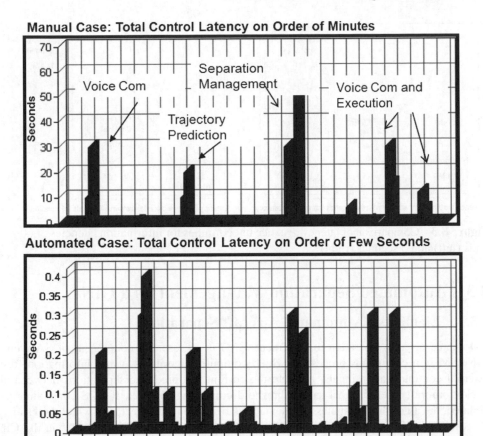

Figure 8.2 Comparison of latency for manual and automated information flows. The horizontal axis represents the segment functions listed in Table 8.2

of estimating future LOS and safe separation deviations. Note that the y-axis in the top graph (manual case) ranges from 0 to 70 seconds whereas the y-axis in the bottom graph (automated case) ranges from 0 to 0.4 seconds. The significant latency peaks for the manual case become negligible for the automated case. These values were derived by estimating the performance of typical HW, SW, and communication performance which suggests that latency on the order of 3 seconds or less is possible with automation.

The information latency, dt, is directly related to the separation distance in which aircraft can be safely controlled. An aircraft with speed v will travel approximately $dt*v$ distance from the time aircraft state information is determined in segment function A (prepare aircraft state information) until it implements any appropriate actions in segment functions S and T (integrate information and initiate control action). In dynamic airspace the aircraft can implement changes in states at any time so that prediction of future states will always have a measure of uncertainty related to this latency. Figure 8.3 shows the relationship between latency distance, uncertainty, and information latency for three aircraft speeds. A desired small latency of 3 seconds associated with automated information flow makes the distance uncertainty negligible when compared with a latency of 60 seconds.

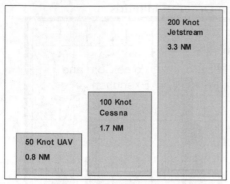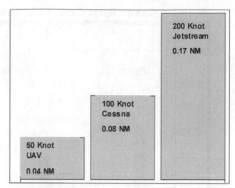

Latency Distance Uncertainty with 60 Seconds Information Latency   Latency Distance Uncertainty with 3 Seconds Information Latency

Figure 8.3   Comparison of latency distance uncertainties for information latencies of 60 seconds (left) and 3 seconds (right)

8.3   Automated Separation Management on a CIN

8.3.1   Elements of Automated Aircraft Separation

Automated aircraft separation management (SM) is an essential component of SAA on a CIN. SM is a continuous process in which computer algorithms, using information from external communication inputs, evaluate the current and predicted separations between all tracked aircraft to ensure that safety threshold constraints are met. If LOS constraints are violated, the algorithm computes safe deviations from planned trajectories which will satisfy all constraints. These trajectory modifications are sent to cooperative aircraft via the CIN and can either be implemented automatically, or presented for a manual oversight approval process.

The key elements of separation management on a CIN are encapsulated in the segment functions defined for Segments 3, 4, and 5 (shown in Table 8.2). Inputs from all users on the CIN are accomplished by segment functions F (aggregate information from all aircraft this control cycle) and G (integrate, filter, and route received information). These two functions prepare the information received from the CIN for processing by the SM algorithm. The corresponding Segment 5 functions determine which communication paths are best able to transmit data resulting from the SM algorithm to users on the CIN.

The segment functions H (estimate aircraft states using current and historical information), I (evaluate separation constraints this cycle over appropriate time windows), and J (determine controls needed to achieve safe separations, perform EMER action, if necessary) are the key elements of the SM algorithm.

There are several different algorithms used to predict future aircraft states using current (including intent information) and historical state data. For controlled airspace, straightforward methods assume that an aircraft simply follows a predefined geometric flight plan at a known speed. For uncontrolled dynamic airspace, flight plan adherence cannot be assumed, so parameter estimation methods can be used to predict future state, such as: linear extrapolation; closed-form extrapolation assuming constant speed and turning radius; data estimation filters; and trajectory prediction using simplified aircraft models. All methods should, ideally, account for the error in the prediction which will inevitably grow as the look-ahead time

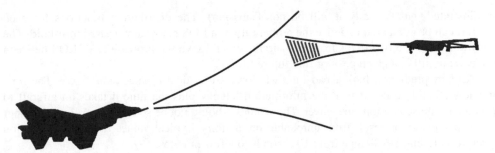

Figure 8.4   Intersection of two maneuver cones

increases. This error depends on several factors including the prediction time window, aircraft speed, maneuverability, weather, state data error, and information latency. A simple representation of all possible trajectories in the near future is illustrated by the maneuver cone, an approximate conical volume of potential future aircraft positions. Figure 8.4 illustrates the case where a fast, highly maneuverable aircraft has a much larger maneuver cone than a slow flying UAS. The intersection of the two cones represents potential LOS scenarios. When planning separation maneuvers the SM is aware of the respective speeds of the conflicting aircraft. In the scenario illustrated, the separation strategy leading to the quickest return to safe separations would be to move the fast flyer away from the trajectory of the UAS, because separation can be achieved in a shorter period of time.

Figure 8.5 shows a probabilistic approach to handling the cone of uncertainty. A probability distribution function (PDF) is estimated for each aircraft position (a Gaussian distribution

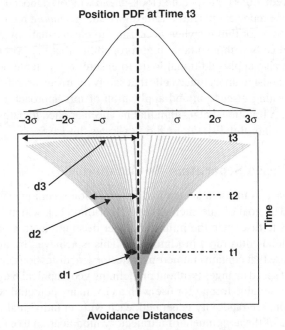

Figure 8.5   PDF describing predicted aircraft positions

is illustrated, but typically it will be non-Gaussian). The distribution functions for each aircraft can be compared to derive the probability of a LOS below any desired threshold. The convolution of the two PDFs from two aircraft would give the probability of LOS between the two aircraft at a given position and future time.

Aircraft prediction look-ahead time windows are critical to successful SAA. The time window should be larger than the information latency plus the time it takes an aircraft to execute a safe separations maneuver. The window should not be so large that a large number of false separations result from unreliable predictions. Typical values for general aviation aircraft including UAS range from 10 seconds to a few minutes.

The components of SM involved with anticipating LOS and computing corrective actions are accomplished within Segment 4 by the segment functions I (evaluate separation constraints this cycle over appropriate time windows) and J (determine controls needed to achieve safe separations). There are many categories of SM algorithms [7]. Since we are restricting ourselves to *centralized control* of aircraft separation, only three categories of automated separation algorithms will be discussed here: grid based, genetic search, and emerging systems.

8.3.2   Grid-Based Separation Automation

An example of a grid-based algorithm could construct a 4-dimensional grid, consisting of three spatial and one time dimension. In this discrete space-time manifold a grid cell at a particular time is either labeled as *occupied* or *not occupied*, implemented by using the values 1 or 0 respectively. More advanced grid-based methods can alternatively populate the grid cells with a probability of cell occupation. The *occupant* of a cell may be considered as a potential threat aircraft or a restricted flight region. Future cell states would be populated as a result of trajectory predictions for a specified look-ahead window. Unoccupied cells within the aircraft maneuver cone represent potential future *safe* and *reachable* regions of airspace. The goal of the grid-based algorithm is, when necessary, to ensure that any cooperative aircraft transverses only grid cells with a suitable high probability that the grid cell is unoccupied. Such algorithms can also apply additional logic to ensure that suitable boundary conditions are applied, such as choosing unoccupied cells that safely return an aircraft back to the original flight path. An interesting discussion and application of this approach can be found in the subject literature [8]. A traditional algorithm that has been applied to path planning problems such as this is the A* algorithm [9]. Figure 8.6 illustrates this case.

8.3.3   Genetic-Based Separation Automation

An alternative approach is to use a search and optimization technique such as a genetic algorithm (GA) [10]. The goal of one such genetic algorithm [11], which has been applied to aircraft separation, is to discretize the infinite number of solutions to a given predicted loss of separation scenario, as illustrated in Figure 8.7. This is achieved by the *a priori* creation of a maneuver database that contains numerous trajectory modifications composed of relative heading, altitude, and speed changes (without predefining geographical waypoints). To encode the LOS scenario in a suitable format for use with a GA, many potential solutions, otherwise known as a population, are randomly generated, each with an individual chromosome. Each chromosome consists of the assignment of a trajectory modification to each aircraft involved in the LOS scenario. Determination of the fitness of each solution is achieved by trajectory

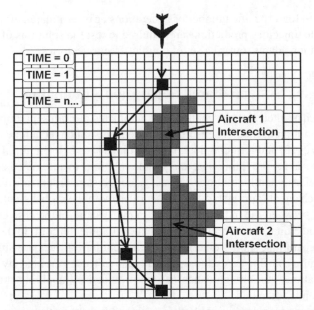

Figure 8.6 Illustration of grid-based aircraft separation management

prediction simulations which test the potential application of the assigned trajectory modification to each aircraft's flight. The fitness of each solution is determined by post-analysis of the
ground tracks recorded during the trajectory predictions. Contributing factors for the fitness
of each solution include the mutual separation distances between each aircraft during the
trajectory prediction, and a measure of the severity that the trajectory modification required
each aircraft to deviate off-track. Every generation, the least fit solutions (close separation
distances and/or a high track deviation component) are culled, while the remaining solutions
are subject to the basic selection, crossover, and mutation genetic algorithm operators. This

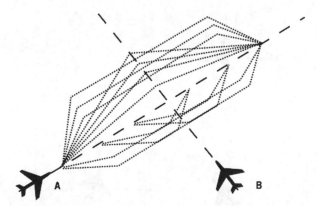

Figure 8.7 Example discretization of the infinite number of solutions to a LOS scenario,
as used by a genetic algorithm for automated aircraft separation. Dashed lines represent the
nominal flight paths while dotted lines represent a subset of collision avoidance maneuvers
for aircraft A

process is repeated for a specific number of generations or is terminated after a set time limit. Although multiple trajectory predictions are required to test the solutions of each generation, the whole population adapts toward a set of conflict-free trajectories.

8.3.4   Emerging Systems-Based Separation Automation

The third approach of interest to the challenge of automated aircraft separation management is based on the inspiration of natural self-organizing systems. Such techniques include artificial potential fields [12–15] and artificial flocking algorithms [16, 17]. On initial inspection, these techniques look ideal for providing neat and simple solutions to automated aircraft separation because they provide dynamic solutions that self-adapt. System order (safe aircraft separation) is achieved by each agent (aircraft) following set rules, with the global solution *emerging* due to the interaction between the individual agents in the system. In the case of potential fields, each aircraft would emit a virtual repulsive force upon all other aircraft. In theory global order, or safe aircraft separation, should emerge as all the repulsive forces are summed relative to each aircraft. Figure 8.8 illustrates the virtual repulsive forces emitted by each aircraft. In some implementations, aircraft guidance can be provided by applying attractive forces to navigation waypoints (diamond).

In flocking methods, rules are applied to each agent to keep the agents together. When applied to aircraft separation, anti-flocking rules must be generated. In their natural forms, these techniques are perhaps too dynamic and unpractical for application to aircraft separation (unrealistic pilot workloads in dense airspace scenarios). Without modification, the systems tend to oscillate and require frequent heading change commands to be sent to each agent, or aircraft. The authors have tested a restricted version of a hybrid potential field/anti-flocking algorithm [18]. This algorithm created avoidance vectors (determined by summing all the avoidance vectors contributed by involved aircraft) to steer aircraft around zones of potential conflict, yet included simple rules of the air to limit solution oscillation, provide predicable

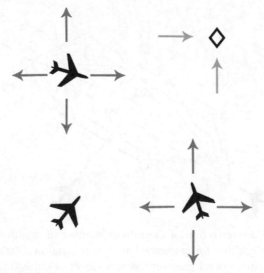

Figure 8.8   Simplification of the potential field approach to aircraft separation

aircraft behavior, and allow overtaking maneuvers. Although results of the algorithm were promising, further testing and integration with typical flight management functions is required.

8.4   Smart Skies Implementation

8.4.1   Smart Skies Background

Sections 8.2 and 8.3 outlined the theory of SAA on a CIN. This section discusses a prototype implementation of SAA on a CIN accomplished as part of the Smart Skies Flight Test Project. The Smart Skies Project (Smart Skies) was a 3-year collaborative research and flight test program, which ran between March 2008 and March 2011. Smart Skies explored future technologies that support the safe and efficient utilization of airspace by both manned and unmanned aircraft. The project brought together specialist researchers from Boeing Research & Technology (BR&T) aided by subcontractors at the University of Sheffield, Boeing Research & Technology Australia (BR&TA), and the Australian Research Centre for Aerospace Automation (ARCAA); a joint venture between the Commonwealth Scientific and Industrial Research Organization (CSIRO) ICT Centre, and Queensland University of Technology (QUT). The project objectives [19, 20] were to explore the development of key airspace automation-enabling technologies, one of which was prototyping an automated SAA system on a CIN.

Central to Smart Skies was a series of integrated flight trials conducted at Burrandowan Homestead near the township of Kingaroy in Queensland (QLD), Australia. These flight trials characterized the performance of SAA as envisioned using a CIN, in uncontrolled airspace under realistic and stressing operating conditions. During this program BR&T implemented several automated dynamic airspace control centers (ADAC) which served as SAA centralized control centers for software development, testing, and flight test support. In addition, ARCAA engineers developed and installed custom *predictive* flight management systems (pFMS) on all test aircraft, which allowed these aircraft to communicate with the ADAC via the CIN. The pFMS accomplishes the key Segment 1 (compiling and transmitting aircraft state) and Segment 7 (receiving and acting on situation awareness and control data) functions of the information flow described in Section 8.2. The pFMS also assists with pilot situational awareness by providing data for cockpit display of traffic information and display of suggested route modifications.

8.4.2   Flight Test Assets

The primary flight test aircraft (illustrated in Figure 8.9) used in the Smart Skies tests included:

1. A Cessna 172R model aircraft, from here on referred to as the Airborne Systems Laboratory (ASL). The custom modified aircraft is fitted with a GPS-INS truth data system, pFMS, custom navigation display (for visualizing flight plans, flight plan deviations, and other information received from the ADAC such as situation awareness data), and a communications management system. The ASL is capable of conventional human-piloted control or an optionally piloted mode (en-route control only).

2. A small autonomous fixed-wing UAS, referred to as the QUT UAS (QUAS). The QUAS has a maximum take-off weight of 20 kg, a payload capacity of 4 kg, and an

Figure 8.9   Flight test assets used during the Smart Skies flight trials. Clockwise from top-left are the ASL, CUAS, flight simulator, QUAS

endurance of approximately 1 hour (full fuel and payload). Onboard systems include: a pFMS; COTS autopilot; UHF, Iridium and 3G communications; and a vision-based sense and avoid payload.

3. A small autonomous helicopter, referred to as the CSIRO UAS (CUAS). The CUAS has a maximum take off weight of 13 kg and endurance of approximately 45 minutes (full fuel and payload). Onboard systems include: a pFMS; custom-designed flight computer and autopilot; UHF communications; Iridium and 3G communications systems located at the CUAS ground control system.

In addition to the real flight test aircraft described, multiple virtual aircraft were also used to increase the number of aircraft involved in a conflict scenario. A pFMS was developed by the University of Sheffield in conjunction with BR&T which allowed autonomous 6 DOF and piloted flight simulations to communicate with the ADAC over the CIN. These virtual aircraft are provided by the piloted engineering flight simulator [21, 22] or a standalone and fully autonomous 6 DOF flight simulation model developed by researchers at the University of Sheffield. The standalone 6 DOF simulations can be run on low-specification personal computers and are networked to the ADAC via the Internet. Each 6 DOF model uses a simple custom script language to initialize and program desired flight plans. The engineering flight simulator can be piloted and connected to the CIN using the Internet or an Iridium transceiver. From the perspective of the ADAC and the separation management algorithms under test, no distinction is made between the different test aircraft (manned or unmanned, real or simulated). Using simulated aircraft in combination with real aircraft and real communications links provides a safe and efficient testing environment for the evaluation of complex potential LOS

Table 8.3 Real and virtual aircraft used for the Smart Skies flight tests

Real aircraft	Autonomous simulations	Piloted *flight simulator*
ASL – Cessna 172	Cessna 172	Cessna 172
CUAS helicopter	Jetstream (twin turboprop)	Jetstream
QUAS Flamingo	Flamingo simulation	
	CUAS simulation	

scenarios. Virtual aircraft can be safely directed to fly toward real aircraft to force the SM to perform SAA over the CIN. Table 8.3 summarizes the test airspace segment assets which have been used in Smart Skies testing.

8.4.3   Communication Architecture

Two independent commercial communication systems were used for the Smart Skies CIN, the Iridium LLC RUDICS system and the Telstra Next G cellular system (NextG). These satellite (Iridium) and 3G cellular (NextG) communication services effectively allowed the ADAC to establish data connections with all aircraft, real and virtual, via TCP/IP Internet connections. The Smart Skies test communication architecture is shown in Figure 8.10.

The use of dual, independent communication channels improves the communication reliability issue for a CIN. Using the well-known formula for combining independent probabilities,

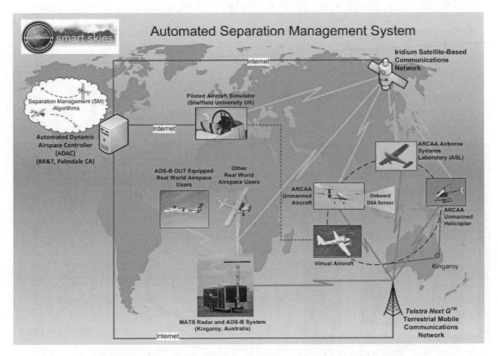

Figure 8.10   Smart Skies test communication architecture

$P(A) + P(B) - P(A \cap B)$, implies that two independent 90% reliable channels, when used in combination, result in a 99% communications reliability.

This architecture allows multiple aircraft to be continually tracked and controlled from an ADAC located anywhere in the world with Internet access [23]. Note that while the real aircraft were flying in Queensland, Australia, the primary flight test ADAC was located in Palmdale, CA. Execution of the virtual aircraft simulations was distributed between Sheffield, UK and in Palmdale, with the simulations offset to produce ground tracks over the test range in Australia. Typically, all test aircraft flew over the test range and with pre-planned flight plans that would cause various LOS scenarios with both virtual and real aircraft.

The ASL test aircraft could be flown either in a cooperative or an uncooperative mode. A mobile airspace tracking system (MATS) was deployed at the test site which could track real aircraft using radar and an ADS-B receiver. The MATS, which is under development by Boeing Research & Technology Australia, is capable of detecting non-cooperative aircraft over short to medium ranges. This information can then be networked with other surveillance sources to provide situational awareness to a ground controller and/or ADAC. This allowed the ADAC to track uncooperative aircraft over the test range [24]. Note that radar was not being used to provide a collision avoidance system (which implies a last-minute function). Rather, it tracks potential targets which the SM algorithms can account for when generating tactical route modifications.

A more detailed illustration of the Smart Skies ADAC architecture is given in Figure 8.11. This figure illustrates how an Iridium satellite connects to the Internet via the Iridium commercial gateway in Tempe, AZ.

The ADAC consists of several computers distributed on a Local Area Network (LAN) performing the tasks of:

1. Interfacing message traffic to and from users outside the ADAC via a messaging system referred to as the *ADAC shell*.

2. Aircraft separation management.

3. Operator situational awareness displays.

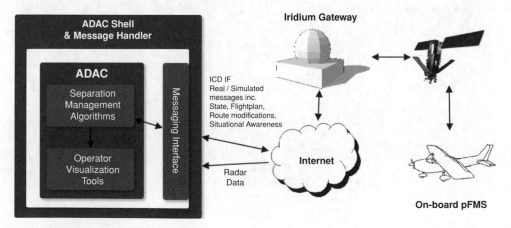

Figure 8.11   ADAC components and connectivity with the Iridium system

8.4.4   Messaging System

The ADAC *shell* was implemented by way of a message handling software component which communicated over the LAN with the SM algorithm. The SM used a network interfacing component (accessed via a dynamic link library) which allowed the development of SM algorithms to be isolated from the details of interfacing with the CIN. The ADAC LAN has a gateway to the Internet, allowing connections with all other nodes on the CIN. Users on the CIN can include both real and simulated aircraft. In this prototype system the ADAC message handling shell provides the role of a messaging server and the cooperative aircraft act as clients. Other ADAC software (such as the SM) can also connect as a client of the messaging server, enabling flexible ADAC configurations. Aircraft join the network by establishing TCP/IP connections with the messaging server and periodically transmitting state vector information and flight plan information. All messages are transmitted with a binary encoding, enabling more efficient usage of data-link bandwidth when compared with pure ASCII messages. In practice, some legacy communication systems require additional encoding of binary data to avoid erroneous insertion of control characters into data streams. This approach has been adopted in this project and all data channels use a consistent encoding scheme to support such legacy systems, regardless of whether the encoding is required by the underlying communication system. In typical use, the airborne pFMS transmits a binary state data message, here termed *trajectory array data set* (TADS), to the ADAC at a nominal rate of 1 Hz (the actual rate is also a parameter varied for particular tests). The TADS message contains aircraft position, speed, and altitude parameters recorded onboard at a given instant. There are approximately 80 bytes per TADS including message overhead.

The ADAC sends two key messages to the pFMS. Periodically, the ADAC will send a *situational awareness* message consisting of the position, heading, and altitude of all known nearby aircraft within a given region. If the SM algorithm determines action is necessary to avoid a LOS, the ADAC will send a flight plan route modification to the pFMS. In this project, we term recommended flight plan modifications issued by the ADAC as *commanded TADS* (CTADS). In the Smart Skies implementation, the following guidance rules were agreed by the Smart Skies researchers for the generation and implementation of the CTADS. Participating real and virtual aircraft conformed to these rules.

1. Route modifications must be terminated at a navigation fix or waypoint coincident with a waypoint on the original flight plan.

2. All waypoints in a CTADS route modification message are considered as *fly-by*.

3. While navigating a set of CTADS waypoints, pilots should maintain their current speed or alter speed to meet any possible waypoint arrival time constraints.

4. Pilots should attempt to maintain standard rate turns when transitioning between the flight plan legs defined by the CTADS.

5. Pilots should attempt to navigate the CTADS waypoints in sequence and navigate each CTADS leg as if it is a *track-to-fix* (TF) leg.

6. CTADS flight plan modifications will arrive in advance of the requirement to perform the first maneuver. The ADAC has to account for communication latencies as well as potential latencies occurring while a pilot or automated system reviews and accepts the flight plan modification.

These rules were implemented so that aircraft are not left *lost* (for example, the creation of flight plan discontinuities, or routing the aircraft far off nominal track) with respect to the ADAC and the onboard aircraft FMS. Although this typically happens when aircraft are vectored by ATC, rejoining with the original flight plan requires manual sequencing of the waypoints. Requiring ADAC-generated flight plan modifications to return aircraft to a point on the original flight plan reduces uncertainty in estimating the location and near-future plan of participating aircraft. With respect to integrating UAS into mixed airspace, reducing such uncertainties contributes to the perception of increasing safety in the system. Restricting aircraft to a particular navigation behavior also reduces uncertainty and improves the accuracy of trajectory predictions computed by the separation management software.

Typically, each aircraft in a flight test maintains an end-to-end connection with the message handling server. Although TCP/IP connections are maintained, the nature of the communications links used in the Smart Skies CIN implementation (Iridium and 3G) means that losses in the connectivity of the wireless data links will tear down the connection with the server. Typically, the physical data-link implementations attempt to keep the connection open even in the event of temporary signal loss. However, the system must compensate for inevitable connection losses. In such situations, the message handling server can be left unaware that a data link has been lost. For this reason, it is the responsibility of the pFMS on each participating aircraft to reconnect with the message handling server. To sense the presence of *zombie* connections on the message handling server, the transmission of a custom *ping* message between the ADAC and the pFMS acts as a connection *heartbeat*. When the server attempts to relay a ping message through a zombie connection, the message handler senses the disconnection and frees any previously allocated resources. The use of a ping message also allows the ADAC to continually monitor CIN latencies with all nodes. Typical latencies measured during flight tests are given in Figure 8.12. The end-to-end information flow latency over the Iridium system is approximately 2.6 seconds, whereas the latency of the cellular NextG was 1.5 seconds. Similar results were observed in each test. Additionally, the standard deviation is greater for Iridium latencies.

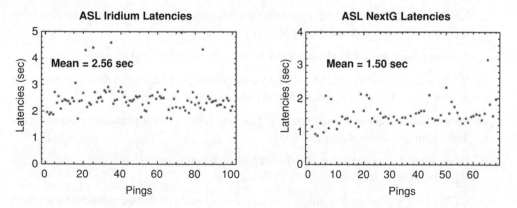

Figure 8.12 Example CIN end-to-end latencies over satellite (Iridium) and cellular (NextG) networks

Table 8.4 Key messages used to communicate between the ADAC and the pFMS

Message type	Length (bytes)*	Transmission frequency (Hz)	Additional notes
TADS (aircraft state data)	80	≥ 1	
Intended flight plan	Variable: 160 for 10 waypoints	Once	Retransmitted if modified by pilot
CTADS (commanded flight plan modifications)	Variable: 128 for 5 waypoints	As needed	Transmitted by the ADAC to cooperative aircraft requiring separation
Ping	30	0.1	Includes response message
Acknowledgment	28	As needed	Transmitted by the ADAC or pFMS in response to receipt of a flight plan or CTADS, respectively
Situation awareness	Variable: 178 for 4 local aircraft	0.1	Transmitted by the ADAC to provide surveillance data to neighboring aircraft

*All message lengths reported include 26 bytes of overhead inclusive of timestamps and aircraft identification.

An ADAC/pFMS/CIN interface control document (ICD) listing in detail the content of these messages was established early in the program. The key messages, lengths, and their typical transmission rates are listed in Table 8.4.

8.4.5 Automated Separation Implementation

During Smart Skies, the primary SM algorithm tested within the ADAC was developed by BR&T scientist Dr Regina Estkowski. The algorithm, a variation of the grid-based method described in Section 8.3, has been named the Virtual Predictive Radar (VPR) algorithm and is considered Boeing proprietary. The algorithm, which was tailored to meet the requirements of the Smart Skies Flight Test program, and modified several times during the project to implement new requirements, successfully managed and separated aircraft LOS scenarios ranging from 2 to 50 aircraft over the test region. Example use of the previously described genetic algorithm conflict resolver (Section 8.3) and potential field algorithms was also briefly fielded but with more limited test and results. However, the ability to swap-in differing aircraft separation algorithms was considered to be beneficial.

8.4.6 Smart Skies Implementation Summary

Table 8.5 lists the segment level information flow decomposition for networked SAA as implemented in Smart Skies.

Table 8.5   Smart Skies implementation of the information flow segments

Segment	Smart Skies implementation
1. Source aircraft data	pFMS collects GPS/INS information and transmits to the CIN through dual communication links. Aircraft include ASL, QUAS, CUAS, and virtual 6 DOFS and hands-on simulators. Dual transmissions for all real cooperative aircraft. Uncooperative aircraft tracked by MATS
2. Communications: aircraft to SAA control center (ADAC)	Cooperative real aircraft: simultaneous satellite (Iridium) and cellular transmission via Internet gateways to ADAC. All real aircraft: radar and/or ADS-B transmissions (if available) over Internet to ADAC. Virtual aircraft: Internet connections to ADAC
3. Estimate aircraft predicted states within airspace	ADAC message handler receives transmissions over dual channels. SM chooses primary (pre-selected) channel if available. Trajectory predictions are made within the SM using a combination of state extrapolation and aircraft intentions (flight plan) information
4. Evaluate separation constraints and perform separation management	Boeing VPR algorithm uses short-term and long-term separation constraints. When violated it issues appropriate CTADS following agreed upon rules if necessary
5. Choose communications path to aircraft	ADAC SM chooses any good communication link (cellular and satellite) and requests the MH to use the selected channel
6. Communications: SAA control center to aircraft	Same paths as Segment 2
7. Aircraft receives SA and control information and takes appropriate action	pFMS receives messages, manned aircraft displays SA data, choice of manually controlling ASL using cues or autonomously controlling heading by lateral autopilot. Real UAS and virtual aircraft autonomously implement commands from ADAC

8.5   Example SAA on a CIN – Flight Test Results

The Smart Skies program consisted of eight separate flight phases, with each phase lasting approximately 3 days, resulting in over 25 flight test days over 2 years. For each flight test with the ADAC, teams were deployed to the test region in Burrandowan, Queensland (ASL, QUAS, CUAS, and MATS teams), Palmdale, CA (ADAC team), and to the Sheffield simulator site in the UK. The early test phases occurring in 2009 had relatively simple

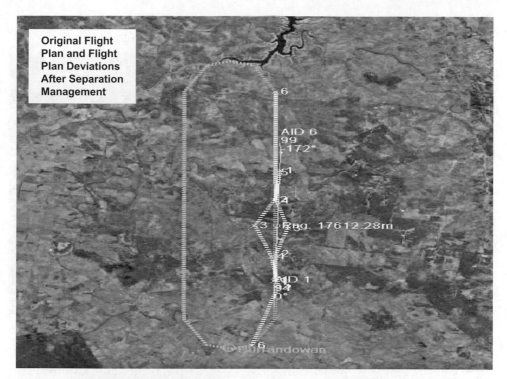

Figure 8.13   Head-on separation test: the ASL and a virtual Cessna

aircraft separation scenarios such as that illustrated in Figure 8.13. This figure shows a screen capture of an experimental ADAC visualization tool (SSBDO) and depicts: two aircraft, the ASL (labeled AID 6) and a virtual Cessna 6 DOF (labeled AID 1); their conflicting oval flight plan trajectories; the CTADS automatically issued by the ADAC to achieve a safe separation. For the illustrated test case, the nominal flight plans were developed so that a head-on LOS would occur on the right-hand side of the oval. For this particular scenario the desired minimum separation threshold was defined as 1 km with both the simulated aircraft and the ASL receiving CTADS, indicating that each aircraft shared the cost of the separation maneuver. The resulting distance of closest approach (DCA) while both aircraft were navigating the CTADS route modifications was approximately 1.7 km and the separation was deemed successful.

More complex and operational scenarios were investigated in 2010. Other scenarios examined SM performance as a function of aircraft type, approach geometry, speed, and data rate [25, 26]. Although all results can't be reported, an example of an operational scenario is shown in Figure 8.14. The left-hand side of the figure illustrates a complex *fire-drop* scenario where the ASL and two additional aircraft are flying trajectories that would enable them to drop fire suppressant on a fire at the lower portion of the figure. Simultaneously, the QUAS (fixed wing) is flying a fire surveillance mission and the CUAS (small helicopter) is simulating a search and rescue mission in the immediate fire zone. The right-hand side of the figure shows a snapshot of the actual test as visualized in SSBDO. In order to further stress

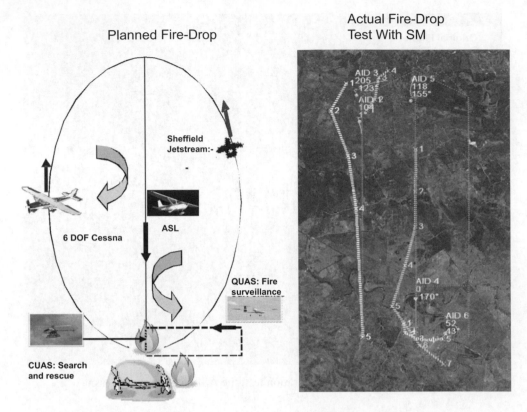

Figure 8.14   Fire-drop with search and rescue scenario; planned and actual

the SM, altitude differences between all aircraft could optionally be ignored, with separation based on horizontal distances only. Most scenarios (including the fire-drop test) were planned to have several near simultaneous LOS events and can be seen by the three active trajectory modifications being executed in the right-hand part of Figure 8.14. In this particular scenario, the planned and actual mutual separation distances of the five aircraft for the first 800 seconds of this flight test are shown in Figure 8.15. Each curve in Figure 8.15 represents the mutual horizontal distance in kilometers versus time for a pair of aircraft involved in the test. In this test there are five aircraft giving rise to ten distinct mutual separation curves. The horizontal dashed lines indicate the upper and lower thresholds for safe separation. In the actual test case the SM resolved all but one of the LOS events. The violation occurred with the CUAS rotorcraft UAS and a 6 DOF simulation of a Cessna aircraft. This result illustrates a problem with predicting the short-term trajectory of a rotorcraft which can hover and change heading in a manner very different from a fixed-wing aircraft. Separations between all fixed-wing aircraft were successful for this test.

Over ten additional complex scenarios were also tested during 2010 [27], ranging from tests involving 2 up to 50 aircraft at varying altitudes. Approximately 10% of the tests involved a mixture of cooperative and uncooperative aircraft; the remaining approximately 90% involved only cooperative aircraft. At the time of this publication researchers are still evaluating test results. High-level descriptions of some of the preliminary results are listed in

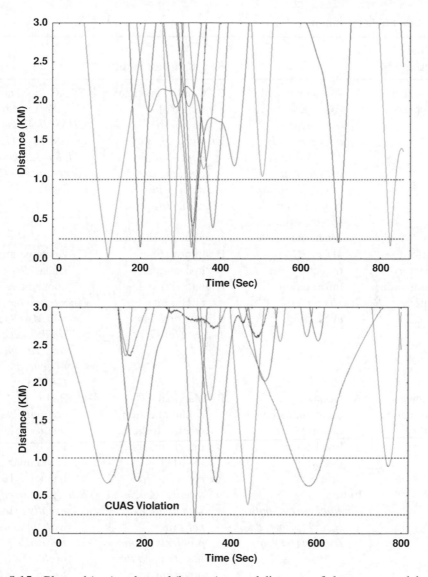

Figure 8.15 Planned (top) and actual (bottom) mutual distances of closest approach between all aircraft plotted against flight trial time for the fire-drop scenario

Table 8.6. Four key SM test parameters were: (i) airspace complexity; (ii) platform information content; (iii) received information quality; (iv) and operator obedience response.

Airspace complexity refers to the types of aircraft, aircraft density, and aircraft maneuverability and separation geometries. *Platform information content* refers to the TRAC flow Segment 1 information characterizing the aircraft state. *Received information quality* refers to Segment 7 information that is received at the aircraft after traversing the CIN. *Operator response obedience* refers to how a pilot or ground controller would use the information such as from SSBDO or SA data to make decisions.

Table 8.6   Flight test summary

Variables impacting SM		Key flight test results	
1. Air space complexity	A. AC density and AC types: *2–50 AC over approximately 15×15 nm² region with four aircraft types: UAS rotor craft, UAS fixed wing, Cessna, and Jetstream*	B. Conflict geometry (angle of approach) *Various angles of approach plus climbing and descending scenarios*	C. AC speed and maneuverability *CUAS: 0–10 knots QUAS: 40–60 knots ASL 80–120 knots Jetstream:150–240 knots*
2. Platform trajectory information content	A. 4D vs 7D trajectory information *Results under evaluation*	B. Information update rate: High (5 Hz) vs low (1 Hz) rates *0.5–1 Hz Iridium 1–2 Hz NextG*	C. Uncooperative AC (radar) vs cooperative *Uncooperative AC and MATS radar tracks successfully tracked and used to generate CTADS*
3. Received information quality	A. Good communication quality: NextG, low latency (<3 s), no dropouts Iridium *Nominal case of both NextG (prime) and Iridium (backup)*	B. Marginal communication quality: Iridium, marginal latency (3–10 s) *Rarely occurred, usually associated with onboard hardware problems. If one link dropped the other link was automatically used*	C. Bad communication quality: Iridium, loss of communication, latency >10 s *Rarely occurred, usually associated with onboard hardware problems. If one link dropped the other link was automatically used*
4. Operator response obedience	A. Automated vs manual *Use of lateral autopilot with CTADS was successful. Easier for pilots vs manual with SA TADS*	B. 4D vs 7D visualization for decisions *4D and 7D BDO displays were captured for further analysis*	C. Mission success vs safety *SM returned AC to flight plan after separation*

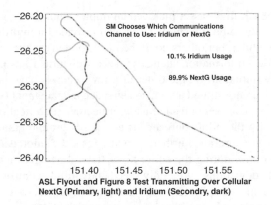

ASL Flyout and Figure 8 Test Transmitting Over Cellular
NextG (Primary, light) and Iridium (Secondry, dark)

Figure 8.16 Example of the ASL connected to the CIN by two communication channels.
Note how the SM decides from which channel to select data

With regard to these four SM test parameters, the SM worked well for various airspace complexities (i). The TADS information set transmitted by cooperative aircraft worked well at data rates between 0.5 Hz and 2 Hz (ii). One research question still being investigated is: under what conditions does having the attitude plus position state information (7D information) provide an advantage over positional information only (4D information)? In general, the information latency on the CIN was less than 3 seconds (iii). When the primary information channel did drop out (manifested either by a large delay in message arrival or a data call disconnection), the SM switched to the backup channel. An example of the SM switching between the primary and secondary CIN communications channels is shown in Figure 8.16. Using commercial communications systems inevitably resulted in instances where poor signal quality communications software interfaces failed onboard the real aircraft, resulting in significant message latencies (>10 s). In general, the ASL test pilots were comfortable using the automated lateral autopilot and permitting the ADAC to directly command the autopilot (through an uplinked and reviewed route modification). One of the prime benefits of permitting the autopilot to guide the aircraft through the ADAC suggested trajectory modifications was that it reduced their workload by reducing head-down concentration while navigating the aircraft in complex scenarios. With regard to the SSBDO operator displays, the belief was that it would reduce ground controller workloads but as yet, this has not been quantified using actual controller inputs (iv).

8.6 Summary and Future Developments

The successful Smart Skies flight test program has demonstrated the feasibility of implementing networked SAA via a CIN with the test architecture described. A description and implementation of a prototype networked SAA system was presented. The Smart Skies flight tests purposely included airspace scenarios, with integrated UAS that were considerably more complicated than one would expect to find in current uncontrolled airspace. Additionally, since UAS were assumed to be fully integrated in the airspace with respect to ATC (as implemented using an ADAC), each type of aircraft was considered as an equal during the airspace management process. The use of COTS communication networks to provide data links was successful. Each flight asset included an implementation of a pFMS that not only enabled aircraft state data to be transmitted, but also provided a multilink interface (in this case Iridium and 3G cellular, but others could be incorporated) to the various autopilots and FMS used in Smart Skies. Information latencies (data acquisition, transmission and reception, and separation control initiation) on the order of a few seconds have been established. The flight tests successfully demonstrated automated airspace management (with integrated UAS) and that aerial platforms can be controlled using trajectory modifications directly from the ADAC, and executed by the pilot (onboard or at a UAS ground station) to avoid loss of separation events between aircraft. Further, in many tests the piloted ASL remained under autopilot and pFMS control during the entire flight (lateral navigation of the nominal flight plan and received route deviations), with only speed and altitude control required by a human pilot. One important achieved milestone was the use of radar data to separate cooperative from uncooperative aircraft and place this information on the CIN.

Although additional research is continuing, it is reasonable to consider and explore issues with networked SAA over CIN for operational use, either exclusively or to complement onboard/direct SAA capabilities. Clearly, any operational system would have to consider key issues such as: affordability; communications reliability with cooperative aircraft; the tracking of uncooperative aircraft; and the multitude of aircraft and FMS/autopilot equipment installed in GA aircraft and UAS. There still remains the need to track uncooperative aircraft and the possibility of not being able to get tracks on uncooperative aircraft if ADS-B or radar information is unavailable. In the case when uncooperative tracks are not available the system would have to rely on current see and avoid procedures.

As mentioned previously, advanced ATM concepts, such as using traffic information services (TIS-B) and ADS-contract from a control center, would represent one implementation of SAA over a CIN. However, questions of affordability for UAS, use within the GA community, and the impact of forcing one solution on all users, have caused much debate on this approach. Ideally, networked SAA on a CIN shouldn't be restrictive to most users, so an option is that UAS should always be suitably equipped and cooperative, while GA and other users can opt out, especially if they are using alternative equipment that can be indirectly placed on the CIN (for example, ADS-B conformant equipment). In such a case, the emphasis would be to deconflict UAS, while other airspace users could opt in. Another approach to allow GA users to opt in, or to provide situational awareness of nearby UAS, is to take advantage of the recent growth in smart phones and smart tablets. Many of these devices have built-in GPS receivers (or the ability to connect to other suitable devices) and attitude sensors and can communicate over cellular and satellite systems with a small cost to the user.

An operational system would also need to resolve the problem of the lack of standardization among the navigational and guidance behavior exhibited by the variety of available FMS. For the Smart Skies implementation, a rigid interface control was adopted by all flight and simulation assets, which would not be possible for an operational system. Currently, a wide variety of FMS/autopilot systems are being used in GA and for UAS navigation, ranging from lateral-only autopilots and simple *direct-to* navigation systems to full 4D trajectory management computers and adaptations of commercial transport aircraft FMS. It is likely that these systems will produce a large variety of navigational guidance responses to flight plan modifications uplinked by future airspace automation systems. Before UAS can be safely integrated into non-segregated airspace managed by a semi- or fully automated CD&R system, the navigation systems and algorithms must produce predictable responses to uplinked commands [28]. One potential option is to enforce UAS to conform to performance-based navigation principles, with additional agreed guidance rules to conform to while executing uplinked suggested route modifications.

A crucial aspect of networked SAA is that information regarding all airspace users in a specific region is available. Boeing Research & Technology is investigating an approach that would amalgamate and translate the many different formats and communication methods into a common interface to address the problem of diverse platforms and formats on the CIN. Communications over multiple networks, satellite and terrestrial, certainly helps the communication reliability problem. Additional research remains to be accomplished in this area, but initial results using cellular and Iridium are very encouraging.

In summary, the computer hardware and communication network advances over the past decade have made it feasible to consider implementing SAA over a CIN in the near future. Such a system would provide the best-known situation awareness to the pilot in the cockpit with a latency of a few seconds. This additional information, while not guaranteed to be complete, would certainly augment the current method of see and avoid and provide an additional margin of safety.

Acknowledgments

The authors would like to acknowledge Mr Ted Whitley for supervising BR&T's effort in the Smart Skies project. Ted developed the concept of the ADAC and has long been a proponent of automated air traffic control. In addition, the many contributions of BR&T scientist Dr Regina Estkowski are acknowledged. Regina developed the primary separation management algorithm used during the Smart Skies project and, as part of the Smart Skies engineering team, offered invaluable advice, insight, and analysis into making this a successful project. We also acknowledge the advice and support of Professor David Allerton at the University of Sheffield. Further, we acknowledge the members of the Boeing Iridium Support and Battlescape development teams for their detailed subject knowledge and technical support. Finally, we wish to acknowledge all ARCAA members of the Smart Skies team. Headed by Professor Rod Walker with assistance from Dr Reece Clothier, the Smart Skies team has accomplished amazing engineering design and development tasks in a short period of time in order to conduct the airborne flight trials. This project has been funded, in part, by the Queensland State Government Smart State Funding scheme.

References

1. S. Temizer, M. J. Kochenderfer, L. P. Kaelbling, T. Lozano-Pérez, and J. K. Kuchar, 'Collision avoidance for unmanned aircraft using Markov decision processes', Proceedings of the American Institute of Aeronautics and Astronautics (AIAA) Guidance, Navigation, and Control Conference, Toronto, Ont., August 2–5, 2010.

2. *2009–2034 Unmanned Systems Integrated Roadmap*, US DOD, 2009.

3. R. C. Matthews, 'Characteristics of U.S. midairs', *FAAviation News*, 40(4), 1, 2001.

4. Joint Planning and Development Office, 'Concept of operations for the next generation air transportation system', Version 3.0, 2009.

5. SESAR Consortium, Deliverable 3 – The ATM target concept, Document DLM-0612-001-02-00, 2007.

6. The Boeing Company, US Patent 7,212,917: *Tracking relay and control information flow analysis process for information based systems*, issued and published May 1, 2007.

7. J. K. Kuchar and L. C. Yang, 'A review of conflict detection and resolution modeling methods', *IEEE Transactions on Intelligent Transportation Systems*, 1(4), 2000.

8. O. Watkins and J. Lygeros, 'Stochastic reachability for discrete time systems: an application to aircraft collision avoidance', IEEE Conference on Decision and Control, Maui, HI, 2003.

9. N. J. Nilsson, *Artificial Intelligence: A New Synthesis*, Morgan Kaufmann, 1998.

10. J. H. Holland, *Adaptation in Natural and Artificial Systems*, University of Michigan Press, 1975.

11. G. T. Spence and D. J. Allerton, 'A genetic approach to automated aircraft separation', CEAS 2009, Manchester, UK.

12. M. Eby and W. Kelly, 'Free flight separation assurance using distributed algorithms', IEEE 1999 Aerospace Conference, March 14–18, 1999.

13. W. Kelly and M. Eby, 'Advances in force field conflict resolution algorithms', AIAA Guidance, Navigation, and Controls Conference, Paper 2000-4360, Denver, CO, August 14–17, 2000.

14. S. Lee and J. Park, 'Cellular robotic collision free path planning', 5th International Conference on Advanced Robotics, Vol. 1, pp. 539–544, 1991.

15. O. Khatib, 'Real-time obstacle avoidance for manipulators and mobile robots', *International Journal of Robotics Research*, 5(1), 90–98, 1986.

16. C. W. Reynolds, 'Flocks, herds and schools: a distributed behavior model', *Computer Graphics*, 21(4), 25–34, 1987.

17. G. W. Flake, *The Computational Beauty of Nature*, The MIT Press, 1998.

18. G. T. Spence, D. J. Allerton, R. Baumeister, and R. Estkowski, 'Real-time simulation of a distributed conflict resolution algorithm', ICAS 2008 26th Congress, Anchorage, September 2008.

19. R. Clothier and R. Walker, The Smart Skies Project, AUVSI North America 2009, Washington, DC.

20. R. Clothier *et al.*, 'The Smart Skies Project', accepted for publication in *IEEE Aerospace and Electronic Systems Magazine*, 2011.

21. D. J. Allerton, 'A distributed approach to the design of a real-time engineering flight simulator', 21st ICAS Congress, September 1998.

22. D. J. Allerton, *Principles of Flight Simulation*, John Wiley & Sons, 2009.

23. The Boeing Company, US Patent 7,457,690: *Systems and methods for representation of a flight vehicle in a controlled environment*, issued and published November 25, 2008.

24. M. Wilson, 'A mobile aircraft tracking system in support of unmanned air vehicle operations', 27th International Congress of the Aeronautical Sciences, Nice, France, 2010.

25. R. Baumeister *et al.*, 'Evaluation of separation management algorithms in class G airspace', AIAA Modeling and Simulation Technologies Conference, Chicago, IL, AIAA-2009-6126, 2009.

26. R. Baumeister *et al.*, 'Test architecture for prototyping automated dynamic airspace control', CEAS European Air and Space Conference, Manchester, UK, 2009.

27. R. Baumeister *et al.*, 'Automated aircraft tracking and control in class G airspace', 27th International Congress of the Aeronautical Sciences, Nice, France, 2010.

28. G. W. Flathers, D. J. Allerton, and G. T. Spence, 'FMS automation issues for future ATM integration', 27th International Congress of the Aeronautical Sciences, Nice, France, 2010.

Part IV

SAA APPLICATIONS

9

AgentFly: Scalable, High-Fidelity Framework for Simulation, Planning and Collision Avoidance of Multiple UAVs

David Šišlák, Přemysl Volf, Štěpán Kopřiva and Michal Pěchouček
Czech Technical University, Prague, Czech Republic

AgentFly is a software prototype providing intelligent algorithms for autonomous unmanned aerial vehicles. AgentFly is implemented as a scalable multi-agent system in JAVA running on the top of the Aglobe platform [1] which provides flexible middle-ware supporting seamless interaction among heterogenous software, hardware and human actors. Thanks to JAVA, AgentFly can easily be hosted on UAVs or computers with different operating systems. The multi-agent approach [2] provides straightforward mapping – each airplane is controlled by one agent. Agents integrate intelligent algorithms providing a coordination-based control for autonomous UAVs. In the presented work, only algorithms which are fully distributed among airplanes are used. Such algorithms provide a real autonomous control for UAVs which do not require any central unit (a ground station or master airplane) controlling a group of UAVs. The main benefit is that the group of UAVs can also operate in situations where the permanent communication link with the central unit or ground operating station is missing. Some of the algorithms presented in this chapter suppose that UAVs are equipped with communication modems which allow them to dynamically establish bi-directional communication channels

Sense and Avoid in UAS: Research and Applications, First Edition. Edited by Plamen Angelov.
© 2012 John Wiley & Sons, Ltd. Published 2012 by John Wiley & Sons, Ltd.

based on their mutual position. Thus, airplanes utilize the mobile ad-hoc wireless network [3] created by their communication modems. These algorithms provide robust control in critical situations: loss of communication, destroyed airplane.

The AgentFly system has been developed over more than five years. It was initially built for simulation-based validation and comparison of various approaches for autonomous collision avoidance algorithms adopting the free-flight concept [4]. Later, AgentFly was extended with a high-level control providing tactical control team coordination. Even though the AgentFly system has been developed primarily for simulation purposes, the same agents and algorithms are also deployed for real UAV platforms. Beside UAV application, the US Federal Aviation Administration (FAA) supports the application of the AgentFly system for simulation and evaluation of the future civilian air-traffic management system which is being studied within the large research program called Next Generation Air Transportation Systems (NGATS) [5]. AgentFly has been extended with high-fidelity models of civilian airplanes and to support a human-based air-traffic control. The current version of AgentFly is suitable for several use-case models: a tool for empirical analysis, an intelligent control for UAVs and hybrid simulations. The hybrid simulation allows us to integrate a real flying UAV platform into a virtual situation and perform initial validation of algorithms in hazardous situations (which could be very expensive while done only with real platforms) and also perform scalability evaluation of intelligent algorithms with thousands of UAVs.

The rest of the chapter is organized as follows. Section 9.1 presents the overall multi-agent architecture of AgentFly. Section 9.2 describes the extensible layered UAV control concept. Section 9.3 briefly introduces algorithms used in the trajectory planning component and their comparison to other existing state-of-the-art methods. Section 9.4 describes the multi-layer collision avoidance framework providing the sense and avoid capability to an airplane. Section 9.5 provides the description of existing high-level coordination algorithms integrated with AgentFly. Section 9.6 presents distribution and scalability of the AgentFly simulation with respect to the number of UAVs. Finally, Section 9.7 documents the deployment of the AgentFly system and included algorithms to the real UAV platform.

9.1 Agent-Based Architecture

All components in AgentFly are implemented as software agents in the multi-agent middleware Aglobe [1]. Aglobe is like an operating system providing run-time environment for agents. It provides agent encapsulation, efficient agent-to-agent communication, high-throughput message passing with both address-determined and content-determined receivers, yellow page services providing the address look-up function, migration support, and also agent life-cycle management. The Aglobe platform has been selected because it outperforms other existing multi-agent platforms with its limited computational resources and very efficient operation. Moreover, Aglobe facilitates modeling of communication inaccessibility and unreliability in ad-hoc networking environments.

The high-level overview of the AgentFly agent-based architecture is shown in Figure 9.1. Basically, there exist three different types of agent in AgentFly: (i) UAV agents, (ii) environmental simulation agents, and (iii) visio agents. When AgentFly is started in the simulation mode, usually all three types of agent are used. On the other hand, when AgentFly is running directly on a real UAV platform, only UAV agents are running (one UAV agent per UAV platform) and an actuator control and sensing perceptions are mapped to real hardware.

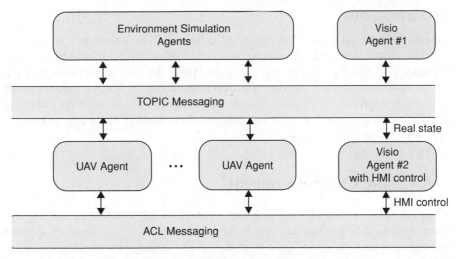

Figure 9.1 AgentFly system structure overview

9.1.1   UAV Agents

Each UAV in AgentFly is represented by one UAV agent. This agent provides the unit control for UAV. Intelligent algorithms for UAVs are integrated in this agent. Based on the configuration, they provide high-level functions like trajectory planning, collision avoidance, see and avoid functionality, and also autonomous coordination of a group of UAVs. AgentFly usually integrates algorithms providing a decentralized control approach. Thus, appropriate parts of algorithms are running in a distributed manner within several UAV agents and they can utilize ACL messaging providing UAV-to-UAV communication channels. If it is required by the experimental setup to use an algorithm which needs some centralized component, another agent can be created which is not tightly bound with any specific UAV.

9.1.2   Environment Simulation Agents

Environment simulation agents are used when AgentFly is started in the simulation mode. These agents are responsible for simulation of a virtual environment in which UAVs are operated. These agents replace actions which normally happen in the real world. They provide simulation of physical behaviors of virtual UAVs (non-real UAV platforms), mutual physical interactions (physical collisions of objects), atmospheric model (weather condition) influencing UAV behavior, communication parameters based on the used wireless simulator, and simulation of non-UAV entities in the scenario (e.g. humans, ground units). Through simulation infrastructure, these agents provide sensing perceptions for UAV agents. Beside simulation, there exist simulation control agents which are responsible for the scenario control (initialization of entities, parameter setups, etc.) and for data acquisition and analysis of configured properties which are studied in a scenario. These agents are created so that they support large-scale simulations that are distributively started over several computers connected by a network, see Section 9.6 for details.

9.1.3   Visio Agents

Visio agents provide real-time visualization of the internal system state in a 3D or 2D environment. Based on a configuration, much UAV-related information can also be displayed in various ways. In AgentFly, several visio agents providing the same or different presentation layers from various perspectives can be connected simultaneously. The architecture of Agent-Fly automatically optimizes data collection and distribution so that the network infrastructure is optimally utilized. A visio agent can be configured to provide HMI for the system, e.g. the user operator can amend the goal for algorithms.

9.2   Airplane Control Concept

The AgentFly UAV control concept uses the layered control architecture, see Figure 9.2. Many common collision avoidance approaches widely used in the research community [6–13] provide control based on a direct change of the appropriate airplane's state, e.g. a heading change control. Such control methods don't provide a straightforward way for complex deliberative UAV control because there is no detailed information about the future flight which is necessary for selection of the suitable solution from several versions. For example, the set of tasks should be fulfilled as soon as possible by a group of UAVs. Due to the lack of detailed flight information, a task controller cannot assign tasks to UAVs with respect to the required time criterion. The method used in AgentFly is based on a complete flight trajectory description. The flight trajectory is the crucial structure which provides a description of future UAV intentions, covering also the uncertainty while they are executed by UAVs. In AgentFly,

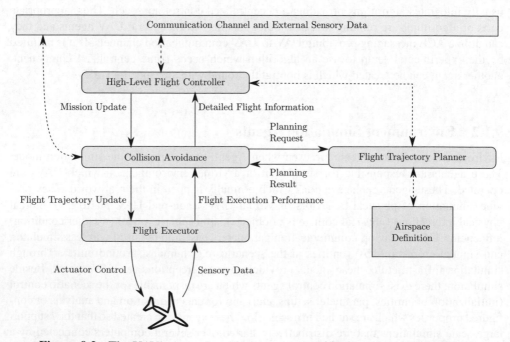

Figure 9.2 The UAV's control architecture: control blocks and their interactions

it is supposed that UAVs operate in a shared limited three-dimensional airspace called the operation space. Additional limits of the operation space are given by separation from the ground surface and by a set of no-flight areas which define a prohibited space for UAVs. No-flight areas are also known as special use airspaces (SUAs) in civilian air-traffic [14]. These no-flight areas can be dynamically changed during the run-time (e.g. there is identified an air defense patrol by an other UAV). Beside the flight trajectory, another crucial structure called the mission is used. The mission is an ordered sequence of waypoints, where each waypoint can specify geographical and altitude constraints and combine optional constraints: time restrictions (e.g. not later than, not earlier than), a fly speed restriction, and an orientation restriction.

The flight control in AgentFly is decomposed into several components, as shown in Figure 9.2:

- *Flight executor* – The flight executor holds the current flight trajectory which is executed (along which the UAV is flying). The flight executor implements an autopilot function in order to track the request intentions in a flight trajectory as precisely as possible. Such intelligent autopilot is known as the flight management system (FMS) [15] in civilian airplanes. The flight executor is connected to many sensors and to all available actuators which are used to control the flight of a UAV platform.

  Based on current aerometric values received from an airplane's flight sensors and the required flight trajectory, the autopilot provides control for UAV actuators. Aerometric data include the position (e.g. from the global positioning system (GPS) or inertial navigation system (INS)), the altitude from a barometric altimeter, the pressure-based airspeed and attitude sensors providing an angular orientation of UAV from gyroscopes. Depending on the UAV construction, the following primary actuators can be used: ailerons (rotation around longitudinal axis control), elevators (rotation around lateral axis control), rudder (compensation of g-forces), and thrust power (engine speed control). Large UAVs can also be equipped with secondary actuators changing their flight characteristics: wing flaps, slats, spoilers, or air brakes.

  The design of the flight executor (autopilot) is tightly coupled with the UAV airframe construction and its parameters. Its design is very complex and is not in the scope of this chapter. The control architecture covers the varying flight characteristics which are primarily affected by changing atmospheric conditions (e.g. the wind direction and speed). The control architecture supposes that the flight trajectory is executed within a defined horizontal and vertical tolerance. These tolerances cover the precision of sensors and also imprecise flight execution. The flight executor is aware of these tolerances and provides the current flight precision in its output. Section 9.7 includes the description of AgentFly deployment to the Procerus UAV platform.

  The current flight execution performance is monitored by layers above the flight executor and if it is out of the predicted one included in the executed flight trajectory, the replanning process is invoked. This is how AgentFly works with the position-based uncertainty present in the real UAV platform.

- *Flight trajectory planner* – The flight trajectory planner is the sole component in the UAV control architecture which is responsible for the creation of all flight trajectories. Planning can be viewed as the process of transformation of a waypoint sequence to the detailed flight intent description considering UAV model restrictions (flight dynamics) and the airspace definition. More information about the flight trajectory planner is provided in Section 9.3.

Beside planning (which is usually used only before the UAV is started), it is also capable of replanning (modification of an existing flight trajectory). In the case of replanning, the planning request contains the unique identifier of the position from which the plan has to be modified. This unique identifier is known as the unchangeable marker. Existence of the unchangeable marker in replanning is necessary for the case when a UAV is already flying and wants to change its currently executed flight trajectory. All intelligent algorithms used for UAV control run in non-zero time based on their complexity. Moreover, while the replanning process is running, the UAV is still flying. The flight trajectory can be changed only in the future, otherwise it is not accepted by the flight executor (as it is not consistent with the current UAV state, including its position). It can happen that for a planning request (an unchangeable marker and waypoint sequence) the planner is not able to provide a flight trajectory (e.g. a waypoint lies outside the operational airspace or it cannot satisfy any optional constraints). In such a case, the flight trajectory planner returns a failure as the planning result.

- *Collision avoidance* – The collision avoidance component is responsible for implementation of the sense and avoid function for a UAV. In AgentFly, the method using additional control waypoints which are inserted into the current mission waypoint sequence is used. These control waypoints are injected so that the final trajectory is collision-free with respect to other UAVs or piloted airplanes operating in the same airspace. The collision avoidance component chooses the appropriate collision modification with respect to the selected airplane preferences and optimization criterion. Each such considered modification is transformed in the flight trajectory utilizing the flight trajectory planner.

  Algorithms for collision avoidance in AgentFly implemented by the Agent Technology Center utilize the decentralized approach based on the free-flight concept [4] – the UAV can fly freely according to its own priority but still respects the separation requirements from others in its neighborhood. This means there is no centralized element responsible for providing collision-free flight trajectories for UAVs operating in the same shared airspace. Intelligent algorithms integrated in this component can utilize a communication channel provided by onboard wireless data modems and sensory data providing information about objects in its surrounding (a large UAV platform can be equipped with an onboard radar system, a smaller one can utilize receivers of transponders' replies or receives radar-like data from a ground/AWACS radar system).

  Collision avoidance implements the selected flight plan by passing the flight trajectory update to the flight executor. Collision avoidance utilizes the flight execution performance (uncertainty in the flight execution) to adjust the algorithm separation used while searching for a collision-free trajectory for the UAV. Collision avoidance monitors modification of the UAV mission coming from the upper layer and also detects execution performance beyond that predicted in the currently executed flight trajectory. In such a case, collision avoidance invokes replanning with new tolerances and conflict detection and separation processes are restarted with the new condition. More details about collision avoidance algorithms in AgentFly are provided in Section 9.4.

- *High-level flight controller* – The high-level flight controller provides a goal-oriented control for UAVs. This component includes intelligent algorithms for the group coordination and team action planning (assignment of the specific task to the particular UAV). Depending on the configured algorithm, the high-level flight controller utilizes the communication channel, sensory perceptions (e.g. preprocessed camera inputs), and the flight trajectory planner to decide which tasks should be done by the UAV. Tasks for the UAV are then formulated as a mission which is passed to the collision avoidance component. During the flight, the high-level flight controller receives updates with the currently executed flight trajectory, including modification caused by collision avoidance. The high-level flight controller can identify that properties of the flight trajectory are no longer sufficient for the current tasks. In such a case, the high-level flight controller invokes algorithms to renegotiate and adjust task allocations for UAVs and thus modify the current UAV mission. Examples of high-level flight control algorithms are given in Section 9.5.

If no high-level algorithm is used, a simple implementation of this component can be made which just provides one initial flight mission for the UAV composed as a sequence of waypoints from a departing position, flight fixes (where the UAV should fly through), and a destination area where it has to land.

9.3   Flight Trajectory Planner

The flight trajectory planner is a very important component in the control architecture presented in the previous section. It is the only component which is responsible for preparation of the flight trajectory for a UAV. Any time when other components require preparing a new version of the flight trajectory they call the planner with an appropriate planning request containing an unchangeable marker and waypoint sequence. The efficiency of this component has influence on the overall system performance. For example, while an intelligent algorithm searches for an optimal solution for an identified collision among UAVs, the planner can be called many times even though finally only one trajectory is selected and applied for its execution. Such a mechanism allows other layers to evaluate various flight trajectories considering their feasibility, quality, and also with respect to the future waypoint constraints. Similarly, the high-level flight controller can call the planner many times while it evaluates task assignments. The planner should be fast enough. There exist many trajectory approaches known from the robotics planning domain. There is a trade-off between optimality of the path planner and its performance.

AgentFly uses the optimization-based flight trajectory planner based on the classical A* algorithm [16]. The planner searches for the valid three-dimensional flight trajectory which has the minimum cost for a given criterion function. The search in the continuous space is transformed to the search within a dynamically generated state space where states are generated by UAV motion elements based on its model. The flight trajectory has to respect all dynamic constraints specified for the UAV model the trajectory has to be smooth (smooth turns), limits and acceleration in flight speed, etc. There is defined a set of possible control modes (flight elements) for the UAV which covers the whole maneuverability of the UAV. The basic set consists of straight, horizontal turn, vertical turn, and spiral elements. These elements can be combined together and are parameterized so that a very rich state space can be generated defining UAV motion in the continuous three-dimensional space. The example of generation of samples in two-dimensional space is shown in Figure 9.3. Us of the original A* algorithm is possible over this dynamically generated state space but its performance is

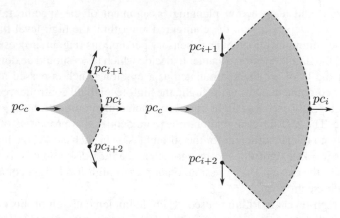

Figure 9.3 An example of generation of new samples from the current planning configuration in two-dimensional space

very limited by the size of a state space rapidly growing with the size of the operation space for the UAV.

In order to improve performance of the search, the *Accelerated A** (AA*) algorithm has been developed [17]. The simplified pseudo-code of the AA* algorithm is provided in Algorithm 9.1. AA* extends the original A* algorithm to be usable in large-scale environments while still providing a certain level of search precision. AA* removes the trade-off between the speed and search precision by the introduction of *adaptive sampling* – parameters used for generating elements are determined based on the distance to the nearest obstacle (operating space boundary or no-flight area). If the state is far from any obstacle, parameters are higher and thus produce longer flight elements and vice versa. This adaptive parameterization is included in each state while it is generated (line 9). Sampling parameterization is then used within the Expand function (line 7). Adaptive sampling leads to variable density of samples, see Figure 9.4 – samples are sparse far from obstacles and denser when they are closer.

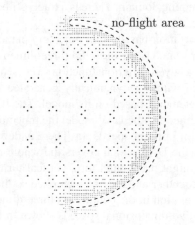

Figure 9.4 An example of adaptive sampling in two-dimensional space

Algorithm 9.1   The AA* algorithm pseudo-code.

```
{1} AASearch
{2}     Initialize OPEN, CLOSED;
{3}     while OPEN ≠ ∅ do
{4}         s_C ← RemoveTheBest (OPEN);
{5}         Insert(s_C, CLOSED);
{6}         if SmoothToEnd (s_C) then return ReconstructPathToEnd(s_C);
{7}         foreach fp_i ∈ Expand(s_C) do
{8}             pc_N ← EndConfiguration(fp_i);
{9}             ξ_N ← DetectSamplingStep (pc_N);
{10}            if Contains( pc_N, CLOSED, ξ_N) then
{11}                continue;
{12}            if not IsValid(fp_i) then continue;
{13}            s_N ← CreateNewState (pc_N, ξ_N, s_C);
{14}            InsertOrReplaceIfBetter (s_N, OPEN);
{15}        end
{16}    end
{17}    return failure;
{18} end
```

Adaptive sampling in AA* requires a similarity test instead of a equality test over *OPEN* and *CLOSED* lists while a new state is generated (lines 11 and 14). Because of the adaptive sampling step, usage of the equality test will lead to a high number of states, leading to a high density of states in areas far from any no-flight area. This is caused by the fact that generation is not the same in the reverse direction because those states typically have different sampling parameters due to different distances to the nearest no-flight areas. AA* uses sampling parameters based on the power of two multiplied by the search precision (minimum sampling parameters). Two states are treated as similar if their distance is less than half the sampling step. The algorithm is more complex as it also consider different orientations within the state, see [17] for details.

The path constructed as a sequence of flight elements can be curved more than is necessary due to the sampling mechanism described above. To remove this undesired feature of the planner, each path candidate generated during the search (line 13) is smoothed. Smoothing in AA* is the process of finding a new parent for the state for which the cost of the path from the start to the current state is lower than the cost of the current path via the original state parent. Such a parent candidate is searched for among all the states from *CLOSED*. The parent replacement can be accepted only if a new trajectory goes only within the UAV operating space and respects all constraints defined by its flight dynamics.

Properties of the AA* algorithm have been studied and compared with existing state-of-the-art planning methods using the modified search domain which is common to many algorithms. The modified domain is known as *any-angle grid-based planning* [18], see Figure 9.5. In such a planning problem, connection of any two vertices is allowed if the connecting line doesn't intersect any blocked cell. This problem is close to the planning described above. Two-dimensional any-angle planning can be viewed as planning for a UAV which is restricted to fly at one altitude and the horizontal turn radius is limited to zero. This means that the UAV can turn (change its horizontal orientation) without any transition. For this reduced planning problem, there is provided a mathematical analysis of AA* sub-optimality. The result

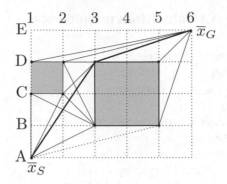

Figure 9.5   Any-angle path planning in the grid

proves that all solutions provided by the AA* algorithm are always within a certain tolerance from the optimal solutions. The sub-optimality range can be uniquely derived from the grid configuration, see [19] for details.

Several existing planning algorithms were selected, and used for comparison with AA*: the original A* algorithm [16] adapted for grid-based any-angle planning, the Theta* algorithm [20], and the very popular rapidly exploring random tree (RRT) techniques with one exploring tree [21] and the dynamic domain RRT algorithm with two exploring trees [22]. Both RRT techniques were combined with post-smoothing applied to their results [23], which removes the unwanted effects in the resulting paths given by the random nature of RRTs. The Field D* algorithm [24] is not included as it has already been shown that Theta* finds shorter paths in less time in similar tests. One experiment provides a comparison of thousands of generated grids with randomly blocked cells. Three different densities of blocked cells were tested: 5%, 10%, 20%, and 30% of blocked cells of the whole grid. Start and goal vertices were selected randomly too. Each different density of obstacles was tested 500 times with newly generated blocked cells, start and goal vertices. Table 9.1 summarizes the results, presenting the average values from all repetitive tests for the same densities. Each planning task was initially validated by the original A* algorithm to ensure that a path between start and goal vertices exists. The same generated planning task was then executed in sequence by Theta*, AA*, and both RRT algorithms.

Table 9.1   Path lengths and run-times (in parentheses) for various densities of blocked cells in randomly generated grids, each row contains average values from 500 generated planning tasks

	Blocked cell density			
Algorithm	5%	10%	20%	30%
A*	54.210 (10.084)	53.190 (11.896)	53.301 (18.476)	53.206 (31.493)
Theta*	54.345 (0.023)	53.428 (0.026)	53.623 (0.036)	53.611 (0.049)
AA*	54.210 (0.079)	53.190 (0.082)	53.301 (0.101)	53.206 (0.129)
RRT	73.834 (0.003)	79.558 (0.013)	85.207 (0.032)	85.344 (0.077)
ddbi-RRT	73.411 (0.0004)	75.896 (0.0015)	78.030 (0.0037)	81.566 (0.0089)

As presented in Table 9.1, the AA* algorithm finds the same shortest paths as the original A* algorithm providing the optimal solution for each planning task. Theta* finds longer paths than the optimal ones but still very close to them – they are about 1% longer than the optimal ones. Paths found by both RRT-based planners are more than 36% longer. The dynamic-domain bi-directional RRT (ddbi-RRT) algorithm provides shorter paths than the uni-directional RRT (RRT) algorithm while it is slightly slower. On the other hand, both RRT-based algorithms are very fast a few milliseconds in comparison to about a hundred milliseconds for AA*. However, AA* is many times faster (from 127 times for 5% density up to 244 times for 30% density) than the original A* algorithm while it provides the optimal path in all 2000 experiments. For higher density of blocked cells, it provides higher acceleration because its run-time has very small dependency on the number of blocked cells. On the other hand, the number of blocked cells has proportional dependency on the run-time of the original A* algorithm. Theta* is about three times faster than AA* but there is no guaranteed sub-optimality limit for Theta*. The acceleration of AA* is primarily gained by the reduction of the number of generated states. Thus, another major benefit of AA* is the lower requirement for memory during its run-time. Comparisons within other grids can be found in [25].

The AA* algorithm described above makes the planning algorithm usable in a large-state environment because it dynamically accelerates search within open areas far from any obstacle (no-flight ares or operating space boundary). The run-time of the search is affected by the number of defined obstacles. Even though tree-based structures are used for indexing, the higher number of obstacles slows down identification of the closest distance to any obstacle and also intersection tests. The AA* algorithm has been deployed to an operation space where there are more than 5000 no-flight areas (each defined as a complex three-dimensional object) defined, and in this case its performance was affected by such high number of obstacles. In AgentFly, there was introduced another extension of the search which is called the *Iterative Accelerated A* (IAA*)* algorithm. IAA* extends AA* in order to be usable in large-scale domains with a high number of complex obstacles. IAA* pushes the limits of fast precise path planning further by running the search iteratively using a limited subset of obstacles. IAA* selects a subset of obstacles which are positioned within a certain limit around a trajectory from the previous iteration. For the first iteration, the subset is selected around the shortest connection from the start to the goal. The set of obstacles used for AA* in subsequent iterations is only extended and no previously considered obstacles are removed. The range around a trajectory in which obstacles are selected is dynamically derived as a configured ratio from the length of the trajectory. After each iteration of IAA*, the resulting path is checked for intersection with all obstacles and if it doesn't intersect any obstacle, the result of this iteration is the result of the planning task. For the other case when any AA* iteration fails (no solution found), this implies that there is no solution for the planning task. Experiments in [26] show that the IAA* approach significantly reduces the number of obstacles and thus the number of computationally expensive operations. IAA* provides exactly the same result as AA* for all 60,000 planning tasks and provides results more than 10 times faster than AA* on average.

9.4   Collision Avoidance

This section contains the description of major algorithms for autonomous collision avoidance in AgentFly. All these algorithms adopt the decentralized autonomous approach based on the free-flight concept [4]. Within the free-flight approach, UAVs can fly freely according to their own priorities but still respect separation from others by implementing the autonomous *sense and avoid* capability. In other words, there is no central element providing collision-free flight paths for UAVs operating in the shared airspace. All algorithms in AgentFly for collision avoidance consider the non-zero time required for the search for collision-free flight trajectories. While a collision avoidance algorithm is running or performs flight trajectory replanning the UAV is still flying and the time for the flight trajectory change is limited.

In AgentFly, the integrated collision avoidance algorithms form two groups: (i) cooperative and (ii) non-cooperative. The *cooperative* collision avoidance algorithm is a process of detection and finding a mutually acceptable collision avoidance maneuver among two or more cooperating flying UAVs. It is supposed that UAVs are equipped with communication modems so that they can establish bi-directional data channels if they are close to each other. UAVs don't have any common knowledge system like a shared blackboard architecture and they can only utilize information which they gather from their own sensors or from negotiation with other UAVs. For the simplification of description in this chapter, we will suppose that UAVs provide fully trusted information. Based on the configuration, they are optimizing their own interests (e.g. fuel costs increase or delays) or the social welfare (the sum of costs of all involved parties) of the whole UAV group. On the other hand, the *non-cooperative* collision avoidance algorithm cannot rely on the bi-directional communication and any background information about the algorithm used by other UAVs. Such an algorithm is used when the communication channel cannot be established due to malfunction of communication modems or due to incompatible cooperative systems of considered UAVs. A non-cooperative algorithm can work only with information provided by sensors providing radar-like data. The rest of the section briefly presents all major algorithms. A detailed formalized description can be found in [7]. Many experiments evaluating properties of these algorithms, and also comparisons to other state-of-the-art methods, are given in [19].

9.4.1   Multi-layer Collision Avoidance Architecture

In AgentFly, the collision avoidance component as presented in Section 9.2 is represented by the complex architecture called the *multi-layer collision avoidance framework*, see Figure 9.6. This architecture is capable of detecting and solving future collisions by means of a combination of variant collision avoidance methods. It provides a robust collision avoidance functionality by the combination of algorithms having different time requirements and providing different qualities of solution. It considers the time aspect of algorithms. Based on the time remaining to the earliest collision, it chooses the appropriate method for its resolution. The architecture is modular and from its nature is domain independent. Therefore, it can be used for deployment on different autonomous vehicles, e.g. UGVs. Collision avoidance algorithms and their detector parts are integrated as plug-ins called collision solver modules.

Beside solver modules, there is also the collision solver manager (CSM) – the main controller responsible for the selection of a solver module that will be used for the specific collision. Each solver module has a detection part which is responsible for the detection of a potential future collision. In the case of a cooperative solver module, this detector uses flight intent information which is shared among UAVs locally using data channels. In the case of a non-cooperative solver module, this detector is based on a prediction of the future trajectory based on observations from radar-like data of its surrounding area. Each potential future collision is then registered within CSM. The same collision can be detected by one or several collision solvers.

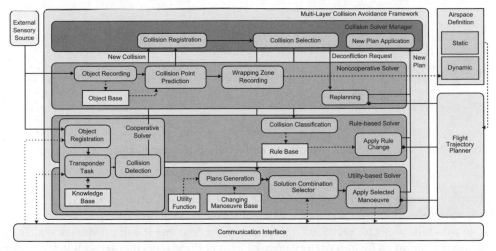

Figure 9.6   The multi-layer collision avoidance architecture

Depending on the configured priority, CSM assigns each registered collision solver a time slot (considering time to collision) that can be used for its resolution by that solver. Usually, the priorities are configured as preset but they can easily be altered during run-time. At any moment, the CSM has complete information about all reported collisions linked to the time axis. Thus, CSM can perform time-oriented switching among various solvers. Sophisticated switching of collision solvers is inevitable as solvers have different properties. Different solvers provide a different quality of collision-free solution and require a different amount of time to find such solution. Specifically, the negotiation-based solvers may provide a better solution than the non-cooperative solvers due to their coordination, but they usually consume more time as they require negotiation through communication modems. The time to collision is a very important parameter in the multi-layer collision avoidance architecture. The solution of an earlier conflict affects the flight trajectory after that conflict and currently identified collisions after the earlier one can be affected by this resolution. The operation of the collision avoidance framework is permanent, anytime when a solver identifies a new more important collision, the resolution of the currently solved one can be terminated.

9.4.2   Cooperative Collision Avoidance

AgentFly integrates three core cooperative negotiation-based collision avoidance algorithms: (i) rule-based, (ii) iterative peer-to-peer, and (iii) multi-party collision avoidance. All these three methods are decentralized, which means that collisions are solved as localized collision avoidance problems. At the same time, there can also be several (different) algorithms running, resolving various conflicts in different local areas. The local problem is restricted by the defined time horizon. In many of our applications we use a 15-minute time horizon, which correlates with the mid-term collision detection and resolution known from civil air-traffic management. Theoretically, this time horizon can be set to a very large value, which implies that algorithms search for a global collision avoidance solution. All three methods use the same detector part, which is based on sharing of UAVs' local intentions. These intentions are shared using the

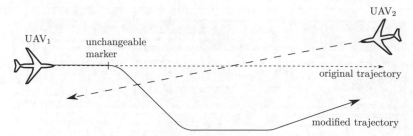

Figure 9.7   Application of a right evasion maneuver

subscribe/advertise protocol and are formed as limited parts of their trajectories. UAVs share flight trajectories from the current time moment for the defined time horizon. By acceptation of the subscribe protocol, each UAV makes a commitment that it will provide an update of this limited part of its flight trajectory once its current flight trajectory is modified (e.g. due to other collision or change in its mission tasks) or the already provided part is not sufficient to cover the defined time horizon. This shared information about flight intention is used only to describe the flight trajectory for that time horizon and doesn't contain any detail about future waypoints and their constraints. So, even though the UAV cooperatively shares its flight intention, it doesn't disclose its mission to others. Using this mechanism, UAVs build a knowledge-base with local intentions of other UAVs in its surrounding. Every flight intention contains information about the flight execution performance (uncertainty), which is then used along with the separation requirement for the detection of conflicting situations (positions of any two UAVs do not satisfy the required separation distance at any moment in time). This mechanism is robust, as cross conflicts are at least checked by two UAVs.

All three algorithms modify the flight path using the trajectory planner by definition of a replanning task. The modification of the flight trajectory is given by a modification of the waypoint sequence in the request. There can be inserted new, modified existing or removed control waypoints. Please note that collision avoidance cannot alter any waypoint which is defined in the UAV mission defined by the high-level flight controller. There is defined a set of evasion maneuvers which can be applied to the specified location with the specified parameterization. Usually, an evasion maneuver is positioned using the time moment in the flight trajectory and its parameterization defines the strength of the applied maneuver. Seven basic evasion maneuvers are defined: left, right (see Figure 9.7), climb-up, descend-down, fly-faster, fly-slower, and leave-plan-as-it-is evasion maneuvers. The name of each maneuver is given by the modification which it produces. Maneuvers can be applied sequentially at the same place so that this basic set can produce all changes. Each UAV can be configured to use only a subset of these maneuvers or can prioritize some of them. For example, a UAV can be configured to solve conflicts only by horizontal changes (no altitude changes). The leave-plan-as-it-is maneuver is included in the set so that it simplifies all algorithms as they can easily consider the unmodified flight trajectory as one of the options.

9.4.2.1   Rule-Based Collision Avoidance (RBCA)

RBCA is motivated by the *visual flight rules* defined by the ICAO [28]. Each UAV applies one predefined collision avoidance maneuver means of the following procedure. First, the type of collision between UAVs is determined. The collision type is identified on the basis of the angle between the direction vectors of UAVs in the conflict time projected to the ground plane. Depending on the collision classification and the defined rule for that collision type, each UAV applies the appropriate conflict resolution. Maneuvers are parameterized to use information about the collision and angle so that the solution is fitted to the identified conflict. Application of the resolution maneuver is done by both involved UAVs. This algorithm is very simple and doesn't require any negotiation during the process of the selection and application of the appropriate collision type solution. The algorithm uses only indirect communication via the updated flight trajectory. Moreover, this algorithm doesn't use explicitly the cost function including the airplane intention during the collision avoidance search, but the UAVs' priorities are already included in the flight planner settings which are used during applications, of predefined evasion maneuvers in the rules. Conflicts of more than two UAVs are solved iteratively and the convergence to the stable solution is given by the used rules.

9.4.2.2   Iterative Peer-to-Peer Collision Avoidance (IPPCA)

IPPCA extends the pairwise conflict resolution with negotiation over a set of combinations considering the cost values computed by the involved UAVs. First, the participating UAVs generate a set of new flight trajectories using the configured evasion maneuvers and their lowest parameterization. Only those which do not cause an earlier collision with any known flight trajectory are generated. Each flight trajectory is evaluated and marked with the cost for its application. Then, a generated set of variants are exchanged (including also original unmodified trajectories). During this exchange, the UAV provides only the limited future parts of the trajectories considering the configured time horizon which is the same as used in the conflict detector. Then, the own set of variants and a received set are used to build combinations of trajectories which can be used to resolve the conflict. It can happen that no such combination is found as some variants are removed, because they cause earlier conflicts with others and the rest do not solve the conflict. In such a case, the UAVs extend their sets of new flight trajectories with modifications using higher parameterizations until some combinations are found. The condition removing variants causing earlier collisions is crucial. Without this criterion the algorithm could iterate in an infinite loop and could also generate new conflicts which are so close that they cannot be resolved.

From the set of suitable combinations of flight trajectories, the UAVs select the best combination based on the configured strategy. UAVs can be configured to optimize the global cost (minimize the sum of costs from both UAVs). Or they can be configured as self-interested UAVs. In such a case, they try to reduce the loss caused by their collision. Then, the best possible collision avoidance pair is identified by a variation of the *monotonic concession protocol* [29] – the protocol for automated agent-to-agent negotiations. Instead of iterative concession on top of the negotiation set, the algorithm can use the extended Zeuthen strategy [30] providing negotiation equilibrium in one step and no agent has an incentive to deviate from the strategy.

9.4.2.3   Multi-party Collision Avoidance (MPCA)

MPCA removes iterations from IPPCA for multi-collision situations – more than two UAVs have mutual future collisions on their flight trajectories. MPCA introduces the multi-party coordinator, who is responsible for the state space expansion and the search for the optimal solution of the multi-collision situation. The multi-party coordinator is an agent whose role is to find a collision-free set of flight trajectories for the whole group of colliding UAVs, considering the limited time horizon. The coordinator keeps information about the group and state space. It chooses which UAV will be invited to the group and requests UAVs to generate modifications of their trajectories. Each state contains one flight trajectory for every UAV in the multi-party group. Initially, the group is constructed from two UAVs which have identified a conflict on their future trajectories. Later, the group is extended with UAVs which have conflicts with them and also which have potential conflicts with any considered flight trajectory in the state space. Similarly to IPPCA, UAVs provide the cost value for the considered collision avoidance maneuver along with its partial future flight trajectory.

The coordinator searches until it finds a state which does not have any collision considering the limited time horizon. Its internal operation can be described as a search loop based on the *OPEN* list and states. States in *OPEN* are ordered depending on the used optimization criterion (e.g. the sum of costs, the lowest is the first). In each loop, the coordinator takes the first state and checks if there are some collisions. If there is no collision, trajectories in this state are the solution. If any trajectory has a conflict with other UAV not included in the multi-party group, the coordinator invites this UAV and extends all states with the original flight trajectory of this invited UAV. Then, it selects a pair of UAVs from that state which have the earliest mutual collision and asks them to provide modifications of flight trajectories so that the collision can be eliminated. This step is similar to IPPCA. From the resulting set of options, new children states are generated.

As described above, the participation of UAVs in one multi-party algorithm is determined dynamically by identified conflicts on their trajectories. Thus, two independent multi-party algorithms can run over disjoint sets of UAVs. A UAV already participating in one multi-party run can be invited into other one. In such a case, the UAV decides which one has higher priority and where it will be active based on the earliest collision which is solved by appropriate multi-party runs. The second one is paused until the first one is resolved. Please note that the run-time of the multi-party algorithm is also monitored by the multi-layer collision avoidance framework. The framework can terminate the participation due to lack of time and can select an other solver to resolve the collision.

Figure 9.8 presents the different results provided by three presented algorithms in the super-conflict scenario [13] – UAVs are evenly spaced on a circle and each UAV has the destination waypoint on the opposite side of the circle so that the initial flight trajectories go through the center of the circle. RBCA was configured to use only predefined rules not to change altitude. While comparing the quality of solutions, MPCA provides the best solution from these three algorithms. On the other hand, considering the time aspect, MPCA requires the largest amount of time searching for a solution. For detailed results and other experiments, see [27].

9.4.3   Non-cooperative Collision Avoidance

AgentFly also integrates the collision avoidance algorithm which doesn't require a bi-directional communication channel for conflict detection and resolution. In such a case, only radar-like information about objects in its surrounding is used. The method used in AgentFly is based on the dynamic creation of no-flight areas, which are then used by the flight trajectory planner to avoid a potential conflict. Such an approach allows us to combine both cooperative and non-cooperative methods at the same time. The dynamically created no-flight areas are

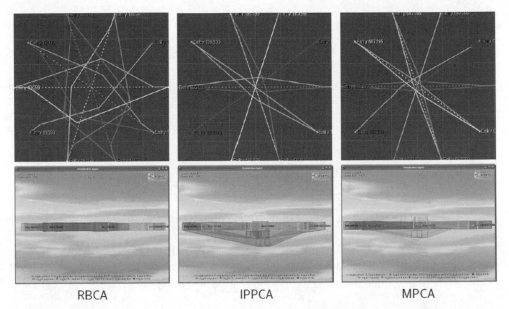

RBCA   IPPCA   MPCA

Figure 9.8 Results of three cooperative collision avoidance methods in the super-conflict scenario

then taken into account when the planner is called by a cooperative solver for application of an evasion maneuver.

Components of the non-cooperative algorithm are displayed in Figure 9.6. The detection part of the algorithm is active permanently. It receives information about positions of objects in the surrounding area. The observation is used to update the algorithm knowledge base. If there is enough history available, the prediction of a potential collision point process is started. The collision point is identified as an intersection of the current UAV's flight trajectory, and the predicted flight trajectory of the object for which the algorithm receives the radar update. Various prediction methods can be integrated in this component: a simple linear prediction estimating the future object trajectory, including the current velocity, which requires two last positions with time information or a more complex tracking and prediction method based on a longer history, which is also able to track a curve trajectory. The result of the collision point prediction is a set of potential conflict points with a probability of conflict. For many cases, it can happen that there is no such collision point found. Then the previously registered conflict within the collision solver manager is canceled.

In the opposite case, the collision points with higher probability than the configured threshold are wrapped by a dynamic no-flight area. The shape of a dynamic no-flight area is derived from the set of possible collision positions. An example of the shape of a dynamic no-flight area is given in Figure 9.9. The predicted dynamic no-flight area is put to the UAV's airspace definition database. Such areas are then used by the trajectory planner while it is called by any other component (cooperative collision avoidance algorithm or high-level flight controller). Information about the predicted future collision is also registered within the collision avoidance manager that will decide when the collision will be solved considering the time to collision. Once the non-cooperative solver is asked to solve the collision, the replanning of the current flight trajectory is invoked by calling the flight trajectory planner

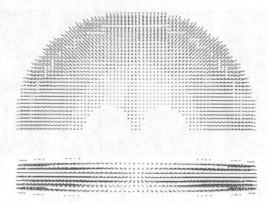

Figure 9.9   An example of the shape of a dynamic no-flight area

which keeps the flight trajectory only within the operation airspace excluding all no-flight areas. The modified flight trajectory is then applied for its execution.

9.5   Team Coordination

In the presented airplane control architecture, algorithms providing coordination of multiple UAVs are integrated in the high-level flight controller part. The output is formulated as a mission specification which is then passed to the collision avoidance module for its execution. Team coordination is the continuous process. Based on the current status, coordination algorithms can amend the previously provided mission. In AgentFly, there were integrated intelligent algorithms for controlling a group of autonomous UAVs performing information collection in support of tactical missions. The emphasis was on accurate modeling of selected key aspects occurring in real-world information collection tasks, in particular physical constraints on UAV trajectories, limited sensor range, and sensor occlusions occurring in spatially complex environments. Specifically, algorithms are aimed at obtaining and maintaining relevant tactical and operational information up-to-date. Algorithms from this domain primarily address the problems of exploration, surveillance, and tracking. The problem of multi-UAV exploration of an unknown environment is to find safe flight trajectories through the environment, share the information about known regions, and find unexplored regions. The result of the exploration is a spatial map of the initially unknown environment. In contrast, the surveillance is a task providing permanent monitoring of the area. Usually sensors' coverage of all UAVs is not sufficient to cover the whole area at the same time and UAVs have to periodically visit all places, minimizing the time between visits to the same region. Finally, the tracking task involves such control of UAVs which do not lose the tracked targets from the field of view of their sensors. There are many variants of the tracking tasks based on the different speed of UAVs and tracked targets, as well as the number of UAVs and targets.

Figure 9.10 shows an example of the scenario which was used for testing of information collection algorithms. In the model, there are more than 300 buildings with different heights and various widths of street in the city. The right top view in Figure 9.10 presents the information known by the group of UAVs once they finished the initial exploration of the unknown area, white buildings are the tallest ones and black ones are the lowest. Initially,

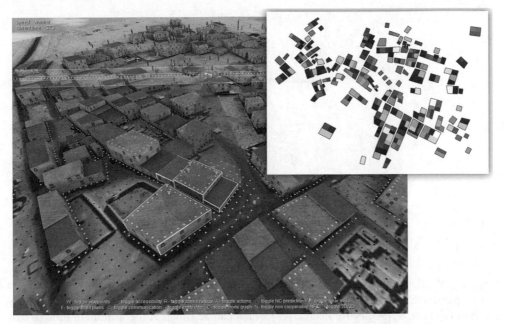

Figure 9.10   A complex urban scenario used for multi-UAV coordination algorithms

UAVs have only planar information about the area taken from a satellite image but they don't have the full three-dimensional information which is required to perform optimized information collection within that area. Each UAV is equipped with an onboard camera sensor which is oriented downwards. Beside UAVs, there are simulated ground entities – people moving within the city. A simplified behavior was created for people, so that they move intelligently within the city. They react to flying UAVs and other actions. The three-dimensional representation of the urban area plays a crucial role in the scenario. While the UAV is flying over the city, its camera sensor can only see non-occluded areas close to building walls. UAVs have to plan their trajectories carefully, so that they can see to every important place in the city while they perform a persistent surveillance task. The abstraction in the simulation supposes that the image processing software provided with the sensor has the capability to extract high-level information from the obtained data. So, intelligent algorithms which were tested are not working with raw images but with information on a symbolic level. It is supposed that the camera sensor is able to detect edges and identify the height of the building and detect people on the ground. The coverage of one UAV camera sensor is visualized in Figure 9.10.

The persistent simulation task can be formulated as an optimization problem, minimizing the objective function. In this case, the objective function is constructed as the sum of information age over the area where the surveillance task should take place. In other words, UAVs try to keep knowledge of any position in the area as recent as possible. The exploration task is just a specific case, where there are some positions in the area with an infinite age, which turns the algorithm to reveal each position by the sensor coverage at least once. During the exploration, UAVs update their unknown information about that area and rebuild a coverage map. A coverage map example is shown in Figure 9.11. The coverage map provides

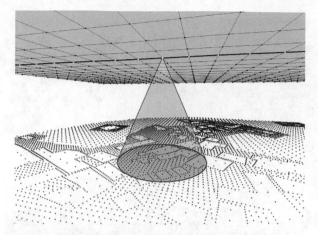

Figure 9.11   A coverage map in the urban area

mapping of all interested ground points to air navigation points (waypoints) considering sensor parameters. The multi-UAV simulation task then searches for the flight trajectory patterns (a cyclic trajectory for each UAV) which can be flown at the minimum time so that all interesting points are covered during those cycles. The intelligent algorithm for persistent surveillance searches for such trajectories and allocates them to available UAVs. The situation becomes more complex when several surveillance areas are defined and airplanes have to be assigned to particular areas or have to cycle between them in order to keep minimized the objective function during the time. Detailed results from the information collection algorithms implemented in AgentFly can be found in [31].

The tracking task requires that a ground target is covered by the UAV's sensor for the whole time. Thus, it should provide an uninterrupted stream of image data about a designated target. Tracking is a challenging task because of the lack of knowledge of a target's future plans and the motion constraints of the UAV. In AgentFly, a circular pattern navigation algorithm is integrated for one UAV which is tracking one target, see Figure 9.12. A virtual

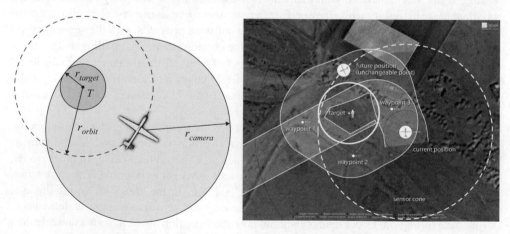

Figure 9.12   The tracking task of a ground entity

circle is constructed above the tracked object – the radius of the circle is derived from motion constraints of the UAV as small as possible to keep the UAV close to the tracked object. The UAV tries to stay on this circular trajectory as this is the best trajectory to keep the tracked object covered by its camera sensor. While the tracked object is moving (in our case it is moving slower than the maximum flight speed of the UAV), the virtual circle moves with it. As described in Section 9.2, the algorithm has to provide a mission for the UAV in the form of a sequence of waypoints. The UAV, considering the current position and the optimal virtual circle, computes positions for waypoints based on the tangent projection from the current position. Then it prepares the next waypoints around the circle. The algorithm is invoked once it gets updated positions of the tracked object. In such a case, the tracking algorithm adjusts the position of computed waypoints if necessary. The algorithm becomes more complex when one UAV is tracking a cluster of ground targets at the same time. In this case, the UAV uses a general orbit trajectory instead of a virtual circle.

Figure 9.13 provides a view of the interface for the interaction with the system. This interface is provided by an agent which implements the hybrid command and control system for mixed information collection tasks. Through this agent, a human operator can specify information collection tasks for the whole group of UAVs and inspect the resulting operation. The agent automatically collects knowledge about the area and presents it to the operator. Tasks are not defined for particular UAVs but are specified as goals for the whole group.

Figure 9.13 Command and control panel in AgentFly

Thus, a multi-agent allocation algorithm is integrated for the optimal allocation of concurrent information collection tasks in the group. There is no centralized element in this system. The command and control system adopts a robust solution which implies that UAVs synchronize and merge partial information together anytime they can communicate together. If any UAV is lost, the group loses only information which was not synchronized within the group and information collection tasks are reorganized so that the remaining UAVs fulfill the requested tasks considering the configured optimization objective function. Similarly, when a new UAV is added to the group, it gets synchronized while it establishes communication with any other UAV in the group. The command and control agent needs a connection only with one UAV in the group.

9.6 Scalable Simulation

Thanks to the agent-based approach used, AgentFly can provide a simulation environment to model and study complex systems with a large number of situated entities, high complexity of their reasoning and interactions. For such a complex system, it is not possible to run the whole simulation system hosted on a single computer. AgentFly provides an effective distribution schema which splits the load over heterogenous computers connected through a fast network infrastructure. Within the AgentFly simulation environment, not only can UAVs or civilian airplanes be simulated, but any type of situated entity. By *situated entity* we mean an entity embedded in a synthetic model of a three-dimensional physical space. AgentFly provides high-fidelity (accurate simulation with respect to the defined physics, not using level of details), scalable (in terms of the number of simulated entities and the virtual world dimension), and fast (produce results for a simulated scenario in the minimum possible time) simulation platform. The development of such functionality has been motivated initially by the need to simulate full civilian air-traffic in the national airspace of the United States, with great level of detail. Further, the same feature has been used to validate intelligent algorithms for UAVs which operate in very complex environments – beside a high number of UAVs there are simulated ground vehicles and running models of people that play a crucial role in the scenario.

The AgentFly simulation supports both passive (their behavior is only defined by their physics) and autonomous (pro-active, goal-oriented actors) situated entities operating and interacting in a realistically modeled large-scale virtual world. Entities could be dynamically introduced or removed during the simulation run-time based on the evaluation scenario needs. Each situated entity carries a state – the component which can be either observable (public) or hidden (private, internal). The fundamental component of the entity's observable state is its position and orientation in the virtual world (e.g. the position of the UAV). The evolution of the entity's state is governed by the defined entity's physics, e.g. the entity's movement is driven by its motion dynamics, typically defined by a set of differential equations which can also refer to the entity's hidden state components. Physics can be divided into intra-entity and inter-entity parts. The intra-entity physics capture those aspects of physical dynamics that can be fully ascribed to a single entity. Although the equations of the intra-entity physics can refer to any state in the simulated world (e.g. weather condition), they only govern the states carried by its respective entity. In contrast, the inter-entity physics captures the dynamics that affects multiple entities simultaneously and cannot be fully decomposed between them (e.g. the effects of a physical collision of two UAVs). Beside physical interactions, autonomous entities can also interact via communication and use sensors to perceive their surrounding virtual environment. Communication between entities can be restricted by the inter-entity physics simulating a required communication medium (e.g. wireless network). Similarly, each sensor has its capabilities defined, which may restrict the observable state it can perceive only to a defined subset, typically based on the distance from the sensor's location (e.g. radar, onboard camera).

Each situated entity (e.g. UAV) in AgentFly is divided into up to three sub-components: (i) body, (ii) reactive control, and (iii) deliberative control. The body encapsulates the entity's intra-physics, which governs the evolution of its state as a function of other, possibly external states (e.g. UAV flight dynamics given by the motion model used affected by the current atmospheric condition). The reactive control component contains a real-time loop-back control for the entity, affecting its states based on the observation of other states (e.g. the flight executor component integrating the autopilot function for the UAV). The deliberative control component contains complex intelligent algorithms employing planning, sensing observations, and communication with others (the flight trajectory planner, collision avoidance framework, and high-level flight controller in the UAV control architecture). The outputs of deliberative control typically feed back into the entity's reactive control module and provide a new control for the entity (the update of the current flight trajectory in the UAV). The body and reactive control components are similar in their loop-back approach and are collectively referred to as the entity's state updater.

As described in Section 9.1, AgentFly is implemented as the multi-agent system with environmental simulation agents and UAV agents which contains intelligent deliberative algorithms. Figure 9.14 presents the decomposition of components of each UAV into software agents. Thanks to the Aglobe multi-agent framework used, each computer node can host one or several agents and an efficient communication infrastructure is available. Within the simulation architecture, the deliberative controller and state updater of each entity are decoupled and integrated within different agents. The respective pair of state updater and deliberative controller is connected by a signal channel through which sensing perceptions (one way) and control commands to the reactive control (the other way) are transmitted.

AgentFly employs a time-stepped simulation approach – the virtual simulation time driving the dynamic processes of the simulation is incremented uniformly by a constant time step in each simulation cycle. All physics and reactive parts of all entities have to be evaluated synchronously in order to provide consistent evolution of the virtual world. Synchronous evaluation means that states are updated once all physics is computed based on the previous state. The virtual time steps and thus state updates can be applied regularly with respect to the external wall clock time or in a fast-as-possible manner. The first mode has to be used when AgentFly is running as a hybrid simulation (the part of the scenario is simulated, some entities are represented by real hardware). Within the simulation mode, the fast-as-possible mode is used in order to get simulation results in the shortest possible time – the next step (simulation cycle) is initiated immediately after the previous one has been completed.

Both environment simulation and deliberative control agents are distributed among multiple computer nodes during the simulation. The whole virtual simulation world is spatially divided into a number of partitions. The partitions are mutually disjoint except for small overlapping areas around partitions' boundaries. Each partition has a uniquely assigned environment simulation agent (ES agent) responsible for updating states (e.g. applying relevant differential equations) corresponding to all entities located within its assigned partition. The number of ES agents running on one computer node is limited by the number of processing

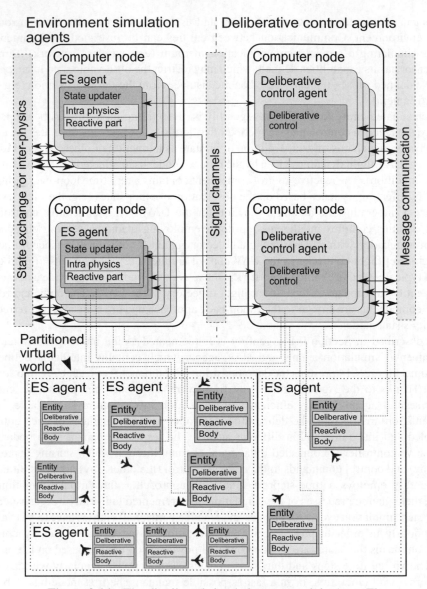

Figure 9.14   The distributed simulation approach in AgentFly

cores at that node. The application of physics may require exchange of state information between multiple partitions, in case affected entities are not located within the same partition. In contrast with the state updaters, entities' deliberative control parts are deployed to computer nodes irrespective of the partitions and corresponding entity's position in the virtual world. The described world partitioning implies that whenever the location of an entity changes so that the entity moves to a new partition, the entity's state updater needs to be migrated to the respective ES agent. The signal channel between the entity's control and its deliberative control modules is transparently reconnected and it is guaranteed that no signal transmission is lost during the migration. The small overlapping areas around partition boundaries are introduced to suppress oscillatory migration of state updaters between neighboring ES agents in case the corresponding entity is moving very close along a partition border.

In order to provide maximum performance throughout the whole simulation, the proposed architecture implements load-balancing approaches applied to both environment simulation agents and deliberative control agents. The virtual world decomposition to partitions is not fixed during the simulation but is dynamically updated depending on the load of computer nodes hosting environment simulation agents. Based on the time required for processing the simulation cycle, the virtual world is repartitioned so that the number of entities belonging to partitions is proportional to the measured times. The ES agent that is faster will be assigned a larger area with more entities and vice versa. Repartitioning is not performed during each simulation cycle but only if the difference in simulation cycle computation by various ES agents exceeds a predefined threshold. It is also triggered whenever a new computer node is allocated for simulation. Similarly, the deliberative components are split among computer nodes based on the load of those computers.

Figure 9.15 presents the results, studying the dependency of the simulation performance on the number of computer nodes. In this experiment, the same number of UAVs was simulated in all configurations. A 1-day scenario was simulated with more than 20,000 UAVs performing various flight operations within that day. There were about 2000 flying UAVs at the same moment in time. The number of simultaneous flying UAVs varied slightly in the scenario.

The left chart in Figure 9.15 presents the dependency of the time to complete the whole scenario on the number of computer nodes available for the environment simulation (thus the number of partitions and ES agents). UAV agents are running in every configuration on the same number of computers. We can observe that there are three utilization regions: (i) overloaded, (ii) optimally loaded, and (ii) under-loaded. In the left part (1, 2, and 3 partitions), the normalized processing time[1] is significantly decreased with more partitions (few ES agents are over-loaded by computing inter-physics). For the mid-part (4 and 5 partitions), the normalized processing time stagnates, which is caused by rising demands for partition-to-partition coordinations. Adding even more computation resources (and thus more partitions) results in a decreasing normalized processing time, the simulation is under-loaded – the overhead for synchronization dominates the parallelized processing. The right chart in Figure 9.15 studies varying computational resources available for deliberative control agents (UAV agents). In this case, the fixed number of partitions was used. Similarly to the previous case, in the left part there is a significant speed-up given by parallelization of execution of heavy-weight agents and for less computational power all CPUs are over-loaded and the simulation is slower. For increasing resources, the time to complete simulation is almost unchanging. In this case, the overall simulation bottle neck is not in the efficiency of UAV agents, but in the limit of the environment simulation. The increasing resources (6, 7, and more) are wasted and thus the normalized processing time is increased due to an increase in the processing time. In contrast to the left chart, addition of more resources for deliberative controller agents doesn't slow down the whole simulation (does not cause coordination overhead for the simulation).

---

[1] The normalized processing time expresses the computational real time (wall clock) per one UAV in the simulation.

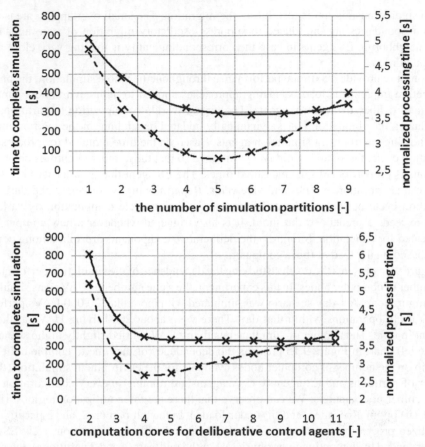

Figure 9.15   The result of varying number of computation nodes available for the simulation

9.7   Deployment to Fixed-Wing UAV

AgentFly is deployed to various UAV platforms including a quad-rotor helicopter. This chapter presents the deployment to the fixed-wing airplane called Procerus UAV. Our Procerus UAV is shown in Figure 9.16. We use this small UAV as we can easily perform experiments and we are not restricted by any regulation. Once AgentFly is able to provide all intelligent algorithms for a small UAV, it can be used successfully for the control of any larger and more equipped one. UAV is based on the Unicorn airframe from EPP foam with 72" wingspan. It is a fully autonomous UAV which has installed four Li-pol batteries, Kestrel autopilot board, data communication modem with antenna, GPS unit, electric motor with necessary power regulator and currency sensor, servos for ailerons, motorized gimbal camera (installed on the bottom side), video transmitter with antenna, and AgentFly CPU board. The weight of the fully loaded UAV platform is about 3 kg. Depending on the number of take-offs, climbs, and usage of the onboard camera system, it can fly up to 60 minutes with a speed from 40 to 65 mph. Now, we primarily use these UAVs for experimentation with described autonomous algorithms for sense and avoid. Thus, the camera system is not connected to the

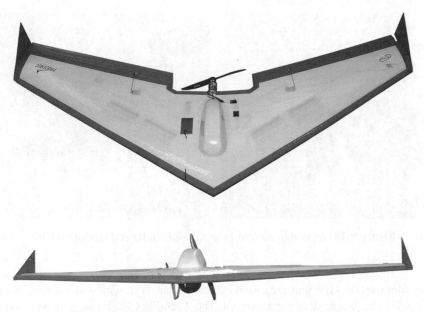

Figure 9.16 Procerus UAV – the fixed-wing airplane

onboard processer but the video is broadcasted to the ground station, see Figure 9.17. The gimbal camera system is electrically retractable (which is usually done before landing), and the operator can control pan, tilt, and zoom of the camera. Even though the communication modem is able to keep connection for several miles the, UAV is able to fly autonomously without connection to the ground station as the AgentFly system is running onboard.

The autopilot board has integrated three-axis gyros, accelerometers, magneto-meter, absolute and differential pressure sensors for altitude/airspeed, and there are also integrated temperature sensors for compensation of sensors' drifts in changing temperature.

Figure 9.17 The screenshot from a camera feed transmitted to the ground station

Figure 9.18   Assisted autonomous take-off (left) and landing (right)

The autopilot uses the GPS unit for its navigation and also estimation of the wind speed and heading, which helps it achieve better control. The AgentFly CPU board is connected through a serial link with the autopilot so that it can read the current flight performance including position data from GPS and provide a control back to the autopilot. The AgentFly CPU board is the Gumstix computer module with ARM Cortex-A8 CPU (RISC architecture) running AgentFly in JAVA. We use the same implementation for the UAV control as is used in the simulation mode. There are only changed interfaces so that sensors, communication channel, and flight executor are mapped to the appropriate hardware. The autopilot is able to control the UAV and navigate it through the provided waypoints. It also supports assisted autonomous take-off (the UAV is thrown from the hand or launched from a starting gun; its autonomous take-off procedure is invoked when the UAV reaches the appropriate speed threshold) and autonomous landing in the defined area, see Figure 9.18. The autopilot is not able to track the flight trajectory intention as described in Section 9.2. Thus, the part of the functionality from the flight executor is running on the AgentFly CPU board. Based on the UAV's parameters, the flight executor converts the flight trajectory intention into a sequence of low-level navigation waypoints which are passed to the autopilot board through the serial connection. These navigation waypoints are selected so that the UAV platform follows the requested trajectory as precisely as possible. On the other hand, the flight executor contains a monitoring module which permanently processes the flight status (including the GPS position and wind estimation) and checks if the flight is executed within the defined tolerance. The monitor module invokes replanning and adjusting the flight performance prediction. For some future version, a module can be integrated that will automatically adjust preconfigured parameters of the UAV model used by the flight trajectory planner. This will minimize the number of replannings when preconfigured parameters don't fit well for the current conditions.

Agents running on the AgentFly CPU board can communicate with other UAVs and also with the ground system through the data communication modem which is connected to the autopilot. Now, we haven't any source for radar-like information about objects in the UAV's surrounding area. We are only working with the cooperative collision avoidance algorithms which utilize negotiation-based conflict identification as described in Section 9.4. All UAVs, both real and simulated, are flying in the same global coordination system.

Acknowledgments

The AgentFly system has been sponsored by the Czech Ministry of Education grant number 6840770038 and by the Federal Aviation Administration (FAA) under project numbers DTFACT-08-C-00033 and DTFACT-10-A-0003. Scalable simulation and trajectory planning have been partially supported by the CTU internal grant number SGS10/191/OHK3/2T/13. The underlying AgentFly system and autonomous collision avoidance methods have been supported by the Air Force Office of Scientific Research, Air Force Material Command, USAF, under grant number FA8655-06-1-3073. Deployment of AgentFly to a fixed-wing UAV has been supported by the Czech Ministry of Defence grant OVCVUT2010001. The views and conclusions contained herein are those of the authors and should not be interpreted as representing the official policies or endorsements, either expressed or implied, of the Federal Aviation Administration, the Air Force Office of Scientific Research, or the US Government.

References

1. M. Šišlák, M. Rehák, M. Pěchouček, M. Rollo, and D. Pavlíček. AGLOBE: Agent development platform with inaccessibility and mobility support. In R. Unland, M. Klusch, and M. Calisti (eds), Software Agent-Based Applications, Platforms and Development Kits, pp. 21–46, Berlin Birkhauser Verlag, 2005.

2. M. Wooldridge. *An Introduction to MultiAgent Systems*. John Wiley & Sons Inc., 2002.

3. P. Santi. *Topology Control in Wireless Ad-hoc and Sensor Networks*. John Wiley & Sons Inc., 2005.

4. R. Schulz, D. Shaner, and Y. Zhao. Free-flight concept. In *Proceedings of the AIAA Guidance, Navigation and Control Conference*, pp. 889–903, New Orleans, LA, 1997.

5. National Research Council Panel on Human Factors in Air Traffic Control Automation. *The future of air traffic control: Human factors and automation*. National Academy Press, 1998.

6. G.J. Pappas, C. Tomlin, and S. Sastry. Conflict resolution in multi-agent hybrid systems. In *Proceedings of the IEEE Conference on Decision and Control*, Vol. 2, pp. 1184–1189, December 1996.

7. K.D. Bilimoria. A geometric optimization approach to aircraft conflict resolution. In *Proceedings of the AIAA Guidance, Navigation, and Control Conference*, Denver, August 2000.

8. J. Gross, R. Rajvanshi, and K. Subbarao. Aircraft conflict detection and resolution using mixed geometric and collision cone approaches. In *Proceedings of the AIAA Guidance, Navigation, and Control Conference*, Rhode Island, 2004.

9. A.C. Manolis and S.G. Kodaxakis. Automatic commercial aircraft-collision avoidance in free flight: The three-dimensional problem. *IEEE Transactions on Intelligent Transportation Systems*, 7(2):242–249, June 2006.

10. J. Hu, M. Prandini, and S. Sastry. Optimal maneuver for multiple aircraft conflict resolution: A braid point of view. In *Proceedings of the 39th IEEE Conference on Decision and Control*, Vol. 4, pp. 4164–4169, 2000.

11. M. Prandini, J. Hu, J. Lygeros, and S. Sastry. A probabilistic approach to aircraft conflict detection. *IEEE Transactions on Intelligent Transportation Systems*, 1(4):199–220, December 2000.

12. A. Bicchi and L. Pallottino. On optimal cooperative conflict resolution for air traffic management systems. *IEEE Transactions on Intelligent Transportations Systems*, 1(4):221–232, December 2000.

13. L. Pallottino, E.M. Feron, and A. Bicchi. Conflict resolution problems for air traffic management systems solved with mixed integer programming. *IEEE Transactions on Intelligent Transportation Systems*, 3(1):3–11, March 2002.

14. Federal Aviation Administration. *Aeronautical Information Manual*. Federal Aviation Administration, US Department of Transportation, 2008.

15. C.R. Spitzer. *Avionics: Elements, Software and Functions*. CRC, 2006.

16. P. Hart, N. Nilsson, and B. Raphael. A formal basis for the heuristic determination of minimum cost paths. *IEEE Transactions on Systems Science and Cybernetics*, (2):100–107, 1968.

17. D. Šišlák, P. Volf, and M. Pěchouček. Flight trajectory path planning. In *Proceedings of the 19th International Conference on Automated Planning & Scheduling (ICAPS)*, pp. 76–83, Menlo Park, CA. AAAI Press, 2009.

18. P. Yap. Grid-based path-finding. In *Proceedings of the Canadian Conference on Artificial Intelligence*, pp. 44–55, 2002.

19. D. Šišlák. *Autonomous Collision Avoidance in Air-Traffic Domain*. PhD thesis, Czech Technical University, Prague, February 2010.

20. A. Nash, K. Daniel, S. Koenig, and A. Felner. Theta*: Any-angle path planning on grids. In *Proceedings of the AAAI Conference on Artificial Intelligence (AAAI)*, pp. 1177–1183, 2007.

21. S.M. LaValle and J.J. Kuffner. Rapidly exploring random trees: Progress and prospects. In B.R. Donald, K.M. Lynch, and D. Rus (eds), *Algorithmic and Computational Robitics: New Directions*, pp. 293–308, A.K. Peters, Wellesley, MA, 2001

22. A. Yershova, L. Jaillet, T. Siméon, and S.M. LaValle. Dynamic-Domain RRTs: Efficient exploration by controlling the sampling domain. In *Proceedings of the IEEE International Conference on Robotics and Automation*, pp. 3867–3872, 2005.

23. A. Botea, M. Müller, and J. Schaeffer. Near optimal hierarchical path-finding. *Journal of Game Development*, 1(1):7–28, 2004.

24. D. Ferguson and A. Stentz. Using interpolation to improve path planning: The field D* algorithm. *Journal of Field Robotics*, 23(1):79–101, 2006.

25. D. Šišlák, P. Volf, and M. Pěchouček. Accelerated A* trajectory planning: Grid-based path planning comparison. In *Proceedings of the 19th International Conference on Automated Planning & Scheduling (ICAPS)*, pp. 74–81, Menlo Park, CA. AAAI Press, 2009.

26. Š. Kopřiva, D. Šišlák, D. Pavlíček, and M. Pěchouček. Iterative accelerated A* path planning. In *Proceedings of 49th IEEE Conference on Decision and Control*, December 2010.

27. D. Šišlák, P. Volf, and M. Pěchouček. Agent-based cooperative decentralized airplane collision avoidance. *IEEE Transactions on Intelligent Transportation Systems*, 12(1):36–46, March 2011.

28. M.S. Nolan. *Fundamentals of Air Traffic Control*, 4th edn. Thomson Brooks/Cole, Belmont, CA, 2004.

29. G. Zlotkin and J.S. Rosenschein. Negotiation and task sharing among autonomous agents in cooperative domains. In *Proceedings of the 11th International Joint Conference on Artificial Intelligence*, pp. 912–917, San Mateo, CA. Morgan Kaufmann, 1989.

30. F.L.B. Zeuthen. *Problems of Monopoly and Economic Warfare*. Routledge and Sons, 1930.

31. E. Semsch, M. Jakob, D. Pavlíček, M. Pěchouček, and D. Šišlák. Autonomous UAV surveillance in complex urban environments. In C. McGann, D.E. Smith, M. Likhachev, and B. Marthi (eds), *Proceedings of ICAPS 2009 Workshop on Bridging the Gap Between Task and Motion Planning*, pp. 63–70, Greece, September 2009.

10

See and Avoid Using Onboard Computer Vision

John Lai, Jason J. Ford, Luis Mejias, Peter O'Shea and Rod Walker
Australian Research Centre for Aerospace Automation, Queensland University of Technology, Australia

10.1   Introduction

10.1.1   Background

The integration of unmanned aircraft into civil airspace is a complex issue. One key question is whether unmanned aircraft can operate just as safely as their manned counterparts. The absence of a human pilot in unmanned aircraft automatically points to a deficiency – that is, the lack of an inherent see-and-avoid capability. To date, regulators have mandated that an 'equivalent level of safety' be demonstrated before UAVs are permitted to routinely operate in civil airspace. This chapter proposes techniques, methods, and hardware integrations that describe a 'sense-and-avoid' system designed to address the lack of a see-and-avoid capability in unmanned aerial vehicles (UAVs).

10.1.2   Outline of the SAA Problem

Non-cooperative collision avoidance (or sense-and-avoid) for UAVs has been identified as one of the most significant challenges facing the integration of unmanned aircraft into the national airspace [1, 2]. Here, the term 'sense' relates to the use of sensor information to automatically detect possible aircraft conflicts, whilst the term 'avoid' relates to the automated control actions used to avoid any detected collisions. Much of the previous research effort on the

Sense and Avoid in UAS: Research and Applications, First Edition. Edited by Plamen Angelov.
© 2012 John Wiley & Sons, Ltd. Published 2012 by John Wiley & Sons, Ltd.

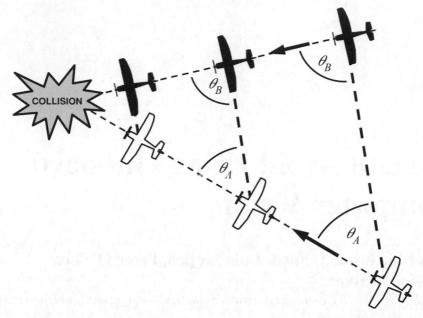

Figure 10.1   Constant bearing between aircraft leads to a collision

sense-and-avoid problem has been focused on the 'sense' or conflict detection aspect of the problem. This is so because, under a crude interpretation of the issues, once a conflict has been 'sensed', the 'avoid' aspects of the problem can be almost routinely achieved through any aircraft manoeuvre that substantially changes heading [3].

10.1.2.1   Collision-Course Geometry

Many authors have reported that a mid-air collision between aircraft travelling with constant velocity occurs when two aircraft are converging with constant bearing [4, 5]. This bearing condition is illustrated in Figure 10.1.

From the perspective of the pilot in the dark aircraft, the light aircraft appears as a stationary feature through the windscreen, and vice versa [6]. This unique dynamic can be exploited by collision warning systems to identify aircraft that are on a potential collision path. For example, a vision-based warning system would perceive objects on a collision course as relatively stationary features on the image plane. Objects that are moving rapidly across the image plane can be discounted as genuine threats. This chapter will focus on target detection, tracking, and avoidance techniques that exploit this constant bearing property of collision-course aircraft.

10.2   State-of-the-Art

An automated sense-and-avoid system is desirable for airborne vehicles to protect them from potential collision with other aircraft. The following review discusses some of the existing technologies that have been used to address this need, as well as providing an overview of emerging techniques that tackle the sense-and-avoid problem through vision-based approaches.

There is a wide variety of possible 'sensing' options, and these options are usually divided into the cooperative and non-cooperative approaches. Cooperative sensing approaches are those involving the mutual sharing of location information as is done in Traffic-alert and Collision Avoidance System (TCAS) transponders [7]. On the other hand, non-cooperative approaches involve directly sensing other aircraft, irrespective of the other aircraft's desire to be sensed. Cooperative approaches such as TCAS are not a completely satisfactory solution to the sense-and-avoid problem because the protection offered by such approaches is dependent on the desire and ability of other aircraft to share information. Since January 2003, the International Civil Aviation Organisation (ICAO) has been progressively mandating the installation of TCAS equipment in various classes of aircraft, including most turbine-engined passenger and cargo aircraft for international commercial air transportation [8]. However, the high cost of TCAS equipment [9] prevents wider uptake by smaller general aviation aircraft, significantly limiting the effectiveness of a TCAS-based sense-and-avoid solution.

Within non-cooperative approaches, schemes that transmit RF energy as part of the sensing (such as radar) are usually called active approaches; conversely, those approaches that do not emit RF energy are called passive sensing approaches [10]. Traditionally, there has been a lot of effort in the areas of active sensing such as radar, but more recently there has been considerable work investigating passive sensors in the sense-and-avoid arena (see [11–13] and references within). This work on passive sensors builds on several decades of research into infrared (IR)-based airborne target detection within the context of missile guidance. Whilst this earlier work on missile guidance does provide some important background information, there are several important differences between the missile guidance problem and the sense-and-avoid problem. In the IR guided missile problem, the target is assumed to occupy tens to hundreds of pixels [14] and, hence, spatial features can be used to assist target tracking. The challenging detection issues relate to maintaining a consistent target track through aspect changes, with advanced decoy/flare rejection achieved through consideration of the target signature characteristics. Conversely, the sense-and-avoid problem, typically, involves attempting to detect conflicts at near sensing limits, when the potential targets have sub-pixel dimensions and have no spatial features to aid target/artefact discrimination [11, 15].

Whilst non-cooperative active sensor approaches such as radar are suitable for many larger platforms, these active sensing solutions are not (yet) suitable on small-to-medium aircraft (including many unmanned aircraft systems (UAS)) [16]. For the above and related reasons, computer vision has emerged as a promising means for addressing the 'sense' and 'detect' aspects of collision avoidance, and is arguably the most feasible non-cooperative solution for general aviation and small-to-medium UAS [17, 18]. As will be seen later in this chapter, however, there are a number of difficulties that must be overcome before the use of computer vision for target detection and tracking becomes routine [19].

Due to the relatively high speeds of aircraft in general, sense-and-avoid systems must, ideally, detect targets while they are still far away; for a vision-based system, this translates to detecting small point-like objects. There has been considerable investigation over the last few decades into computer vision techniques for detecting dim small-sized targets from image data – both visual spectrum and IR imagery [20–23]. The techniques that have been proposed are all designed to enhance potential target features and, at the same time, suppress background noise and clutter. Within this body of literature, two distinct approaches have emerged: (i) intra-frame enhancement and (ii) inter-frame enhancement.

Intra-frame processing techniques operate on individual image frames. They are, therefore, suited to exploiting the instantaneous qualities of the target that differentiate it from noise/clutter (e.g. size, shape, brightness of the target in a specific frame). Max-mean and max-median subtraction filters are examples of intra-frame image enhancement tools that have been applied to the small target detection problem [24]. Another class of intra-frame filtering tools that has shown great promise in the detection of dim small-sized targets has its basis in mathematical morphology [25]. Numerous morphology-based filters have been proposed for the detection of small targets in IR images [26–29] and visual range images [11, 22, 30].

In contrast to intra-frame techniques, inter-frame processing methods are designed to operate over a sequence of image frames. They exploit the temporal or dynamic qualities of the target that may differentiate it from noise/clutter (e.g. the change in size, shape, position, brightness of the target over time). Two particular inter-frame or temporal filtering approaches have received much attention in the literature: recursive ad-hoc Viterbi-based approaches [20, 21, 23, 31, 32] and Bayesian-based approaches [31, 33–35]. As the name suggests, many ad-hoc Viterbi-based approaches have characteristics that resemble certain features of the standard Viterbi tracking algorithm, a dynamic programming approach for efficiently determining an optimal target path without explicit enumeration of all path possibilities [36]. On the other hand, Bayesian filtering approaches are based on well-established probability theory formalisms that allow target detection properties and uncertainties to be propagated in time via probability distributions.

10.3   Visual-EO Airborne Collision Detection

While intra-frame and inter-frame processing are both powerful in their own right, they are even more powerful when used synergistically. Accordingly, there are many target detection schemes which combine intra-frame and inter-frame image processing techniques to enhance detection performance [11, 23, 30].

The authors, through work at the Australian Research Centre for Aerospace Automation (ARCAA), have already completed significant pilot activity on the passive sense-and-avoid problem. Between 2009 and 2011, they have investigated automated aircraft separation management technology and visual-electro-optical (EO)-based airborne collision detection technology [11, 15, 37].

This visual-EO-based collision detection research has provided some important insights into the challenges of detecting other aircraft using airborne imaging sensors in a realistic sensing environment. Specifically, it has highlighted difficulties in differentiating collision threats from within a cluttered background (see also [13]); difficulties stabilizing the image to facilitate target detection via inter-frame processing techniques; and difficulties brought on by the variability and unique propagation characteristics of light.

Despite these challenges, the collision detection research has led to the development of visual-EO-based warning technology capable of detecting real-time conflicts at distances suitable for collision avoidance [11]. The basic components of the proposed sense-and-avoid system are as shown in Figure 10.2. The system incorporates an image capture device, an image stabilization process, a target detection and tracking system, and an avoidance control algorithm.

The various components of the system are discussed in the sections which follow.

10.3.1   Image Capture

A number of different image capture systems have been used by the authors for digitizing and recording image measurements in the field. The answer to the question, 'Which image capture system should be employed in the sense-and-avoid system?' is strongly influenced by

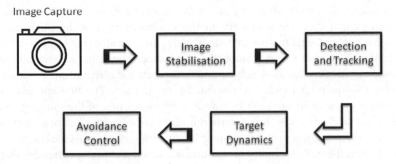

Image Capture

Figure 10.2   Components of a computer vision-based sense-and-avoid system

the aircraft platform being used. Some UAV platforms, for example, have a significant amount of inbuilt hardware for capturing images. Accordingly, it is appropriate to describe the image capture system within the context of the rest of the system hardware. Such a description is provided later in this chapter within Section 10.7.

To assist with the processing of digitized image measurements, a model for relating 3D scene elements to their representation on a 2D image has been used. Details of this camera model are discussed next.

10.3.2   Camera Model

The optical sensor is modelled using a first-order approximation of the mapping from a 3D scene to a 2D image, i.e. a pinhole camera model [38, 39]. This model is appropriate in most cases provided that (1) a suitable calibration that accounts for distortion models is known and (2) suitable coordinate transformations can be applied to the image. Other effects that are sufficiently small can be neglected if a high-quality imaging device is used.

Using a pinhole camera model, a point $P(X, Y, Z)$ in 3D space referenced to the camera coordinate frame can be projected onto a point $p(x, y)$ in a 2D image plane using the following relationship:

$$\begin{bmatrix} x \\ y \\ 1 \end{bmatrix} = \frac{f}{Z} \begin{bmatrix} X \\ Y \\ Z \end{bmatrix}, \tag{10.1}$$

where $f > 0$ is the focal length.

More sophisticated camera models could take into consideration all the camera intrinsic parameters, such as the coordinates of the principal point and the ratio of pixel dimension. The use of these more complex models is recommended if millimeter or sub-pixel accuracy is a major concern.

10.4   Image Stabilization

10.4.1   Image Jitter

Image jitter is an undesirable effect caused by the motion of the imaging sensor relative to objects in the scene. As a result, when imaging sensors are mounted on moving platforms, the observed image jitter can be largely attributed to platform motion.

In the presence of image jitter, objects in the camera field-of-view can appear to have motion when in fact they are stationary in the environment. For detection and tracking algorithms that exploit target motion dynamics in the image frame, the motion distortion that image jitter introduces can severely impact performance. The issue is especially significant with narrow field-of-view lenses which tend to exacerbate the effect of image jitter.

There are two main approaches to combating image jitter. The first approach addresses the fundamental cause of image jitter by minimizing the motion of the imaging sensor itself through physical mechanisms such as passive motion-dampening devices or actively stabilized mounts. However, image jitter cannot be completely eliminated by this means, particularly in airborne platforms that are constantly in motion and are subject to unpredictable disturbances such as wind gusts. An alternative/complementary approach is to apply image processing techniques that attempt to realign jitter-affected image frames based on image features or direct measurements of the platform motion.

10.4.2 Jitter Compensation Techniques

Jitter compensation is the process of generating a compensated image sequence where any and all unwanted camera motion is subtracted from the original input. The jitter compensation process can be separated into two components: (a) motion estimation and (b) motion correction. Motion estimation is the main component of an image-based compensation system. Jitter compensation systems may be evaluated based on the performance of the motion estimation module alone, in which case one could use synthetic or calibrated sequences where the interframe motions are known. Two distinct approaches for motion estimation are presented in the literature: (a) feature-based motion estimation [40, 41] and (b) global intensity-based motion estimation [42, 43]. The effectiveness of jitter compensation is closely tied to the accuracy of detecting the local motion vectors in order to produce the right global motion vector. Here, three stabilization techniques that have been used in the sense-and-avoid system are presented.

10.4.2.1 Optical Flow

The optical flow technique obtains local velocity vectors of each pixel in the current image frame. These vectors are then used to determine the global translational and rotational motions with the assistance of a motion dynamics model and least squares estimation. A detailed description of the optical flow technique can be found in [44]. The output of optical flow is a velocity field, $V(i, j) = (u(i, j), v(i, j))$, of each pixel at position (i, j). This is the local motion field that is used to compute the global rotational and translational motions. Consider the case where the image frame sequence is purely rotated about a particular rotational centre, (i_0, j_0), by an angular velocity, ω. The rotational velocity vector, $V_r(i, j) = (u_r(i, j), v_r(i, j))$, that describes motion at (i, j) about centre point (i_0, j_0) can be decomposed into

$$u_r = |V_r| \cos \theta = \omega r \, \cos \theta = \omega(j - j_0);$$
$$v_r = -|V_r| \sin \theta = -\omega r \, \sin \theta = -\omega(i - i_0);$$

leading to

$$(u_r, v_r) = (\omega(j - j_0), -\omega(i - i_0)). \tag{10.2}$$

Here, θ is the angle of the vector that joins (i_0, j_0) and (i, j) with respect to a horizontal reference axis. For the case where both translational and rotational motions are present, let the coordinate frame at the rotational centre have translational velocity $\acute{u}$ and $\acute{v}$, in the vertical and horizontal directions, respectively. Then the velocity $(u(i, j), v(i, j))$ at any point (i, j), including translational and rotational components, will be given by

$$u(i, j) = \acute{u} + \omega(j - j_0) \quad \text{and} \quad v(i, j) = \acute{v} - \omega(i - i_0). \tag{10.3}$$

To determine the global velocities from many local velocity estimates as per equation (10.3), least squares estimation is used. Once the motion is estimated, the correction step consists of displacing the pixel location with a value that is proportional to the estimated translation and rotation.

10.4.2.2 Image Projection Correlation

The simplicity of the projection correlation (PrC) algorithm makes it an attractive option for real-time image stabilization, especially when compared to more computationally intensive block-matching methods [45]. The PrC algorithm seeks to characterize a 2D image frame by simpler 1D signals known as the image's row and column projections. The row projection is formed by summing the grey-scale pixel values of each row of the image frame; similarly the column projection is formed from a summation of the image frame columns, as illustrated in Figure 10.3. The translational displacement between two image frames can then be determined from the cross-correlation peak between the projections: row projections are compared to estimate vertical motion and column projections are used to estimate horizontal motion. Enhancements and variations to the basic technique outlined above have been proposed, including methods to improve precision. These variations include passing the projections through a raised cosine filter before correlating and modifications to allow estimation of image rotation [46]. Hybrid techniques that combine image projection and block-matching methods have also been proposed [45].

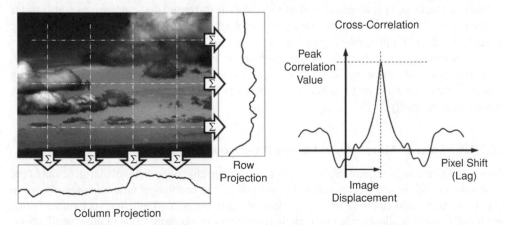

Figure 10.3 Illustration of the projection correlation technique

10.4.2.3   Inertial Measurement

In contrast to the previous two image-based compensation techniques, the inertial-based method is robust to featureless 'blue-sky' conditions that may be encountered in an airborne environment. Inertial-based image stabilization compensates the image sequence by employing motion sensors (typically, gyroscopes and accelerometers packaged in an inertial measurement unit (IMU)) to detect the camera movement. Measurements of camera motion can be translated to equivalent pixel displacements, which can then be used to shift image frames into alignment. This type of image stabilization is hardware-dependent and requires accurate timing and correlation between IMU measurements and captured image frames.

Motion measured with the IMU is translated to motion in pixels based on the following. Let f denote the camera focal length and let $\Delta\phi_k = \phi_k - \phi_0$ denote the pitch angle displacement at time k, based on the difference between the instantaneous IMU pitch measurement, ϕ_k, and a fixed reference angle, ϕ_0. The vertical pixel displacement caused by a pitching camera motion is then given by $\rho_{i,k} = f \tan \Delta\phi_k$. A similar relationship exists for yawing camera motion; that is, $\rho_{j,k} = f \tan \Delta\psi_k$, where $\Delta\psi_k$ is the heading angle displacement at time k. The image frame is corrected for camera rotational motion by shifting the image vertically and horizontally a specific number of pixels, proportional to the values of $\rho_{i,k}$ and $\rho_{j,k}$, respectively. (Note that the constant of proportionality depends on camera parameters such as image resolution, field-of-view, etc.). Camera roll motion is compensated directly by applying a basic geometric transformation that rotates the image frame $\Delta\xi_k = \xi_k - \xi_0$ degrees, where ξ_k denotes the instantaneous IMU roll measurement, and ξ_0 denotes a fixed reference angle.

10.5   Detection and Tracking

10.5.1   Two-Stage Detection Approach

A two-stage detection paradigm has become popular over the last few decades for detection of dim, pixel-sized collision targets [11, 20–23]. This detection paradigm focuses on the fact that collision threats tend to be small objects without spatial extent, and they are persistent or slowly moving in the image frame. These two characteristics separately lend themselves to different types of signal processing, and hence motivate a two-stage processing approach. These two stages are: (1) an image processing stage (intra-frame) that emphasizes point targets without spatial extent (often incorporating morphological filtering) and (2) a temporal filtering stage (inter-frame) that emphasizes features that are persistent in the scene.

As observed earlier, intra-frame and inter-frame processing stages can work in concert to emphasize and encourage detection of persistent pixel-sized features, whilst rejecting features that either have larger spatial extent (such as cloud artefacts) or features that are only observed temporarily.

10.5.1.1   Stage 1: Morphological Image Pre-processing

Morphological image processing is an intra-frame image enhancement tool that arose out of the seminal work of Georges Matheron and Jean Serra on the analysis of mineral compositions in thin geological cross-sections [47]. They derived a number of mathematical techniques which eventually found application in practical image processing scenarios, one of these applications being aircraft detection.

Image morphology techniques help to discriminate genuine intruder aircraft from 'target-like' image artefacts that can cause false alarms. Popular morphological filtering techniques include the top-hat, bottom-hat, and close-minus-open transformations [48, 49]. In general, a top-hat approach can be used to identify positive contrast features (features brighter than the local background), whereas a bottom-hat approach can be used to highlight negative contrast features (features darker than the local background). A close-minus-open (CMO) approach combines the power of both the top-hat and bottom-hat operators to simultaneously highlight both positive and negative contrast features.

Analysis of aircraft image data captured by machine vision sensors has shown that distant aircraft predominantly manifest as negative contrast features, suggesting that it is the shadow of the aircraft (rather than the reflected light) that is responsible for the visible contrast [50]. Hence, a bottom-hat filtering approach is particularly suited to identifying distant collision-course aircraft in a sense-and-avoid application.

Let $Y \oplus S$ and $Y \ominus S$ denote the dilation and erosion respectively of a greyscale image Y by a morphological structuring element S (see [25, 51] for more details about the dilation and erosion operations). The structuring element S acts like a cut-off parameter for filtering out features that are too large to be of interest. The bottom-hat transformation is then defined as $BH(Y, S) = [(Y \oplus S) \ominus S] - Y$. Figure 10.4 shows an example case where an intruder aircraft is highlighted, whilst an image artefact is suppressed via a bottom-hat filtering approach.

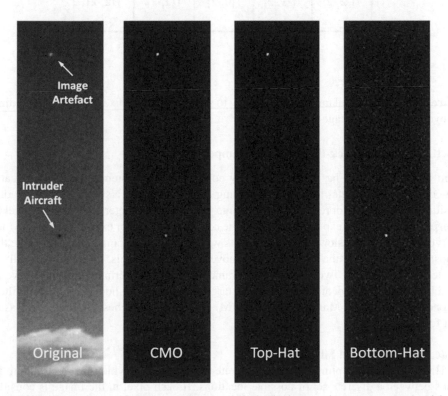

Figure 10.4 Illustration of three common morphological operations on a sample airborne image

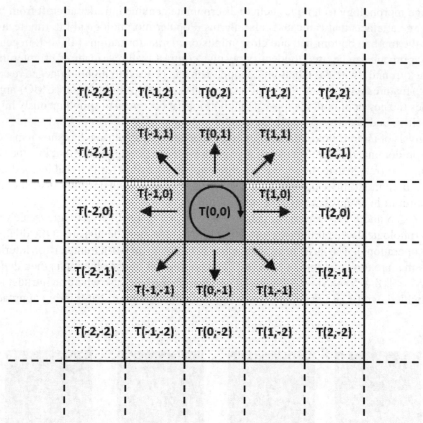

Figure 10.5   A potential target is most likely to stay in the same pixel or move to an adjacent one over consecutive image frames

10.5.1.2   Stage 2: Track-Before-Detect Temporal Filtering

Given that image jitter has been adequately compensated, a potential collision threat can be modelled as a discrete-time process moving across the camera's image plane, gradually transiting across pixels (or remaining at the same pixel). The characteristics of these pixel transitions are related to the expected motion of the collision threat. For example, collision threats will typically appear as slowly moving objects within the image frame, and, hence, the threat is most likely to stay at the same pixel, or move to adjacent pixels. This is graphically represented in Figure 10.5. Two types of inter-frame or temporal filtering approaches have been found to be useful for identifying persistent targets: recursive ad-hoc Viterbi-based filtering approaches and hidden Markov model (HMM) filtering approaches. Both are discussed in what follows.

Hidden Markov Model Filtering

The HMM filter is an optimal filter for discrete-time processes which involve random transitions between a discrete set of possible locations. If each pixel in the image is considered to be a possible location at which the collision threat could reside, and the target follows a random walk across pixel locations, then the HMM filter can be used to track (and detect) the target's motion in the image [11].

   The two key filter design characteristics impacting on detection performance are (1) the model of pixel intensity (in the morphological output) when a target is present in that pixel, denoted $B(\cdot)$ and (2) the model of how the target transitions between pixels, denoted $A(\cdot)$.

   The model of how the target transitions between pixels can be represented by a one-step target motion patch illustrated in Figure 10.5. $T(a, b)$ represents one possible transition where the target moves a pixels in the vertical direction and b pixels in the horizontal direction (note that $-2 \leq a,b \leq 2$ in this particular patch). Given a motion patch, the mean and variance of the expected target motion between frames can be evaluated. The mean expected motion $[\mu_a, \mu_b] = \sum_{\text{all } a,b} [a, b] P(T(a, b))$, and the variance equals $\sum_{\text{all } a,b} ([a, b] - [\mu_a, \mu_b])^2 P(T(a, b))$.

   Under the assumption that a target exists, the HMM filter calculates the conditional mean estimate of the target location based on a measurement sequence. (See [52] for a detailed presentation on HMMs.) However, in the collision detection problem, knowledge of target location is arguably secondary to the initial concern of: 'Is there a collision threat present?' An intermediate normalization factor within the standard HMM filter is proportional to the probability of the target being present, and hence this factor can be used as the basis of a detection test statistic. That is, the probability P(target in image frame | measurements) can be evaluated and serve as the metric for a threshold test that reflects a particular collision risk tolerance.

   Consider input image frames H pixels in height and W pixels in width, where the image at time step k is denoted by Y_k. The output of the HMM filter is another image. Let $\alpha_k(i, j)$ denote the output at pixel (i, j) at time step k. The basic HMM filter equations can now be written as:

Algorithm 10.1: HMM filter.

For $1 \leq i \leq H$, $1 \leq j \leq W$ and all k,

   1. **Initialization:** At step 0, $\alpha_0(i, j) = 1/(HW)$ (assuming no *a priori* target information).
   2a. **Recursion:**

$$\bar{\alpha}_k(i, j) = \left[\sum_{m=1}^{H} \sum_{n=1}^{W} \alpha_{k-1}(m, n) A((i, j)|(m, n)) \right] B(Y_k|(i, j)),$$

   where $A((i, j)|(m, n)) = P(T(i - m, \ j - n))$ and $B(Y_k|(i, j))$ is the probability of observing measurement Y_k given the target is at pixel (i, j).
   2b. **Normalization:**

$$\alpha_k(i, j) = N_k \bar{\alpha}_k(i, j),$$

   where $N_k = 1/\sum_{i=1}^{H} \sum_{j=1}^{W} \bar{\alpha}_k(i, j)$.
   3. **Test statistic:**

$$\gamma_k = \beta \gamma_{k-1} + (1 - \beta) \ln(1/N_k),$$

   where β is a scalar weighting coefficient between zero and one. Note that $\gamma_0 = 0$.

The test statistic γ_k for declaring the presence of a target is in the form of an exponentially weighted moving average with weighting coefficient β (experimentally, a weighting of $\beta = 0.9$ has been found to produce good detection results). When γ_k exceeds a predefined threshold (corresponding to a certain probability that a target is present), the HMM filter algorithm considers the target to be present in the image frame. Note that the detection threshold can be selected to achieve a specific design trade-off between detection probability and false alarm rate (a false alarm event occurs when the filter incorrectly declares a target to be present; that is, the test statistic γ_k crosses the threshold, but there is no target). Higher thresholds reduce the incidence of false alarms, but also lower detection probabilities. The system design objective is to select $A(\cdot)$ and $B(\cdot)$ so as to maximize the detection probability for a given false alarm rate (or, equivalently, to minimize the false alarm rate for a given detection probability).

Extensive implementation details for HMM filters are provided in [11, 15].

Ad-hoc Viterbi-Based Filtering

One difficult feature of the collision detection problem is that the detection filter must be able to detect collision threats with any heading in the image plane. However, any particular collision threat is likely to have almost constant heading. Thus, if a HMM filter is designed to detect targets with any possible heading, then its detection performance is degraded compared to a HMM filter design with knowledge of the target's specific heading (i.e. with a patch choice corresponding to the target's actual heading and having small variance in heading direction).

For this (and possibly other reasons), several researchers have proposed a different filtering approach that starts from the premise that any specific collision threat can be approximated as a slow-moving target with constant heading. The basic philosophy behind this alternative detection approach is that the uncertainty about target direction can be handled by using a bank of filtering branches (one branch for each of the four compass directions). In this way, if a target is present then it must be, at least partially, detected in one of the filter branches. If the filter branch output is combined in a suitable way, then detection of a target with any heading can be achieved.

It is interesting to note that in this ad-hoc approach, the set of filtering branches replaces the role of transition probability patches in describing the range of possible target headings.

Unfortunately, unlike the HMM filter, there is no simple connection between filter output and the probability of a collision threat being present. However, intuition suggests that the strength of filter returns is closely linked to the likelihood of a target being present, and hence filter outputs can again be used as a test statistic (even if the connection to collision risk is not straightforward).

Let $\alpha_k^r(i, j)$ denote the output at pixel (i, j) of filter branch r, and let $Y_k(i, j)$ denote the greyscale level of pixel (i, j) in the input image. Then the basic ad-hoc Viterbi filter equations are:

Algorithm 10.2: Ad-hoc Viterbi filter.

For $1 \leq i \leq H$, $1 \leq j \leq W$, $1 \leq r \leq 4$ and all k,

  1. **Initialization:** At step 0, $\alpha_0^r(i, j) = 0$.

  2a. **Recursion:**

$$\alpha_k^r(i, j) = \beta \max_{1 \leq m \leq H, 1 \leq n \leq W} \left[\alpha_{k-1}^r(m, n) A^r((i, j)|(m, n)) \right] + (1 - \beta) Y_k(i, j)$$

    where $A^r((i, j)|(m, n))$ is a branch-specific pixel transition function that is either 1 or 0 to indicate if a transition from pixel (m, n) to pixel (i, j) is allowed, and β is a scalar 'forgetting' factor between zero and one.

  2b. **Branch combination:**

$$\alpha_k(i, j) = \max_{1 \leq r \leq 4} \left[\alpha_k^r(i, j) \right].$$

3. **Test statistic:**

$$\gamma_k = \max_{1 \leq i \leq H, 1 \leq j \leq W} [\alpha_k(i, j)].$$

When γ_k exceeds a predefined threshold, the ad-hoc Viterbi filter algorithm considers the target to be present in the image frame. Experimentally, a forgetting factor of $\beta = 0.75$ has been found to produce reasonable detection results [30].

Filter Bank Approach

A key deficiency of the ad-hoc Viterbi-based filter is that there is no systematic way to tune filter parameters. Recently, the authors proposed a track-before-detection technique that combined the best features of the ad-hoc Viterbi-based and HMM approaches. In [15], new HMM filter bank techniques were presented that allowed filtering branches to be optimized to a set of distinct behaviours, while the overall filter bank system could be designed to optimize detection performance. For example, one can design a HMM filter bank with four branches, with each branch being a HMM filter with a unique transition model A. In this way, all of the branches can be designed to represent motion in a particular direction (in a systematic manner that is more flexible and better performing than the ad-hoc Viterbi approach). A test statistic can also be devised with this approach which is tightly connected to conflict risk.

The basic HMM filter bank equations are:

Algorithm 10.3: HMM filter bank.

For $1 \leq i \leq H$, $1 \leq j \leq W$, $1 \leq r \leq 4$ and all k,

1. **Initialization:** At step 0, $\alpha_0^r(i, j) = 1/(HW)$ (assuming no *a priori* target information).

2a. **Recursion:**

$$\bar{\alpha}_k^r(i, j) = \left[\sum_{m=1}^{H} \sum_{n=1}^{W} \alpha_{k-1}^r(m, n) A^r((i, j)|(m, n)) \right] B(Y_k|(i, j)),$$

where $A^r((i, j) \mid (m, n))$ is the branch-specific transition probability from pixel (m, n) to pixel (i, j) and $B(Y_k|(i, j))$ is the probability of observing measurement Y_k given the target is at pixel (i, j).

2b. **Normalization:**

$$\alpha_k^r(i, j) = N_k^r \bar{\alpha}_k^r(i, j),$$

where $N_k^r = 1/ \sum_{i=1}^{H} \sum_{j=1}^{W} \bar{\alpha}_k^r(i, j)$.

3. **Test statistic:**

$$\gamma_k = \max_{1 \leq r \leq 4} [\gamma_k^r],$$

where $\gamma_k^r = \beta \gamma_{k-1}^r + (1 - \beta) \ln(1/N_k^r)$, and β is a scalar weighting coefficient between zero and one. Note that $\gamma_0^r = 0$.

Studies have shown that HMM filter bank systems offer superior dim target detection performance compared to other HMM filters. That is, they have higher detection probabilities for a specific false alarm rate [15]. Moreover, studies on sample target image sequences suggest that HMM filter banks have better false alarm rejection than ad-hoc Viterbi filtering approaches (although they may be more sensitive to image jitter) [11].

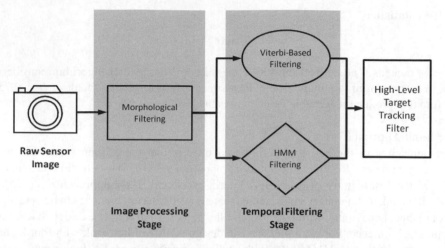

Figure 10.6 Computer vision-based detection and tracking process

10.5.2 Target Tracking

After detection has occurred, target position estimates are then passed to a high-level target tracking filter (such as an extended Kalman filter), as illustrated in Figure 10.6. Target tracking is a well-researched field with a rich history, and there are numerous candidate tracking approaches that could be applied at this stage of the sense-and-avoid problem. Hence, specific target tracking approaches will not be discussed in detail in this chapter, but more information can be found in [53].

To handle multiple detected targets, one possible approach would be to have a dedicated extended Kalman filter for each potential target, and a track file manager to solve the data association problem [54]. The data association process addresses the issue of whether a newly detected target corresponds to a new target or an existing target. In the latter case, another decision is required if there are multiple existing targets. The track file manager could even assist with higher-level decisions about which objects are genuine collision threats. If a potential target is considered a genuine collision threat, the target image positions from the corresponding track file are then used in the next phase of the sense-and-avoid task – namely, characterization of target dynamics and avoidance control.

10.6 Target Dynamics and Avoidance Control

10.6.1 Estimation of Target Bearing

As discussed in Section 10.3.1, the image must be first transformed from 3D space (characterized by an X,Y,Z coordinate system) into 2D space (represented by just x, y coordinates). Based on the geometry depicted in Figure 10.7, two important parameters of the target with regard to the camera can be extracted; namely the target bearing λ and elevation δ. The bearing represents the angle formed by the vector q_{xz} (the projection of q in the x–z plane) with the z-axis, and the elevation is the angle formed by the vector q_{yz} (the projection of q in the y–z plane) with the z-axis.

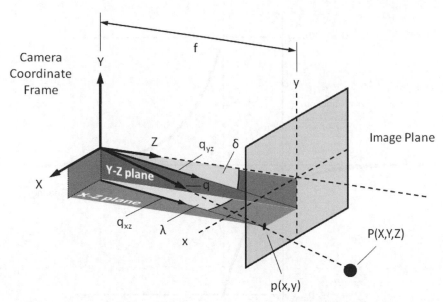

Figure 10.7 Camera model and target geometry used for bearing and elevation estimation

The relative bearing and elevation of the target are estimated as follows:

$$\lambda = \tan^{-1}\left(\frac{x}{f}\right) \quad \text{and} \quad \delta = \tan^{-1}\left(\frac{y}{f}\right),$$
(10.4)

where $f > 0$ is the camera focal length.

It is possible to infer target motion in the image plane by exploiting the bearing and elevation rates $\dot{\lambda}$ and $\dot{\delta}$, respectively. This type of information can be useful in determining whether the target represents a likely collision threat (zero or low angular rates indicate a target is on collision course).

10.6.2   Bearing-Based Avoidance Control

The underlying principle behind the proposed avoidance strategy is to move the actuator (camera/aircraft) away from the features (target). This is achieved through a combination of 2D and 3D vision-based control [55, 56]. Let $\Phi_{\max}$ denote the maximum heading command; let λ represent the current target bearing; let $\hat{\lambda}$ denote the *least* desired target bearing; and let c correspond to a positive gain. Then an exponential error function of the form $\Gamma = [\Gamma_\lambda, \Gamma_\delta]'$ can be defined, where

$$\Gamma_\lambda = \begin{cases} -\Phi_{\max}e^{c(\lambda-\hat{\lambda})}, & \lambda - \hat{\lambda} < 0 \\ \Phi_{\max}e^{-c(\lambda-\hat{\lambda})}, & \lambda - \hat{\lambda} \geq 0 \end{cases} \quad \text{and} \quad \Gamma_\delta = 0.$$
(10.5)

This error function will be maximum when $\lambda = \hat{\lambda}$, and will decrease in magnitude in an exponential manner away from $\hat{\lambda}$, as illustrated in Figure 10.8.

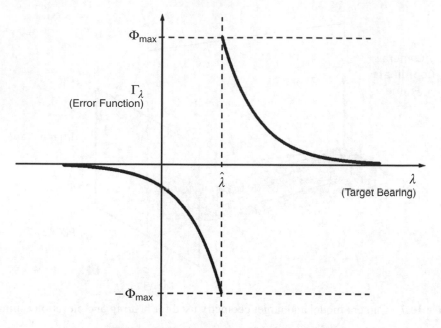

Figure 10.8   Exponential error function

Letting $\hat{\lambda} = 0$ corresponds to a control strategy that tries to drive the target to the left or right edge of the image frame (or keeps the target away from the centre of the image).

Based on the error function Γ, one can develop a control law $\Omega = -\eta L^{+}\Gamma$, where η is a positive gain and L^{+} is the pseudo-inverse of an interaction matrix L that relates velocities in 3D space to motion on the 2D image plane. L is dependent on camera intrinsic parameters [57]. The above control law can be used to achieve avoidance behaviour as illustrated in Figure 10.9, where upon detection of an intruder aircraft, an avoidance waypoint at bearing angle $\kappa \propto \Omega$ can be generated and tracked to avert a collision.

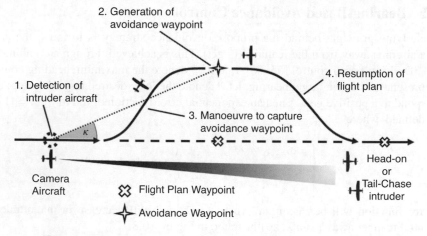

Figure 10.9   Illustration of collision avoidance control strategy

10.7   Hardware Technology and Platform Integration

Avionics and sensor hardware were integrated onto various aircraft platforms to facilitate testing and evaluation of the proposed sense-and-avoid system. An important aspect of this activity involved the development of target/intruder platforms that play the role of a non-cooperative collision-course aircraft, as well as camera platforms that capture image data for either online or offline post-processing. The key principles that guided the design of the platform architectures included making subsystems modular, reusable, and that exploit commercial-off-the-shelf (COTS) components where possible.

The key innovations in the platform architectures are the approaches to (1) precisely associate captured image data with the corresponding aircraft state information measured at the time of image capture and (2) real-time image processing.

10.7.1   Target / Intruder Platforms

The main function of the target platform is to act as the 'aircraft to avoid' in collision scenarios and to precisely log its own state information. Two different types of target platforms have been employed: (i) a Boomerang UAV and (ii) a piloted Cessna 182 light aircraft.

10.7.1.1   Boomerang UAV

The UAV target platform is a Boomerang 60 model airplane manufactured by Phoneix Models. A photograph of the platform is shown in Figure 10.10. The model airplane measures 1.5 m from nose to tail and has a wingspan of 2.1 m. It is powered by an OS 90 FX engine driving a 15 by 8 inch propeller.

System Architecture
The Boomerang carries the highly modular UAV base system architecture illustrated in Figure 10.11. It relies on the MicroPilot 2128's autopilot and suite of onboard sensors for

Figure 10.10   Boomerang UAV target platform

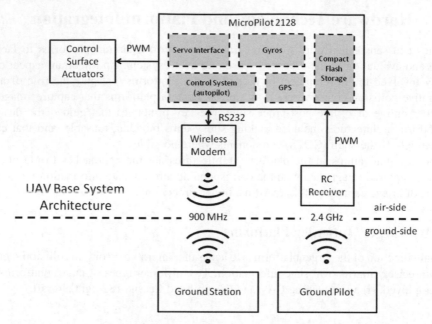

Figure 10.11   UAV base system architecture

flight control and navigation. The UAV can be operated autonomously or flown manually under radio control (RC). The Boomerang UAV provides basic unmanned flight capabilities, and a detailed breakdown of the system components is given in Table 10.1.

10.7.1.2   Piloted Cessna 182 Light Aircraft

The light aircraft target platform is a standard Cessna 182 aeroplane. During flight tests it carries a NovAtel DL-V3 GNSS receiver for logging aircraft state information.

10.7.2   Camera Platforms

The camera platforms have two distinct roles: (1) image data collection and (2) onboard real-time sense-and-avoid processing. The UAV camera platform is equipped to perform only the data collection role, whereas the light aircraft camera platform is capable of full in-flight data collection and closed-loop sense-and-avoid testing.

Table 10.1   Boomerang UAV system configuration

System component	Hardware selection
Inertial measurement sensor	MicroPilot® MP2128 gyros
Flight controller	MicroPilot® MP2128 control system
GPS sensor	MicroPilot® MP2128 GPS navigation
Communications with ground station	Microhard Systems Spectra 920A wireless modem
Communications with ground pilot	Spektrum AR9000 SM2 9-channel RC receiver

Figure 10.12   Flamingo UAV camera platform

10.7.2.1 Flamingo UAV

The UAV camera platform is a Flamingo UAV manufactured by Silvertone [58]. It is powered by a 26-cc two-stroke Zenoah engine driving a 16 by 6 inch propeller and is shown in Figure 10.12.

System Architecture

The Flamingo system design exploits the base system architecture of the Boomerang target platform for general flight control. In addition, it has a separate and independent vision payload system for data capturing (but no onboard image processing capability), as illustrated in Figure 10.13. The vision payload system employs dedicated high-quality inertial and

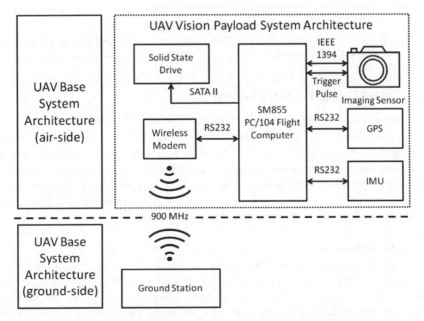

Figure 10.13   Flamingo UAV system architecture

position sensors to provide timely (high update rate) and precise state information critical for image stabilization. A custom real-time operating system ensures that recorded image frames are associated with aircraft state data at precisely the point when the camera is triggered. In particular, a multi-data source synchronization process was developed around a global triggering pulse to coordinate the simultaneous capture of image, global positioning system (GPS), and IMU data. The system has the capacity to record 1024 by 768 pixel image frames (at 8 bits per pixel bit depth) and associated state data to a solid state drive at a rate of up to 15 frames per second (approximately 12 Mb/s sustained writing to disk). Furthermore, the vision payload system can be activated/deactivated remotely from the ground station. A detailed breakdown of the system components is given in Table 10.2.

System Configuration

Table 10.2 Flamingo UAV system configuration

System component	Hardware selection
Vision sensor	Basler Vision Technologies Scout Series scA1300-32fc area scan camera
Inertial measurement sensor	Atlantic Inertial Systems SilMU04®
Flight controller	MicroPilot® MP2128 control system
GPS sensor	NovAtel OEMV-1
Communications with ground station	Microhard Systems Spectra 920A wireless modem (2x)
Communications with ground pilot	Spektrum AR9000 SM2 9-channel RC receiver
Flight computer	Digital-Logic® SM855 PC/104; Intel® Pentium® M 1.8 GHz processor; 1 GB SODIMM DDR RAM; Linux Debian operating system with customized kernel for real-time processing
Image data storage	OCZ Technology SATA II 2.5" Solid State Drive 120 GB

10.7.2.2 Piloted Cessna 172 Light Aircraft

The light aircraft camera platform is a custom-fitted Cessna 172 light aircraft, which is shown in Figure 10.14. The design, manufacture, maintenance, and operation of this cost-effective flight testing platform are detailed in [59]. The Cessna 172 has been equipped with onboard data capturing and real-time image processing capabilities to produce a system suitable for complete closed-loop sense-and-avoid testing; that is, a system capable of automatically (1) detecting intruder aircraft; (2) issuing avoidance control commands; and (3) executing control commands, all without external pilot or ground station interaction.

Basic System Architecture

The data capturing system onboard the Cessna 172 is based on the vision payload system of the Flamingo. The introduction of graphic processing unit (GPU) hardware, as illustrated in Figure 10.15, provides a real-time image processing capability that was absent in the Flamingo. Computationally intensive tasks such as image stabilization and target detection

Figure 10.14   Light aircraft camera platform

are handled entirely by the GPU, allowing 1024 by 768 pixel image frames (at 8 bit per pixel bit depth) to be processed at a rate of up to 15 frames per second. Other processing tasks are distributed across two flight computers, with one computer connected directly to the aircraft flight controller for automated avoidance control. A tightly coupled GNSS and INS sensor suite provides high-quality aircraft state information, and the overall sense-and-avoid system can be managed and monitored via a compact personal digital assistant (PDA) interface. The image sensor is mounted onto the aircraft wing strut using a certified custom-made bracket. A detailed breakdown of the system components is given in Table 10.3.

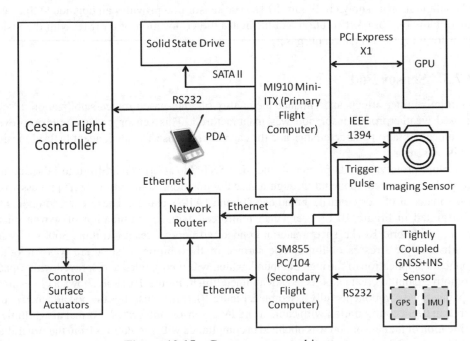

Figure 10.15   Cessna system architecture

Basic System Configuration

Table 10.3   Piloted Cessna 172 light aircraft system configuration

System component	Hardware selection
Vision sensor	Basler Vision Technologies Scout Series scA1300-32fc area scan camera
Inertial measurement sensor	iMAR IMU-FSAS*
GPS sensor	NovAtel OEMV-3*
Primary flight computer	Backplane Systems Technology MI910 Mini-ITX; Intel® Core 2 Duo 2.4 GHz processor; 2 GB SDRAM; Linux Debian operating system
Secondary flight computer	Digital-Logic® SM855 PC/104; Intel® Pentium® M 1.8 GHz processor; 1 GB SODIMM DDR RAM; Linux Debian operating system with customized kernel for real-time processing
Image processing GPU	Gigabyte™ NVIDIA® GeForce® 9600 GT; 512 MB GDDM1R3 RAM
Image data storage	OCZ Technology SATA II 2.5″ Solid State Drive 120 GB

*This sensor is part of a NovAtel SPAN (Synchronized Position, Attitude and Navigation) product (tightly coupled GNSS+INS sensor) in a ProPack-V3 enclosure.

Vision Sensor Pod Architecture
A variation of the basic system architecture has been developed to accommodate a sensor pod configuration as shown in Figure 10.16. The sensor pod provides an upgrade to the basic camera mounting bracket solution and facilitates the co-location of an IMU with the vision sensor for image stabilization purposes.

10.7.3   Sensor Pod

To minimize jitter effects and enhance the quality of state-based image stabilization, a self-enclosed weatherproof sensor pod was manufactured. This sensor pod featured improved mechanical vibration characteristics and the capacity to house an independent IMU alongside the vision sensor.

A rapid prototyping 3D printer (Dimension SST 786 [60]) was utilized to fabricate the core pod structure components through a fused deposition modelling (FDM) process from a base material of acrylonitrile butadiene styrene (ABS) plastic. These core components, as illustrated in Figure 10.17, were then reinforced with a combination of woven glass, carbon, and hybrid Kevlar sheet material bonded with epoxy resin (Araldite 3600) in order to withstand the stresses of flight. The surface of the reinforced structure was smoothed through the application of fairing compound, which was then sanded and sprayed with 2-pack automotive paint to create a polished streamlined finish. Figure 10.18(a) shows a close-up of the painted pod components fully assembled. Figure 10.18(b) illustrates the pod attached to the aircraft in flight configuration with camera and IMU sensors integrated. Formal airworthiness certification of the sensor pod was obtained in compliance with regulations from the Australian Civil Aviation Safety Authority.

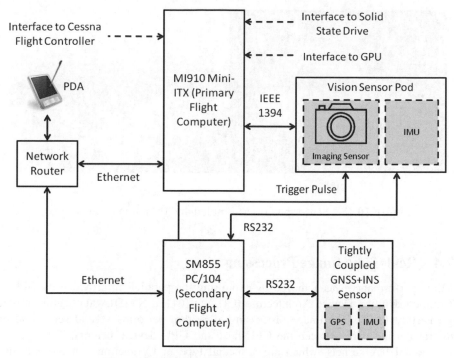

Figure 10.16   Sensor pod system architecture

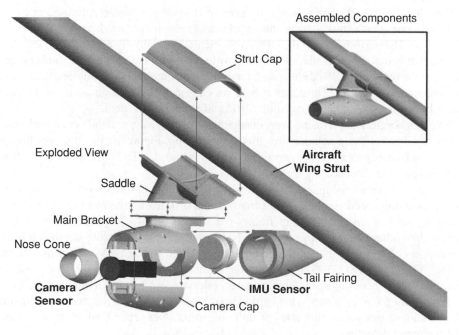

Figure 10.17   Sensor pod components

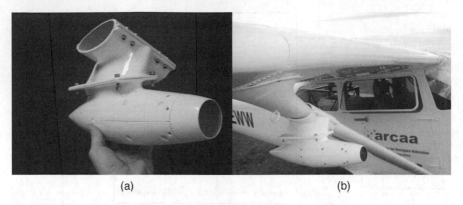

(a)   (b)

Figure 10.18   Sensor pod (a) assembled and (b) mounted on aircraft

10.7.4   Real-Time Image Processing

Real-time performance was achieved by exploiting the parallelism provided by GPUs and the Compute Unified Device Architecture (CUDA) [61], a NVIDIA application programming interface (API). The image processing algorithm performs several sequential operations transferring data between the CPU host and GPU device memory. The implementation uses CUDA kernels, which are a special type of C function that are executed N times in parallel by N different CUDA threads. Threads are grouped into blocks, and communicate only with other threads in the same block using quick access L1 cache type memory.

The block size, and, therefore, the number of threads per block, is limited and can be optimized to suit (1) the task, (2) the amount of cache memory required, and (3) the particular GPU device. The performance of the GPU implementation is closely related to (1) the number of under-utilized warps; (2) the number of multiprocessors and blocks per multiprocessor specific to the particular GPU device; and, finally, (3) the number of threads per block (ideally always a multiple of 32). The latter should be chosen to be as high as possible, limited obviously by the GPU compute capability [61] and available registers.

Several laboratory experiments were conducted to evaluate the scalability and performance of various GPU hardware solutions for image processing. Figure 10.19 illustrates the speed (in terms of frames processed per second) at which various COTS NVIDIA GPU cards were able to execute the detection algorithm.

A near-linear relationship was found between the number of microprocessors and the raw image processing speed (excluding data transfer and disk read/write overheads). Substantial processing rates of up to 150 Hz or frames per second (1024×768 pixel image; 8 bit per pixel bit depth) have been achieved using a NVIDIA GeForce GTX 280, but the high power consumption (236 W) of this card precludes its practical use in UAV platforms. Instead, a NVIDIA GeForce 9600 GT GPU (low-power version) has been tested which is considered to have the best processing speed/power consumption trade-off. It has 8 microprocessors and consumes only 59 W of power. This card is used in the flight-ready hardware configuration and can achieve raw processing rates of up to 30 frames per second, which is sufficient for real-time target detection.

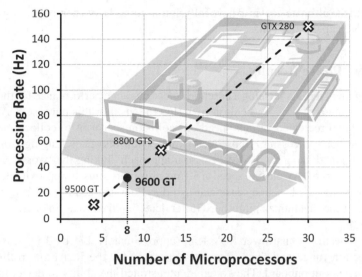

Figure 10.19   Frame processing rate of detection algorithm vs. number of microprocessors in candidate GPU cards

10.8   Flight Testing

Flight tests were undertaken to collect collision-course image data and to evaluate the performance of the proposed sense-and-avoid system under realistic operating conditions in phases of progressively increasing complexity and autonomy. This testing philosophy is reflected in the flight testing schedule shown in Table 10.4. The first three phases of testing have been

Table 10.4   Flight testing schedule

Testing phase	Platforms	Data processing	Avoidance control
Phase I	Boomerang and Flamingo UAVs	Offline post-processing	None
Phase II	Cessna 182 and Cessna 172 light aircraft	Real-time onboard processing	Autopilot command generation
Phase III	Cessna 182 and Cessna 172 light aircraft	Real-time onboard processing	Full closed-loop
Phase IV	Boomerang and Flamingo UAVs	Real-time onboard processing	Autopilot command generation
Phase V	Boomerang and Flamingo UAVs	Real-time onboard processing	Full close-loop

completed, and the ultimate goal is to have fully autonomous UAVs demonstrating closed-loop sense-and-avoid capabilities (Phase V).

10.8.1   Test Phase Results

Phase I testing involved establishing a baseline detection range performance for the sense-and-avoid system using a 51° by 40° field-of-view (FOV) lens. For this purpose, UAV platforms were deployed and recreated various 'head-on' collision scenarios to collect relevant image data for offline post-processing. Subsequently, detection distances ranging from 400 m to 900 m were obtained using the HMM detection approach. Taking the 'worst-case' scenario and approximating the UAV closing speed at 50 m/s, this represents an approximate 8 s warning ahead of impact that the baseline sense-and-avoid system can achieve. This approaches the 12.5 s response time that human pilots need (after detecting the threat) in order to safely avoid a collision [62].

In Phase II testing, a narrower FOV lens (approximately 17° by 13°) was selected to improve detection range and the system was operated for the first time in-flight with all processing carried out onboard. The system demonstrated the ability to detect targets out to distances ranging from 3 km to 5 km. Even with the increased closing speeds (approximately 100 m/s) of the light aircraft platforms, the detection distances represent timely warnings ahead of impact that exceed the minimum response time recommended for human pilots.

Finally, in Phase III testing the system ultimately demonstrated full closed-loop sense-and-avoid functionality. Future test phases will work towards reproducing current closed-loop capabilities onboard UAV platforms.

10.9   Future Work

One of the key impediments to practical realization of computer vision-based sense-and-avoid systems with existing technology is the incidence of false alarms that could cause unnecessary avoidance responses. It is imperative, therefore, to refine existing algorithms (or find new ones) which can minimize false alarm events. There may be alternatives to conventional morphological filtering front-ends, for example, which can provide the necessary improvements. Adaptation of the morphological processing to suit the application at hand is another avenue which merits further investigation. Adaptation might also prove useful within the HMM filtering domain. For example, one could bring additional filter branches online when needed, and discard them when not needed. This added flexibility might reduce the computational burden and further enhance detection performance.

There is also room for improvement in the way the image processing and control portions of the sense-and-avoid system are implemented. One might be able to make performance gains by implementing the processing algorithms using the OpenCL programming framework/language. This could also make the code more portable to other GPUs, since the CUDA-based designs used for realization of the proposed sense-and-avoid system are only supported on hardware from NVIDIA.

Finally, it is important to point out that a key limitation of UAVs is their inability to accommodate large, heavy, or high-power payloads. A promising area of future work, then, is the miniaturization of an entire closed-loop sense-and-avoid system, such that it can fit inside a relatively small UAV.

10.10   Conclusions

This chapter has outlined a number of key advantages to realizing a sense-and-avoid system based on computer vision in the optical range. Optical camera-based sensing systems are relatively low in cost, volume, power, and weight compared to alternatives such as radar and TCAS. Vision-based systems also do not have to rely on (possibly non-existent) cooperation from other aircraft.

Although optical sense-and-avoid systems have a number of key advantages, they also bring with them an array of challenges. Since aircraft generally travel at relatively high speeds, one has to detect targets when they are a long distance away and which occupy only a minute fraction of the image frame. Under such challenging circumstances, accurate and timely detection of targets is difficult enough, even without factoring in the dynamic and unpredictable airborne environment. This environment is characterized by an abundance of interfering elements such as clouds and other weather-dependent phenomena that conspire to hide genuine targets and introduce target-like artefacts. A system that does not account for these factors is likely to demonstrate an unacceptably high incidence of false alarms. Finally, optical sense-and-avoid systems also rely on sophisticated image processing methods which are, generally, computationally intensive.

This chapter has shown that significant progress has been made towards overcoming the challenges associated with using computer vision for sense-and-avoid. Elegant new processing techniques involving morphological filtering and hidden Markov model filter banks are particularly promising. Tests using realistic collision-course image data have shown that these techniques are effective in sensing airborne targets if the cloud and background clutter is not excessive. Furthermore, flight trials have demonstrated that real-time processing can be achieved and that closed-loop sense-and-avoid functionality is possible. The existing technology does, however, suffer from a moderate number of false alarms, which would trigger unnecessary avoidance actions. It is believed that if false alarm events associated with cloud artefacts could be eliminated, then the resulting system performance would be acceptable for the purposes of routine automated sense-and-avoid.

Acknowledgements

This chapter was supported under the Australian Research Council's Linkage Projects funding scheme (Project Number LP100100302) and the Smart Skies Project, which is funded, in part, by the Queensland State Government Smart State Funding Scheme.

References

1. M. DeGarmo, 'Issues concerning integration of unmanned aerial vehicles in civil airspace', 2004.
2. US Army UAS Center of Excellence, ' "Eyes of the Army" U.S. Army Roadmap for Unmanned Aircraft Systems 2010–2035', 2010.
3. W. Graham and R. H. Orr, 'Separation of air traffic by visual means: an estimate of the effectiveness of the see-and-avoid doctrine', *Proceedings of the IEEE*, pp. 337–361, March 1970.
4. P. Zarchan, *Tactical and Strategic Missile Guidance*, 4th edn. American Institute of Aeronautics and Astronautics, Reston, VA, 2002.

5. N. Shneydor, *Missile Guidance and Pursuit: Kinematics, Dynamics and Control*, Horwood Publishing, Chichester, 1998.

6. Australian Transport Safety Bureau, 'Limitations of the See-and-Avoid Principle', 1991.

7. T. Williamson and N. Spencer, 'Development and operation of the Traffic Alert and Collision Avoidance System (TCAS)', *Proceedings of the IEEE*, 77(11), 1735–1744, 1989.

8. International Civil Aviation Organization, 'Annex 6 to the Convention on International Civil Aviation – Operation of Aircraft – Part I – International Commercial Air Transport – Aeroplanes', in Annexes to the Convention on International Civil Aviation, pp. 6–10, 2001.

9. Forecast International [Online], www.forecastinternational.com/Archive/es/es0077.doc, January 2007.

10. Office of the Secretary of Defense, 'Airspace Integration Plan for Unmanned Aviation', 2004.

11. J. Lai, L. Mejias, and J. J. Ford, 'Airborne vision-based collision-detection system', *Journal of Field Robotics*, 28(2), 137–157, 2011.

12. C. Geyer, S. Singh, and L. Chamberlain, 'Avoiding Collisions Between Aircraft: State of the Art and Requirements for UAVs operating in Civilian Airspace', Robotics Institute, 2008.

13. G. Fasano *et al.*, 'Multi-sensor-based fully autonomous non-cooperative collision avoidance system for unmanned air vehicles', *Journal of Aerospace Computing, Information, and Communication*, 5, 338–360, 2008.

14. Y. Bar-Shalom, H. M. Shertukde, and K. R. Pattipati, 'Use of measurements from an imaging sensor for precision target tracking', *IEEE Transactions on Aerospace and Electronic Systems*, 25(6), 863–872, 1989.

15. J. Lai and J. J. Ford, 'Relative entropy rate based multiple hidden Markov model approximation', *IEEE Transactions on Signal Processing*, 58(1), 165–174, 2010.

16. J. Keller, Military and Aerospace Electronics [Online], http://www.militaryaerospace.com/index/display/mae-defense-executive-article-display/0220059175/articles/military-aerospace-electronics/executive-watch-2/2010/10/sierra-nevada_corp.html, October 2010.

17. B. Karhoff, J. Limb, S. Oravsky, and A. Shephard, 'Eyes in the domestic sky: an assessment of sense and avoid technology for the army's "Warrior" unmanned aerial vehicle', in Proceedings of the IEEE Systems and Information Engineering Design Symposium, Charlottesville, VA, pp. 36–42, 2006.

18. D. Maroney, R. Bolling, M. Heffron, and G. Flathers, 'Experimental platforms for evaluating sensor technology for UAS collision avoidance', in Proceedings of the IEEE/AIAA Digital Avionics Systems Conference, Dallas, TX, pp. 5C1-1–5C1-9, 2007.

19. W. Rosenkrans, 'Detect, sense and avoid', AeroSafety World, pp. 34–39, July 2008.

20. J. Arnold, S. Shaw, and H. Pasternack, 'Efficient target tracking using dynamic programming', *IEEE Transactions on Aerospace and Electronic Systems*, 29(1), pp. 44–56, 1993.

21. Y. Barniv, 'Dynamic programming solution for detecting dim moving targets', *IEEE Transactions on Aerospace and Electronic Systems*, AES-21(1), 144–156, 1985.

22. T. Gandhi *et al.*, 'Detection of obstacles in the flight path of an aircraft', *IEEE Transactions on Aerospace and Electronic Systems*, AES-39(1), 176–191, 2003.

23. T. Gandhi *et al.*, 'Performance characterization of the dynamic programming obstacle detection algorithm', *IEEE Transactions on Image Processing*, 15(5), 1202–1214, 2006.

24. S. D. Deshpande, M. H. Er, R. Venkateswarlu, and P. Chan, 'Max-mean and max-median filters for detection of small targets', in Proceedings of the Signal and Data Processing of Small Targets, Denver, CO, pp. 74–83, 1999.

25. E. R. Dougherty and R. A. Lotufo, *Hands-on Morphological Image Processing*, SPIE Optical Engineering Press, Bellingham, MA, 2003.

26. L. JiCheng, S. ZhengKang, and L. Tao, 'Detection of spot target in infrared cluster with morpho-logical filter', in Proceedings of the IEEE National Aerospace and Electronics Conference, Dayton, OH, pp. 168–172, 1996.

27. Z. Zhu, Z. Li, H. Liang, B. Song, and A. Pan, 'Grayscale morphological filter for small target detection', in Proceedings of the SPIE International Symposium on Optical Science and Technology: Infrared Technology and Applications, San Diego, CA, pp. 28–34, 2000.

28. M. Zeng, J. Li, and Z. Peng, 'The design of top-hat morphological filter and application to infrared target detection', *Infrared Physics & Technology*, 48(1), 67–76, 2006.

29. N. Yu, H. Wu, C. Wu, and Y. Li, 'Automatic target detection by optimal morphological filters', *Journal of Computer Science and Technology*, 18(1), 29–40, 2003.

30. R. Carnie, R. Walker, and P. Corke, 'Image processing algorithms for UAV "Sense and Avoid"', in Proceedings of the IEEE International Conference on Robotics and Automation, Orlando, FL, pp. 2848–2853, 2006.

31. S. J. Davey, M. G. Rutten, and B. Cheung, 'A comparison of detection performance for several track-before-detect algorithms', *EURASIP Journal on Advances in Signal Processing*, 2008, 1–10, 2008.

32. S. M. Tonissen and R. J. Evans, 'Performance of dynamic programming techniques for track-before-detect', *IEEE Transactions on Aerospace and Electronic Systems*, 32(4), 1440–1451, 1996.

33. M. G. S. Bruno, 'Bayesian methods for multiaspect target tracking in image sequences', *IEEE Transactions on Image Processing*, 52(7), 1848–1861, 2004.

34. M. G. S. Bruno and J. M. F. Moura, 'Multiframe detector/tracker: optimal performance', *IEEE Transactions on Aerospace and Electronic Systems*, 37(3), 925–945, 2001.

35. M. G. S. Bruno and J. M. F. Moura, 'The optimal 2D multiframe detector/tracker', *International Journal of Electronics and Communications*, 53(6), 1–17, 1999.

36. G. D. Forney Jr., 'The Viterbi algorithm', *Proceedings of the IEEE*, 61(3), 268–278, 1973.

37. L. Mejias, S. McNamara, and J. Lai, 'Vision-based detection and tracking of aerial targets for UAV collision avoidance', in Proceedings of the IEEE/RSJ International Conference on Intelligent Robots and Systems, Taipei, 2010.

38. R. Hartley and A. Zisserman, *Multiple View Geometry in Computer Vision*, 2nd edn, Cambridge University Press, New York, 2004.

39. Y. Ma, S. Soatto, J. Kosecka, and S. S. Sastry, *An Invitation to 3-D Vision: From Images to Geometric Models*, Springer-Verlag, New York, 2004.

40. J.-Y. Chang, W.-F. Hu, M.-H. Cheng, and B.-S. Chang, 'Digital image translational and rotational motion stabilization using optical flow technique', *IEEE Transactions on Consumer Electronics*, 108–115, 2002.

41. A. Censi, A. Fusiello, and V. Roberto, 'Image stabilization by features tracking', in Proceedings of the International Conference on Image Analysis and Processing, Venice, pp. 665–667, 1999.

42. V.-N. Dang, A.-R. Mansouri, and J. Konrad, 'Motion estimation for region-based video coding', in Proceedings of the International Conference on Image Processing, Washington, DC, pp. 189–192, 1995.

43. S. Erturk and T. J. Dennis, 'Image sequence stabilisation based on DFT filtering', *IEE Proceedings on Vision, Image and Signal Processing*, 147 (2), 95–102, 2000.

44. B. K. P. Horn and B. G. Schunck, 'Determining optical flow', *Artificial Intelligence*, 17(1–3), 185–203, 1981.

45. G. Ren, P. Li, and G. Wang, 'A novel hybrid coarse-to-fine digital image stabilization algorithm', *Information Technology Journal*, 9(7), 1390–1396, 2010.

46. S. B. Balakirsky and R. Chellappa, 'Performance characterization of image stabilization algorithms', *Real-Time Imaging*, 2(5), 297–313, 1996.

47. J.-P. Serra, *Image Analysis and Mathematical Morphology*, Academic Press, New York, 1982.

48. R. C. Gonzalez, R. E. Woods, and S. L. Eddins, 'Morphological image processing', in *Digital Image Processing Using MATLAB*, Pearson-Prentice Hall, Upper Saddle River, NJ, pp. 334–377, 2004.

49. D. Casasent and A. Ye, 'Detection filters and algorithm fusion for ATR', *IEEE Transactions on Image Processing*, 6(1), 114–125, 1997.

50. C. Geyer, D. Dey, and S. Singh, 'Prototype Sense-and-Avoid System for UAVs', Robotics Institute, 2009.

51. P. Soille, 'Opening and closing', in *Morphological Image Analysis: Principles and Applications*, Springer, Berlin, pp. 105–137, 2003.

52. R. J. Elliott, L. Aggoun, and J. B. Moore, in B. Rozovskii and G. Grimmett (eds), *Hidden Markov Models: Estimation and Control*, Springer, Berlin, 1995.

53. Y. Bar-Shalom, X.-R. Li, and T. Kirubarajan, *Estimation with Applications to Tracking and Navigation: Theory, Algorithms and Software*, John Wiley & Sons, New York, 2001.

54. Y. Bar-Shalom and T. E. Fortmann, *Tracking and Data Association*, Academic Press, Boston, 1988.

55. A. C. Sanderson and L. E. Weiss, 'Adaptive visual servo control of robots', in A. Pugh (ed.), *Robot Vision*, IFS Publications, pp. 107–116, 1983.

56. M. W. Spong, S. Hutchinson, and M. Vidyasagar, *Robot Modeling and Control*, John Wiley & Sons, Hoboken, NJ, 2006.

57. F. Chaumette and S. Hutchinson, 'Visual servoing and visual tracking', in B. Siciliano and O. Khatib (eds), *Handbook of Robotics*, Springer, Berlin, pp. 563–582, 2008.

58. B. Young, Silvertone UAV [Online], http://www.silvertoneuav.com, March 2011.

59. D. Greer, R. Mudford, D. Dusha, and R. Walker, 'Airborne systems laboratory for automation research', in Proceedings of the International Congress of the Aeronautical Sciences, Nice, pp. 1–9, 2010.

60. Stratasys, 'Dimension BST 768/SST 768 User Guide', 2006.

61. NVIDIA, 'NVIDIA CUDA C Programming Guide Version 3.2', 2010.

62. Federal Aviation Administration, 'FAA Advisory Circular: Pilots' role in collision avoidance', 1983.

11

The Use of Low-Cost Mobile Radar Systems for Small UAS Sense and Avoid

Michael Wilson
Boeing Research & Technology-Australia

11.1 Introduction

The overarching goal of the unmanned aircraft systems (UAS)[1] community is to achieve routine, regular and safe integration of UAS into the national airspace system (NAS). Access to airspace, however, is often restricted to small geographic areas and the time required for gaining approval for this access can often be measured in months. Thus, for time-critical applications such as bushfire monitoring, the current situation is unacceptable.

UAS have demonstrated their ability to fly, navigate and perform useful missions. The challenge is to provide UAS with a capability that replicates a human function: the ability to *see and avoid*. The inability to meet this challenge is holding back the integration of UAS into the NAS.

In order to realise the benefits of UAS for commercial and civilian applications, it is necessary to achieve a greater degree of operational freedom within the NAS. To gain this

---

[1] The International Civil Aviation Organisation (ICAO) has adopted UAS instead of unmanned aerial vehicle (UAV) [59].

Sense and Avoid in UAS: Research and Applications, First Edition. Edited by Plamen Angelov.
© 2012 John Wiley & Sons, Ltd. Published 2012 by John Wiley & Sons, Ltd.

freedom, however, there is an overarching requirement for UAS to have a level of safety that is at least as good as manned aviation. Thus, until *sense and avoid* (SAA) for UAS reaches an equivalent capability to that of *see and avoid*, the operation of UAS within the NAS will continue to be restricted.

This chapter focuses on the use of low-cost mobile radar systems for small UAS sense and avoid. These systems are relatively low cost when compared with air traffic control radars or military radar systems. One aim of these systems is to support UAS operations at any required location; therefore, the system must be mobile.

Small unmanned aircraft (UA) are highlighted because of their limitations in carrying additional sensors and systems. There is currently no standard definition of a small UA. One recommendation is for small UA to have takeoff weights that are less than 25 kg [1]. Another definition limits the takeoff weight of small UA to less than 150 kg [2]. Many UA within this weight range have already proven their value. The ScanEagle, for example, achieved 500,000 operational hours during June 2011. On-board SAA solutions for small UAS, however, are difficult to implement due to restrictions on the space, weight and power (SWAP) available on-board the UA. One alternative solution is to use off-board sensors and systems to perform the SAA function.

The Mobile Aircraft Tracking System (MATS) is a mobile, network-enabled air traffic surveillance system. The aim of the MATS is to support the operation of UAS in non-segregated civilian airspace. The MATS does this by detecting the other airspace users with a ground-based primary surveillance radar (PSR). Aircraft fitted with Automatic Dependent Surveillance – Broadcast (ADS-B) may also be independently detected by a co-located ADS-B receiver.

The main function of the MATS is to provide information to the UAS pilot, enabling a ground-based sense and avoid (GBSAA) capability where the pilot is in the control loop. In this case the MATS provides the 'sense' function and the UAS pilot provides the 'avoid' function by manoeuvring the UA. The MATS also enables an automated SAA capability for UAS operations. In this case, the MATS acts as an aircraft sensor that forms part of a larger aircraft tracking and control network.

The MATS system has been undergoing initial demonstrations and characterisation trials as part of the Smart Skies project [3]. Smart Skies explored the development of technologies that support the greater utilisation of the NAS by both manned and unmanned aircraft. One important feature of the Smart Skies flight trials was the use of an aircraft that accurately logged its position and attitude during experiments. This aircraft was also a valuable calibration target for the MATS.

This chapter begins with an overview of the UAS operating environment, a review of SAA and GBSAA systems and an overview of the Smart Skies project. Smart Skies provided the opportunity to flight test a number of radar-based SAA scenarios and the results of this testing are presented. The tracking of two aircraft that simulated a midair collision, during computer vision SAA experiments, is also included. Examples of the radar tracking a variety of manned and unmanned aircraft are also shown. Weather has always been important to aviation, and an example of the MATS providing information about the location and movement of storms is also provided.

The ultimate aim for UAS operations is to 'file and fly': file a flight plan and fly – all within the same day. GBSAA systems enable a pathway for UAS operations in the NAS and these systems are available now.

11.2   The UAS Operating Environment

11.2.1   Why Use a UAS?

In general, missions that are 'dull, dirty or dangerous' are thought to be well suited to the application of UAS [4]. A wide variety of homeland security missions, civilian government applications, scientific missions and commercial applications have been identified [5].

Unmanned aircraft have a long history of being used by the military in segregated airspace. Segregating or isolating the UA from unexpected air traffic greatly simplifies the problem of see and avoid. Segregated airspace, of course, does not provide the freedom of movement that is required for many civilian and commercial UAS applications.

11.2.2   Airspace and Radio Carriage

Many countries have adopted the International Civil Aviation Organisation (ICAO) airspace classes [6]. Each country then selects the airspace classes that are appropriate for their needs. The airspace classes can be either *controlled* or *non-controlled*.

Air traffic control (ATC) services are provided in controlled airspace. Class A airspace, for example, is high-level en-route airspace that is typically used by high-performance jets and turbo-prop aircraft. Separation services are provided by ATC and there are no speed restrictions. Visual flight rules (VFR) flights are not permitted within Class A airspace.

In non-controlled airspace both instrument flight rules (IFR) flights and VFR flights are permitted. Speeds are generally restricted to 250 knots below 10,000 feet (ft) above mean sea level (AMSL).[2] Only IFR flights are required to have continuous two-way communication with ATC.

In Australia, Class G airspace is used wherever the airspace is not designated as being one of the controlled airspace classes. As such, Class G airspace accounts for the majority of Australian airspace and often covers the region from the surface of the earth to the base of the overlying controlled airspace layer. Radio carriage is only required for VFR flights above 5000 ft AMSL and at aerodromes where the carriage and use of radio is required.

The main problem of operating a UAS in Class G airspace is that the location and intent of the other airspace users are unknown. Traditionally, the final defence against a midair collision in Class G airspace is see and avoid, where the flight crew of each aircraft maintains a vigilance so as to see and then avoid other aircraft.

11.2.3   See and Avoid

Aviation regulations require pilots to see and avoid aircraft and other objects while flying in visual meteorological conditions. While see and avoid prevents many collisions, the principle is far from reliable. Numerous limitations, including those of the human visual system, the demands of cockpit tasks and various physical and environmental conditions, combine to make see and avoid an uncertain method of traffic separation [7, 8].

---

[2] Speed limitations are not applicable to military aircraft.

How good are pilots at seeing other aircraft? A series of flight tests to measure pilot air-to-air visual acquisition performance for *unalerted* visual searches has been conducted [9]. Visual acquisition was achieved in only 56% of encounters, with a median acquisition range of 0.99 nautical miles (NM).

Another experiment also recorded the time when pilots were first able to detect approaching aircraft [10]. Even though the pilots were *alerted* to the threat, and they knew where to look for the oncoming traffic, they typically could not see the aircraft until it was within a range of 1 to 1.5 NM. The high closing speeds of some of the scenarios meant that the pilots thought they would have been very challenged to avoid the collision.

The unalerted air traffic detection ranges of inexperienced pilots have also been investigated [11]. Each subject pilot was presented with two aircraft conflicts. The first conflict was set up with the target aircraft crossing the flight path of the test aircraft at a 90° angle. The average detection range was found to be 1.3 NM. The second conflict involved the traffic aircraft approaching the test aircraft from straight ahead. The average detection range for this scenario was 0.9 NM. The head-on conflict is thought to be the most difficult scenario in which to detect traffic. This is because of the illusion that the traffic aircraft is not moving. The problem is compounded by empty field myopia where, in the absence of visual cues, the eyes focus at a resting distance of around half a metre [7].

An Advisory Circular has noted that it takes 12.5 seconds for a pilot to recognise and react to a potential collision threat [12]. For a 200 knot closing speed, which is at the lower end of the possible speeds in Class G airspace, this represents a distance of 0.7 NM, which is only slightly less than the average detection range of pilots.

Another study investigated the visual scanning patterns used by pilots [13]. Unfortunately, these patterns did not resemble the prescribed systematic scanning pattern. It was also found that the pilots participating in the study spent more time looking inside the cockpit than outside, and that the average scanning performance of the participants would make them vulnerable to not detecting aircraft conflicts quickly enough to avoid a collision. An important conclusion from the study was that the relatively low rate of midair collisions, in general aviation, is as much a result of the 'big sky' as it is of effective visual scanning.

11.2.4   Midair Collisions

It is worth understanding the locations and causes of midair collisions between manned aircraft. One of the primary fears of operating an unmanned aircraft is a collision with a manned aircraft. This fear has driven the requirement for unmanned aircraft to have a SAA capability that is at least as good as the see and avoid function performed by pilots.

Midair collisions involving large commercial aircraft are now extremely rare. This is thought to be due to the on-board Traffic-alert and Collision Avoidance System (TCAS) [14] and the requirement for aircraft to be equipped with transponders in busy airspace [15].

General aviation (GA) refers to a range of civilian aviation activities and businesses, primarily using smaller aircraft and secondary airports. The term 'general aviation' is sometimes used to describe all civilian aviation activities other than those involved in scheduled public air transport services.

The characteristics of GA midair collisions in the United States, from 1983 through August 2000, have been investigated [15]. The average risk of a midair collision was found to be one per 800,000 flight hours. Accident reports show that approximately 88% of pilots involved in midair collisions do not see the other aircraft in time to initiate evasive manoeuvres. Another important point is that the failure to see and avoid other aircraft is not strongly correlated with closing speeds. Most midair collisions involve relatively low closing speeds, as one aircraft usually strikes the other from the rear, from above, or from a quartering angle.

Traffic density is also known to be a major factor in midair collisions. The typical midair collision occurs at low altitude on approach and landing or, less frequently, on take-off and climb-out. Thus, most midair collisions occur near airports, especially non-towered aerodromes.

A review of midair collisions that involved GA aircraft in Australia, between 1961 and 2003, has also been conducted [16]. Midair collisions account for approximately 3% of the fatal accidents involving general aviation aircraft, and 0.4% of all accidents involving general aviation aircraft.

The study found that 78% of the midair collisions have occurred in or near the circuit (traffic pattern) area. This reflects the higher traffic density in this area. A high proportion of the collisions occurred on the final approach or on the base-to-final turn. Most of the collisions involved one aircraft colliding with another from behind, or both aircraft converging from a similar direction.

The study also found that, in general, the characteristics and contributing factors of midair collisions in Australia appear to be similar to those observed in other countries such as the United States, France and Canada.

A 2005 review noted the failure to see and avoid, inadequate visual lookout, or a failure to maintain visual and physical clearance as the probable causes in 94% of midair collisions [8].

A recent study has highlighted that midair collisions result from a combination of proximity risk and see and avoid failure [17]. The study concluded that to perform as well as a human pilot, and thus comply with the regulations, a SAA system only needs to have a failure rate on the order of 10^{-2} to 10^{-3} per flight hour. Thus, the SAA system does not need to be a flight critical system, which requires a failure rate of less than 10^{-7} per flight hour. The required failure rate will have a direct influence on the final cost of any SAA system.

11.2.5   Summary

Initial UAS operations are likely to occur away from populous areas. This means that a typical environment could be a non-towered aerodrome in Class G airspace. The operating environment could also include a variety of aircraft performing a variety of operations, which include training, currency flights and joy flights.

What are the implications for operating a UAS? The environment may contain a variety of aircraft and only some of these aircraft will carry transponders or provide VHF radio reports. The risk of a midair collision is likely to be higher when operating from a non-towered aerodrome. The field of view of any SAA system will need to be carefully considered because midair collisions do not tend to be head-on.

A number of studies have shown the limitations of pilots performing see and avoid. The good news is that the SAA technology developed for unmanned aircraft may also benefit the manned aviation community.

11.3   Sense and Avoid and Collision Avoidance

11.3.1   A Layered Approach to Avoiding Collisions

A layered approach is used to avoid collisions between aircraft in civilian airspace [18]. The idea is that failures would need to occur at multiple layers to cause a system failure that results in a collision. At the strategic level there is airspace structure, procedures and equipment to manage the airspace.

At a more tactical level, separation management is provided by air traffic control and VHF-radio location and intention reports. The aim of this level is to keep aircraft separated by at least a prescribed minimum distance and, in general, to avoid dangerous situations. The goal of this layer has been stated as 'don't scare others' [19]. The intent is to manage the airspace by small deviations from the desired flight plan. Remaining 'well clear' from other aircraft is the primary aim of the self-separation function of SAA [20].

The collision avoidance layer is activated when self-separation has failed. The aim of this level is to escape dangerous situations. This may involve last-minute and potentially large changes from the desired flight plan. The goal of this layer has been stated as 'don't scrape paint' [19]. This collision avoidance function of SAA involves the last-minute manoeuvring to avoid a collision [20].

Intruding aircraft need to be acquired in time to perform the SAA subfunctions: Detect, Track, Evaluate, Prioritise, Declare Threat, Determine Action, Command, and Execute [20, 21]. There can be some consideration of trading sensor acquisition range for the severity of the avoidance manoeuvre. Separation management needs the intruder acquisition to occur at greater ranges to achieve its goal using more gradual evasive manoeuvres. Intruder detection may occur at shorter ranges for collision avoidance at the cost of prompt and vigorous evasive manoeuvres.

The various SAA options may be considered in terms of a detection technology trade space [22]. This trade space may be divided into active or passive SAA systems and then into cooperative or non-cooperative intruder aircraft.

Active systems transmit a signal to receive information about other aircraft. *Passive* systems do not transmit a signal but use sensor measurements to detect the other aircraft. *Cooperative* aircraft have an electronic means of identification on-board that is operating (e.g., a transponder) [23]. Thus, to be cooperative, an aircraft is required to carry specific avionics and have this equipment operating normally. The cooperative SAA solution is ideal if *every* aircraft is fitted with this technology, although this is not the case today. *Non-cooperative* aircraft do not have an electronic means of identification on-board or the required equipment is not operational due to a malfunction or deliberate action [23].

A recent review recommended that the Federal Aviation Administration (FAA) mandate that all presently non-cooperative aircraft be installed with and operate a short-range, low-power and lightweight electronic means of identification [24]. Another study, using 2007 estimates, noted that it would cost approximately US$58 million to equip the remaining United States fleet of aircraft with ADS-B OUT (transmit only) [25].

11.3.2   SAA Technologies

A literature review of the cooperative and non-cooperative SAA technologies for UAS has been conducted [26]. The review presented a summary of the various technologies but concluded that one single approach may not be adequate for the SAA requirement on a UAS.

A second review also concluded that no single on-board sensor is sufficient to satisfy all the SAA requirements for all UAS [24]. One recommendation was to mandate that all presently non-cooperative aircraft should be installed with and operate a short-range, low-power and lightweight electronic means of identification; that is, all aircraft should be made coopera-tive. The report noted that continuing the development of on-board sensors to detect non-cooperative aircraft may cost US$2 billion and take more than ten years to achieve a solution.

The need for an on-board capability to SAA other aircraft has been noted as one of the most daunting challenges regarding the routine integration of UAS into the NAS [25]. This study provided a summary of the SAA alternatives and an estimate of the expected time scales: small UAS visual line-of-sight regulations (+2 years), GBSAA dedicated sensors (1–2 years), GBSAA repurposed sensors (2–3 years), cooperative airborne-based sense and avoid (ABSAA) (10+ years) and non-cooperative ABSAA (12+ years).

The following sections present an overview of some of the technologies used to provide UAS with a SAA capability.

11.3.2.1   ABSAA Systems

A number of on-board technological solutions to the SAA problem are being explored, including conventional radar systems [10, 27, 28], multiple-input multiple-output (MIMO) radar systems [29, 30] and passive vision-based systems [11, 31, 32].

ABSAA systems have the following advantages and disadvantages.

Advantages

- Provide surveillance for the UA throughout its mission.

- Provide a pathway for fully autonomous UA.

- May perform better than the pilots of manned aircraft.

Disadvantages

- Are still being developed.

- Are limited by the space, weight and power available on-board.

- Will reduce the UA payload carrying capability.

- The cost is per airframe.

Although an international consensus has not been reached, the ASTM Standard F2411 provides an example of the requirements for airborne SAA systems [33]. The requirements include a field of regard of ±110° in azimuth and ±15° in elevation. The nature of manned midair collisions, however, suggests that a 360° azimuth coverage may be required.

One study used a systems engineering approach to evaluate a number of on-board SAA technologies for a specific UAS [34]. In general, the most appropriate technology will depend on the airframe, the mission and the operational area.

11.3.2.2   GBSAA Systems

The traditional method of coping with non-cooperative aircraft in the NAS is by using a PSR. A PSR detects the presence of a target based on the radar signal scattered from the target.

Traditionally, PSRs tend to be located at busier airports. A busy airport, however, is not an ideal environment to start the integration of UAS into the NAS. A mobile PSR, however, can support UAS operations at any desired location.

GBSAA systems have the following advantages and disadvantages.

Advantages

- Ideal for supporting small UA operations.
- The technology is available now.
- Do not require modifications to the UA.
- The cost is per location (or system), not per airframe.

Disadvantages

- Provide a static surveillance volume, which may be smaller than the maximum operating range of the UA.
- Local terrain may reduce coverage.
- May detect false targets such as ground vehicles, weather, birds, etc.
- SAA-related manoeuvres depend on a data link to control the UA.

A study of the surveillance technologies that are currently used within the air traffic management system listed the strengths and weaknesses of primary and secondary radar, ADS-B, multilateration systems and ADS-C [35]. Primary radars remain as the only technology in the air traffic management system capable of detecting non-cooperative aircraft.

11.3.2.3   Example Ground-Based Sense and Avoid Systems

Table 11.1 provides an overview of three ground-based radar systems. The Sense-and-Avoid Display System (SAVDS) uses the Sentinel AN/MPQ-64 air defence radar to support UAS operations.[3] The Star 2000 is a modern air traffic control radar.[4]

The instrumented range is shown for each radar system. Ideally, detection ranges for a standard radar cross section (RCS) target would be shown, but this would require detailed information about each system, which is beyond the scope of this study [36]. The RCS is a measure of how detectable an object is with a radar. In general, the RCS of an object is a complex function of the structure and constituent materials of the object, the radar frequency, the radar configuration (monostatic or bistatic) and the aspect angle of the object with respect to the radar. Some radar manufacturers provide theoretical radar coverage information for a standard RCS target. In general, targets that have a larger RCS can be detected at longer ranges than targets with a smaller RCS.

Most ATC radars today are two-dimensional (2-D) systems, which provide range and azimuth information but not elevation information and, therefore, not altitude information [35]. The MATS is also a 2-D system, although a three-dimensional (3-D) height-finding system

[3] www.thalesraytheon.com
[4] www.thalesatm.com

Table 11.1   Overview of three ground-based radar systems

Specification	MATS	AN/MPQ-64 Sentinel (SAVDS)	Thales Star 2000
		Radar	
Frequency band	9.41 GHz X-band	9.37–9.99 GHz X-band	2.7–2.9 GHz S-band
Peak power	25 kW	23 kW	28 kW
Instrumented range	54 NM	40 NM	100 NM
Range resolution	180 m	150 m	230 m
Elevation information	No (2-D)	Yes (3-D)	No (2-D)
Elevation beam width	20°	2° (Coverage = 20°)	30°
Azimuth beam width	0.95°	1.8°	1.4°
Rotation rate	24 rpm	30 rpm	15 rpm
Update rate	2.5 s	2 s	4 s
Portable	Yes	Yes	No

is being developed [37]. Some military radar systems, like the one detailed in Table 11.1, do provide elevation and altitude information.

Altitude information is an innate part of aviation, so is a 3-D radar required? The factors to consider include the frequency and nature of the UA operations, the complexity of the airspace and the density of the air traffic in the area of operation. In remote areas, where the traffic density is low, a 2-D radar may be sufficient, especially if cooperative systems are able to provide accurate altitude information to supplement the radar data.

ATC radar systems are engineered for high levels of availability and, as such, tend to be expensive. A cost guide for terminal manoeuvring area (TMA) radars is AUD$8 million [35]. High-technology military systems are also known to be expensive. The MATS provides a relatively low-cost solution to supporting UAS operations. Low cost is particularly important for commercial UAS applications because of the competition with manned aircraft solutions. McGeer provides a sobering view on UAS economics and notes 'On price, robotic aircraft have a long way to go' [38].

11.3.3   The UA Operating Volume

For each GBSAA system a number of airspace volumes may be defined, as shown in Figure 11.1. There are a number of similarities to the volumes defined for airborne SAA [20].

The surveillance volume describes the effective limits of the surveillance system. Aircraft that have a low RCS, particularly small aircraft, may not be detected between the surveillance volume and the detection and tracking volume. Aircraft that have a large RCS, however, may be detected and tracked in this region.

The detection and tracking volume aims to provide a minimum level of detection and tracking performance for an aircraft with a nominated minimum RCS. One example of a

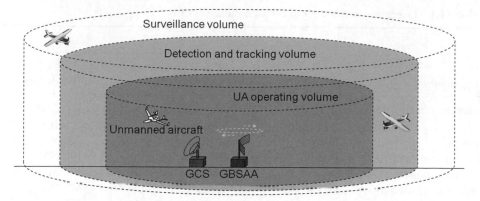

Figure 11.1   The airspace volumes associated with operating a GBSAA system. The GBSAA system may be located separately from the ground control station (GCS)

boundary specification is where the system detects a 2 m^2 Swerling type 1 target with a probability of detection of 80% and a false alarm rate of 10^{-6} [36].

The UA operating volume is the volume of airspace where the UA conducts operations. This volume is smaller than the detection and tracking volume in order to provide a temporal and spatial buffer between any intruder aircraft and the operational UA. An *intruder* has been defined as an aircraft that is within the surveillance volume [20].

The performance characteristics of a radar system will set the dimensions of each airspace volume. One important variable to consider is the speed range of the intruding aircraft. The speed of the aircraft and the range to the UA sets the warning time for the SAA timeline [20]. Thus, greater radar tracking ranges will be required for operating environments that contain high-speed aircraft.

11.3.4   Situation Awareness

One of the most widely used formal definitions of situation awareness is '... the perception of the elements in the environment within a volume of time and space, the comprehension of their meaning and the projection of their status in the near future' [39]. Put another way, this means '... being aware of what is happening around you and understanding what that information means to you now and in the future' [40]. These definitions highlight that a person cannot be given situation awareness. GBSAA systems aim to provide information to UAS pilots and, as a result, enhance the pilot's situation awareness.

Maintaining a high level of situation awareness is essential for effective decision-making. Effective decision-making is a key ingredient of SAA, especially while UAS pilots remain in direct control of the UA.

11.3.5   Summary

A wide range of SAA technologies has been developed and each has its own strengths and weaknesses. The diversity of UAS sizes and missions means that a 'one size fits all' solution is unlikely.

Small UAS, in particular, have a limited capacity to carry additional SAA avionics. Visual line-of-sight observers have been proposed to provide a collision avoidance capability [23]. Operating with observers is quite restrictive. The observers are to be positioned no greater than one nautical mile laterally and 3000 ft vertically from the UA. Another option, which is available today, is to use GBSAA systems.

A layered approach is used to avoid collisions within the NAS. GBSAA systems are able to provide an additional layer of surveillance, a layer which can assist the integration of UAS into the NAS.

GBSAA systems form part of the strategy for the military to gain access to the NAS [41]. GBSAA systems can also provide a pathway for civilian and commercial applications of UAS.

11.4   Case Study: The Smart Skies Project

11.4.1   Introduction

One goal of the UAS community is to achieve routine, regular and safe integration of UAS into the NAS. To achieve this goal there is a need to research, develop and flight test the technologies that will facilitate this integration. To meet this goal a cooperative research project was initiated: the Smart Skies project.

The aim of the Smart Skies project was to develop technologies that facilitate the greater utilisation of the national airspace system by both manned and unmanned aircraft [3]. These technologies include:

- an automated separation management system capable of providing separation assurance in complex airspace environments;

- sense and avoid systems capable of collision avoidance with dynamic and static obstacles; and

- a MATS that utilises a cost-effective primary radar and cooperative surveillance systems.

An important objective of Smart Skies was to integrate, demonstrate and validate the performance of these technologies through a series of eight integrated flight-test activities. The aim of these flight tests was to characterise the performance of the developed technologies under realistic operating conditions.

An overview of the Smart Skies project, including the MATS, is provided in the following sections.

11.4.2   Smart Skies Architecture

The Smart Skies system architecture diagram is shown in Figure 11.2. The system includes manned and unmanned aircraft, virtual aircraft, public mobile data and Iridium communication links, an Automated Dynamic Airspace Controller (ADAC) and the MATS.

This architecture enables a diverse range of experiments, which extend from the testing and characterisation of individual technologies to automated SAA experiments that involve a system-of-systems.

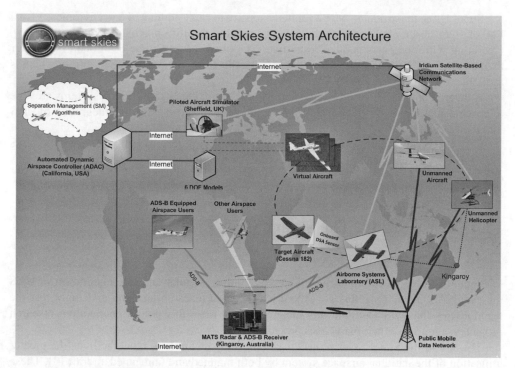

Figure 11.2   The Smart Skies system architecture

An important point to note about the architecture is that the manned and unmanned aircraft flight tests occurred near Kingaroy in Australia, but the SAA control system, the ADAC, was located in California in the United States.

All of the systems within the Smart Skies architecture are linked to the Internet by the Iridium satellite communication network and a third-generation (3G) public mobile data network. The Iridium satellite communication system provided global coverage but low bandwidth communication [42, 43]. The 3G public mobile data network provided higher bandwidth links but with a relatively limited geographic coverage [44]. The Smart Skies architecture is not limited to using these two communication links. The aim was to demonstrate a variety of Smart Skies concepts using data links that had different coverage and bandwidth capabilities.

The Smart Skies architecture enables a mixture of centralised and decentralised automated separation management. In the centralised mode of operation the ADAC provides separation assurance to aircraft. In the decentralised mode an aircraft can use the information from the local sensors and systems on-board to remain well clear of other aircraft.

Figure 11.2 shows two methods for simulating aircraft: (a) six degree of freedom (6 DOF) models and (b) the Sheffield engineering simulator [45]. A variety of aircraft could be simulated with the models. These virtual aircraft were important for testing the Smart Skies architecture and also provided a facility to load the architecture with aircraft.

An overview of the MATS, ADAC, Airborne Systems Laboratory (ASL) and Flamingo UAS are found in the following sections. The Smart Skies overview paper provides further information about the autonomous unmanned helicopter and other aspects of the system [3].

11.4.3   The Mobile Aircraft Tracking System

11.4.3.1   Mission

The primary mission of the MATS is to detect and track aircraft that may intrude into the operational area of the UAS. Detecting the operational UAS is secondary as the ground control station (GCS) often tracks the unmanned aircraft via a telemetry link.

The MATS has, however, tracked a number of unmanned aircraft. While not its primary mission, the MATS can provide an independent source of information about the location of an unmanned aircraft. This independence means that the MATS could be used as a secondary navigation system. This capability may be valuable if the UAS navigation system fails or if navigation using the global positioning system (GPS) fails for any reason.

The main function of the MATS is to provide information to the UAS pilot located in the GCS. Due to the differing operating requirements of each system, the MATS may be positioned in a different location to the GCS, as shown schematically in Figure 11.1. In this configuration the MATS provides its information about the local airspace users via a network link.

11.4.3.2   Architecture

Figure 11.3 shows the current architecture of the MATS, which consists of the following subsystems:

- a primary surveillance radar system;

- an ADS-B receiver;

- a VHF voice transceiver; and

- a server that performs data fusion and communications management.

The UAS flight crew may be located inside the MATS or inside a remote GCS.

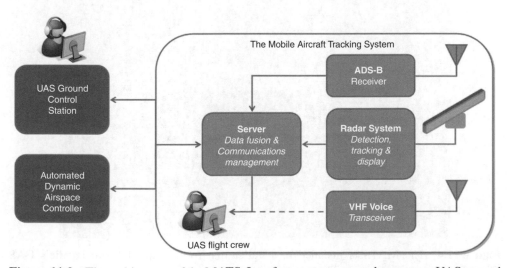

Figure 11.3   The architecture of the MATS. Interfaces to two external systems, a UAS ground control station and an ADAC, are also shown

Figure 11.3 also shows interfaces to two external systems. The two interfaces provide the means for the MATS data to be used by external systems. These interfaces also enable two different methods of using the MATS for SAA.

One interface enables the MATS to provide information to the UAS pilot. The radar's TCP/IP data networking capability allow tracks and plots to be sent to a remote TrackViewer Workstation (TVW) [46], which may be located in the GCS. In this case the radar and the UAS pilot form part of a pilot-in-the-loop SAA system, where the MATS provides the 'sense' function and the UAS pilot provides the 'avoid' function by manoeuvring the UA.

The second interface is used by the MATS to provide information about cooperative and non-cooperative aircraft to external systems, such as the ADAC [47]. In this case the SAA system may be automated, where the MATS provides the 'sense' function but the 'avoid' function is automated. The automation involves the ADAC assessing the airspace situation and then providing updated flight plans to the cooperative aircraft in order to avoid any conflicts.

11.4.3.3   The MATS Radar System

The MATS is installed in a trailer that forms part of the UAS Flight Demonstration System operated by Insitu Pacific Limited (IPL), as shown in Figure 11.4.

A key part of the MATS is the primary surveillance radar system. The radar consists of a commercial off-the-shelf marine radar 'front end' and a 'back end' that performs the detection, tracking and display functions.

The 'front end' of the radar is a non-coherent marine radar: a Furuno FAR-2127-BB. A summary of the key characteristics of the Furuno radar is shown in Table 11.2. The radar uses an eight-foot-long slotted waveguide array antenna. This standard Furuno antenna generates a vertical fan antenna pattern and, as a result, no elevation information is available.

Figure 11.4   The MATS is installed in a trailer that forms part of Insitu Pacific's UAS Flight Demonstration System. The marine radar is shown mounted on a mast. The Accipiter® detection and tracking system is located inside the trailer

Table 11.2 Key characteristics of the Furuno FAR-2127 radar

Frequency	9.410 GHz (X-band)
Output power	25 kW
Pulse length, range resolution, PRF[†], Rmax[‡]	(1) 0.07 µs, 10.5 m, 3000 Hz, 27 NM
	(2) 0.3 µs, 45 m, 1500 Hz, 54 NM
	(3) 1.2 µs, 180 m, 600 Hz, 135 NM
Antenna rotation rate	24 rpm
Beam width (horizontal)	0.95°
Beam width (vertical)	20° (±10°)

[†]PRF – pulse repetition frequency
[‡]Rmax – maximum unambiguous range

The three selectable pulse lengths of the radar provide three different range resolution options. In general, the short pulse-length option is used for short-range applications to provide 10.5 m range resolution. The long pulse-length option provides 180 m range resolution and is typically used for long-range applications, such as intruder detection and tracking. The medium pulse length provides a compromise option.

The Accipiter® detection and tracking system forms the 'back end' or 'brain' of the MATS radar. The radar's performance has been enhanced by replacing the standard marine radar processing with a powerful, software-definable radar processor and tracker [48].

Accipiter's multi-target tracker is designed to manage many dynamic and manoeuvring targets. The system employs a multiple-hypothesis-testing (MHT) interacting-multiple-models (IMM) tracker that enables the system to detect and track manoeuvring targets that have a low radar cross section.

The radar operator is able to set the parameters for the detection and tracking algorithms, allowing the operator to optimise the radar's settings for specific surveillance scenarios.

The MATS provides a variety of display options. The detections from each radar scan may be displayed; these are called plots. Confirmed tracks, with estimated speeds and headings, are usually displayed. The background clutter level may also be selected for display. All of this radar information can be displayed with background maps to provide a geographic context.

11.4.3.4   The MATS ADS-B Receiver

The ADS-B receiver currently used by the MATS is an SBS-1 from Kinetic Avionic Products Limited.[5] The SBS-1 is a portable and low-cost 1090 MHz ADS-B receiver [49]. The SBS-1 provides the capability to track and log information about ADS-B equipped aircraft. This information includes the latitude, longitude, altitude, speed, heading and identity.

One advantage of ADS-B is that the aircraft information is transmitted to the receiver. The accuracy of the information is not imposed by the receiver. This must be contrasted with radar systems where the information about the aircraft is measured and, therefore, the accuracy of these measurements is range-dependent.

---

[5] http://www.kinetic-avionics.co.uk

Figure 11.5   The ASL is a Cessna 172R

AirServices Australia provides an overview of ADS-B and the Australian ADS-B network.[6]

11.4.4   The Airborne Systems Laboratory

Figure 11.5 shows the ASL, which is a Cessna 172R. The ASL has been equipped with a Novatel SPAN integrated GPS–INS navigation system to provide real-time 'truth' data about the aircraft's state. This data includes the aircraft's three-dimensional position, velocity and attitude [50].

A certified roll-steering converter was fitted to the ASL to provide a digital interface to the existing Honeywell KAP140 autopilot. This interface allowed the aircraft's flight management system to command the aircraft's autopilot directly. This capability allows flight plans to be followed autonomously during the cruise phases of flight.

The ASL is also fitted with an Iridium satellite transceiver and a 3G public mobile data network modem. These two communication systems provide the main communication links to the Smart Skies network.

The ASL is fitted with a 1090 MHz extended squitter (ES) ADS-B transmission system. This system provided an excellent means of independently monitoring the ASL during flight trials.

From the MATS point of view the important features of the ASL are its ability to follow predetermined flight plans and its ability to provide accurate information about the aircraft's position and attitude.

---

[6] http://www.airservicesaustralia.com/projects/ads-b/

Figure 11.6   The Flamingo fixed-wing UAS

The ASL is a typical GA aircraft, which makes it ideal for radar characterisation studies. The ASL also represents a typical 'intruder' aircraft. Thus, the ASL is ideally suited to demonstrating how the MATS can support UAS operations in Class G airspace.

11.4.5   The Flamingo UAS

The primary fixed-wing UAS used during the Smart Skies flight trials was based on the Flamingo radio control airframe [3], which is shown in Figure 11.6. The Flamingo system was developed to provide an autonomous unmanned aircraft that could fly beyond visual range. The Flamingo has a maximum takeoff weight of approximately 20 kg, a 4 m wingspan and an endurance of approximately 1 hour.

The avionics payload of the Flamingo included a MicroPilot® MP2128g autopilot, Microhard radio modems, an inertial measurement unit (IMU) and a customised PC104 mission computer. The Flamingo achieved over 70 hours of autonomous flight during the Smart Skies project.

The Flamingo has also served as the image acquisition platform during vision-based collision-detection experiments [32].

11.4.6   Automated Dynamic Airspace Controller

The ADAC provides a capability for automated air traffic control [47]. The ADAC uses existing communication infrastructure, currently the Iridium and 3G public mobile data networks, to control both manned and unmanned aircraft.

The ADAC exchanges custom messages with a data-linked 'predictive' flight management system (pFMS), which is on board each manned and unmanned Smart Skies aircraft [47]. These cooperative aircraft periodically send their current and future state information, such as time, position, velocity and attitude, to the ADAC. Each aircraft may also send a message to the ADAC that contains its desired flight plan. The predictive component of the pFMS aims to take into account communication and other latencies so that the ADAC can make latency-compensated decisions.

The ADAC uses the aircraft data to estimate the current and future airspace situation. If the ADAC identifies that a conflict between aircraft may occur, it is able to send a recommended trajectory modification to each aircraft, in the form of a short-term flight plan, to maintain adequate separation between all aircraft. After an aircraft completes the avoidance manoeuvre it returns to its original flight plan.

The ADAC allows the control of cooperative manned and unmanned platforms from any location around the globe that has access to the Internet. During the Smart Skies flight trials this meant that the aircraft flying in Australia were controlled from the United States [47, 51, 52]. A recent addition to the ADAC is a capability to utilise the information about non-cooperative aircraft that is provided by the MATS.

11.4.7   Summary

This section provided an overview of the elements of the Smart Skies architecture. This architecture provides a comprehensive and flexible system for testing a wide variety of manned and unmanned technologies and concepts.

The MATS used the Smart Skies infrastructure for a variety of ground-based SAA experiments, which are discussed in the following sections.

11.5   Case Study: Flight Test Results

The Smart Skies project provided the opportunity to test the MATS as a ground-based SAA system. The following sections discuss the characterisation of the performance of the MATS, its use in SAA experiments and the radar's ability to detect a variety of aircraft.

11.5.1   Radar Characterisation Experiments

11.5.1.1   Introduction

The aim of the initial series of flight trials was to characterise the performance of the MATS using the ASL. For these tests the ASL was provided with a variety of flight plans to test different aspects of the radar's performance.

There are a large number of variables when testing the detection performance of a radar system. A target's maximum detection range is one of the key metrics when assessing the performance of a radar. The detection range depends on the target's RCS, which is a measure of how detectable an object is with a radar. A target's RCS is a complex function that varies with the radar's 'view' of a target. For example, an aircraft flying directly towards the radar will present a different RCS from an aircraft flying directly away from the radar. Thus, the RCS that a target presents to the radar has a direct influence on the maximum range where the target can be detected. The ASL is a typical GA aircraft. This makes the ASL ideal for testing the radar's detection performance because it has a RCS that is typical of an intruder aircraft.

The radar's performance is also influenced by the local environment. Terrain and structures such as buildings can have an impact on detection performance because targets are detected against the associated clutter from the local environment. In these situations target detection depends on the signal-to-clutter ratio and not simply the signal-to-noise ratio.

Some of the initial flight trials were aimed at understanding the performance of the MATS primary radar, including the influence of the background clutter on detection performance. This understanding of the radar's performance was important for the subsequent SAA experiments.

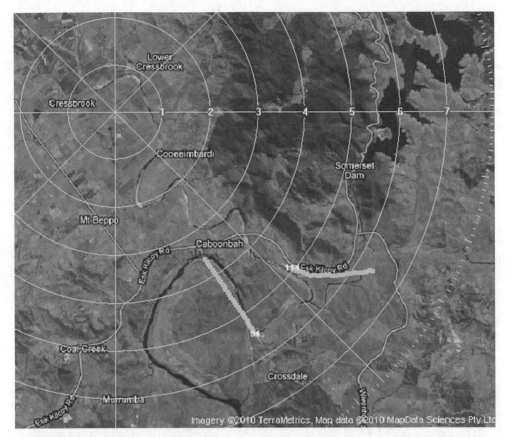

Figure 11.7   Two MATS radar tracks from targets of opportunity in the vicinity of the Watts Bridge Memorial Airfield. Range rings, in one nautical mile increments from Watts Bridge, are also shown. The aircraft speed, in knots, is shown on each track

11.5.1.2   The Operating Environment – Watts Bridge

A number of the MATS characterisation flight trials were carried out at Watts Bridge Memorial Airfield, Queensland, Australia (27° 05′ 54″S, 152° 27′ 36″E). The airfield has three grass runways: two parallel runways and one cross strip. Mount Brisbane (2244 ft) is located approximately 4 NM to the east of the airfield. Intensive skydiving can often occur at 5 NM to the north-west of the airfield.

Insitu Pacific Limited currently uses the airfield for flight training with the ScanEagle UAS.[7] Therefore, the airfield represents a realistic environment for testing the MATS.

Many different types of aircraft use the Watts Bridge airfield. These aircraft provide 'targets of opportunity' for testing the MATS radar. An example of the radar tracks from two targets of opportunity is shown in Figure 11.7. The figure shows an aircraft that had departed from Watts Bridge airfield and was travelling at 94 knots and an aircraft approaching Watts Bridge airfield at 114 knots.

---

[7] http://www.insitu.com/scaneagle

Targets of opportunity may be used to test the ability of the MATS to track a variety of aircraft. Aircraft may be tracked approaching, departing and performing circuits near the airfield, although the altitude and attitude of the targets of opportunity are unknown and these aircraft also tend to follow their own flight plans. Thus, with the large number of unknown variables involved, it is difficult to get meaningful quantitative results from targets of opportunity alone. The ASL, however, is an ideal aircraft for radar characterisation experiments. It represents a typical GA aircraft and has an accurate on-board position and attitude monitoring system.

11.5.1.3   Circular Flight Paths

For these experiments the ASL was provided with circular flight paths at a number of ranges from Watts Bridge airfield. The circular flight paths meant that the ASL flew at a fixed range from the radar and presented a constant RCS to the radar. Thus, the main variable was the background clutter from the local environment.

Figure 11.8 shows the radar tracks when the radar's long (1.2 µs) pulse was used. The figure shows the tracks from circular flight paths with radii of 2.7 NM (5 km), 3.2 NM (6 km), 4.3 NM (8 km), 6 NM, 10 NM and 14 NM. These results are from flight trials held on 6 May 2010 and 13 July 2010.

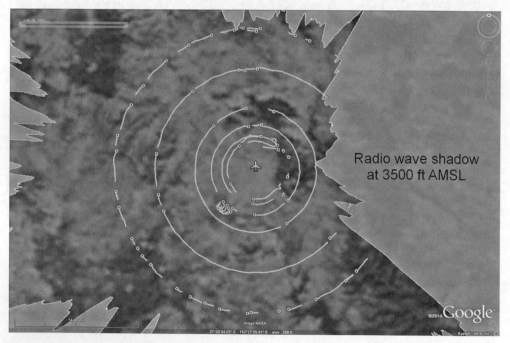

Figure 11.8 The MATS radar tracks of the ASL when it flew circular flight paths around Watts Bridge airfield. The flight paths have radii of 2.7, 3.2, 4.3, 6, 10 and 14 NM. The grey shaded regions show the radio wave shadow at 3500 ft AMSL

It should be noted that the individual tracks that make up the circular paths have been extracted and plotted in Google Earth.[8] Tracks from other aircraft and short-lived tracks have not been included. The gaps in the circular paths represent regions where the ASL was not tracked.

The radio wave shadow regions at 3500 ft AMSL are also shown in Figure 11.8. This map was generated using Global Mapper's Viewshed function.[9] NASA's Shuttle Radar Topographic Mission (SRTM) digital elevation data, which provided 90 m resolution, was used in the Viewshed analysis. The large shadow region to the east of Watts Bridge is caused by Mount Brisbane. Airspace restrictions prevented the ASL flying above this shadow region and staying within radio line-of-sight of the MATS. Thus, the ASL was not tracked in this region.

Figure 11.9 shows the background clutter environment from Watts Bridge when the radar used its long (1.2 µs) pulse. The figure shows the significant background clutter that results from Mount Brisbane at ranges of 2–4 NM from the north north-east to the south-east. The figure also shows that the terrain provides high background clutter at other locations (e.g., 5 NM to the north of the radar).

Figure 11.10 shows a map of the areas of the local terrain that are visible to the radar. This figure provides a simple model of where the ground clutter may affect target detection by the primary radar. This map was generated using Global Mapper's Viewshed function. This function highlights all areas that have a clear line-of-sight to the transmitter. An option to show only those regions that have at least a 60% first Fresnel zone clearance was selected.

It should be noted that this is a simple model and does not take into account vegetation or man-made structures. Also, at this stage, no attempt is made to predict the intensity of the backscatter at each location.

Figure 11.8 shows the radar tracks when the ASL flew a number of circular flight paths at different ranges. Whilst some of the longer-range circles are almost continuous, the figure shows that large sections of the shorter-range circular tracks are missing. The background clutter environment shown in Figure 11.9 provides an explanation. The high-clutter regions shown in Figure 11.9 produce low signal-to-clutter ratios, which hamper the detection and tracking of aircraft.

Despite the simplicity of the model, Figure 11.10 confirms that the areas of strong background clutter are produced by terrain that is within radio line-of-sight of the radar.

The high background clutter levels may be mitigated by modifying or replacing the COTS marine radar antenna. The presence of Mount Brisbane, however, highlights that the clutter may be minimised but not completely removed.

Figure 11.8 shows that the 4.3 NM and 6 NM tracks are continuous in the direction of Mount Brisbane. This is because the ASL stayed within radio line-of-sight of the radar. These areas are also located in low-clutter regions, which enhance the detection of aircraft.

Figure 11.8 also highlights the problem of trying to assign a single detection range for the radar. The ASL is almost continuously tracked at 10 NM, outside the terrain shadow region, but the tracks at 14 NM are more intermittent.

In general, aircraft that have a large RCS, such as passenger jet aircraft, will be detected and tracked at longer ranges while aircraft that have a smaller RCS, including small UAS, will only be tracked at shorter ranges.

---

[8] Available from http://earth.google.com/
[9] Available from www.globalmapper.com

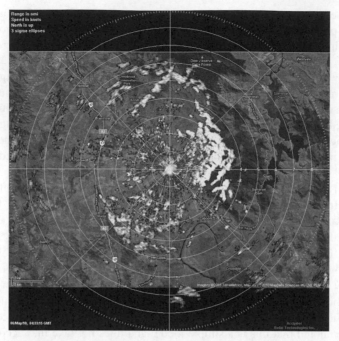

Figure 11.9 The clutter environment observed by the MATS radar. The clutter signal strength is displayed using a greyscale, where white indicates the strongest clutter. Range rings, in one nautical mile increments, are also shown

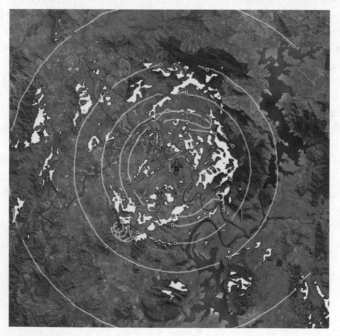

Figure 11.10 A map showing areas of the local terrain that are visible to the radar (lighter areas). This map provides a simple method of predicting where the radar will experience high background clutter. The outer circle shows the 10 NM tracks from Figure 11.8

The flight test results show that the clutter environment leads to both a range and an angular dependence on detection performance. The modelling results show that the poor detection regions are predictable, even with a simple model.

Finally, aircraft are not expected to always fly circular paths. The circular flight paths are, however, a means of keeping many of the detection-related variables constant, thus providing a method of examining the influence of the local environment on the performance of the radar.

11.5.1.4   Diamond Flight Paths

For the flowing flight tests the MATS was located on a farm at Burrandowan, Queensland, Australia (26° 27′ 36.78″S, 151° 24′ 15.66″E). The flight tests occurred during November and December 2010.

In the following experiments diamond flight plans were used to test the radar. In these examples the range and the radar–target geometry changes with time. The geometry change means that the RCS presented by the target to the radar also changes.

Figure 11.11 shows the radar tracks when the ASL flew diamond flight plans. Each side of a diamond could represent a flight path where the target is transiting the area near the radar. The diamond flight paths provide a convenient and reproducible means of sampling the large number of possible flight paths.

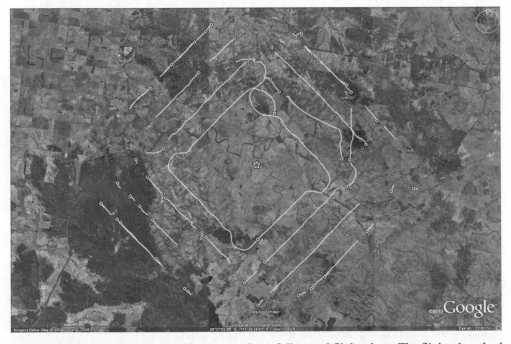

Figure 11.11   The ASL's tracks from a number of diamond flight plans. The flight plans had apices that were 4.3, 6, 8 and 10 NM, respectively, from the local grass runway. The star symbol shows the location of the MATS

The diamond flight plans had apices that were 4.3, 6, 8 and 10 NM, respectively, from the local grass runway, which was located approximately 0.6 NM to the south-west of the MATS.

For each side of the diamond flight plan the range from the MATS to the ASL initially decreases and then increases as the aircraft flies by the MATS. The RCS that the ASL presents to the radar also changes along this path, along with the background clutter. Thus, although the flight plans are relatively simple, there are a number of factors that simultaneously influence the detection performance of the radar.

The collection of tracks produced by the diamond flight plans also shows the effect of increasing range from the MATS. At the longer ranges the tracks are consistently lost near the corners of the diamond. In this region the aircraft turns and the RCS presented to the radar changes the most – from an oblique tail-on view of the aircraft to an oblique head-on view.

The ground clutter map measured at Burrandowan did not display a strong azimuth dependence, unlike the measurements at Watts Bridge. This meant the tracks were more consistent with azimuth. As a result, the main variables that influenced the tracking performance, at this location, were the range and the RCS – as expected.

11.5.1.5 Summary

This section presented the results from testing the MATS using the ASL. The ASL and its 'truth' system provided a means of quantifying the performance of the MATS because the aircraft's location and state information were independently measured. The ASL could also autonomously follow predefined flight paths and, as a result, provided a capability to gather repeatable results.

Circular flight paths were used at Watts Bridge to gain an appreciation of the influence that the environment has on detection and tracking performance. The results show that high-clutter regions can produce a momentary loss of tracking in some areas.

The high-clutter regions are predictable and many of these areas could be reproduced using a Viewshed analysis. Plans are in place to improve the radar's antenna system. The aim is to reduce the effects of ground clutter.

Tests using diamond flight paths, at a different location, showed the expected range and RCS dependence.

The flight trial results show that, for an ASL-sized aircraft, initial tracking may occur at ranges of approximated 14 NM. Consistent tracking is likely to occur at shorter ranges, approximately 6 to 10 NM. These ranges are greater than what can be expected from the visual observers that are currently required to support UAS operations.

The results also show that the UA operating volume, as shown in Figure 11.1, may be restricted to a range of 6 to 10 NM. The aim is to provide a spatial buffer between the areas where an intruder may be detected and where the UA is being operated. The speed of the intruder and the distance to the UA may be converted into a warning time, which may be used to manoeuvre the UA well clear of the intruder.

Figure 11.8 also suggests that a buffer may be required around any local terrain features that cause a radio wave shadow. Shadow regions reduce the detection range of intruding aircraft, and hence the warning time. This means that the UA operating volume may not be as symmetrical, in practice, as the one shown in Figure 11.1.

What if longer-range radar coverage is required? There are two general options. The first uses a primary radar that provides the desired range coverage. The second option uses a network of low-cost radars [46]. The logistics and costs of this option can be weighed against those of the single-radar option.

11.5.2   Sense and Avoid Experiments

11.5.2.1   Introduction

The aim of the MATS is to support UAS operations. A typical place to operate a UAS is a small non-towered aerodrome in Class G airspace. At this type of aerodrome it is usual to have a variety of aircraft approaching and departing. If these aircraft carry VHF radio then their position and intent are easy to obtain. In general, this radio-based coordination works well.

A percentage of the aircraft, however, will not carry VHF radio. Therefore, coordinating UAS activities with these aircraft will not be possible. The result is that the primary radar of the MATS is one of the only ways to track these aircraft. Non-cooperative aircraft that do not carry VHF radio represent the most challenging class of airspace user – for both manned and unmanned aircraft. This is also the scenario where the primary surveillance radar comes into its own.

There is an important question to be answered: if a non-cooperative intruder aircraft is detected by a radar then what actions should the UAS pilot take to avoid conflicts with the aircraft? Some action may be required if the intruder aircraft penetrates the UA operating volume, as shown in Figure 11.1.

The aim of this section is to explore the operation of a UAS in non-segregated airspace where intruder aircraft are also part of the environment.

11.5.2.2   Intruder Scenario

The general scenario being considered is:

- a small UA is operating near a non-towered aerodrome;

- UA operations are supported by a primary radar whose role is to detect aircraft beyond the visual range of any observers at the aerodrome;

- intruder aircraft are non-cooperative and are not carrying VHF radio; and

- intruder aircraft may arrive at any time from any direction.

The intentions of the intruder aircraft are likely to be unknown. One option is for the intruder to overfly the airfield, join the circuit (traffic pattern) and land. Another option is that the intruder will transit the area. The intruder may also have a less-predictable flight path, such as general sight-seeing. As such the UA may need to move to a new location or change altitude to mitigate the risk of collision.

11.5.2.3   Speeds and Distances

It is quite common for an intruding GA aircraft to have a speed advantage over a small UA. This means the UA will not, in general, be able to outrun the intruder.

The speed advantage of the intruder limits the separation options of the UA. If the intruder is detected at 6 NM and is flying at 100 knots, then it will reach the aerodrome in 216 seconds (1 NM in 36 seconds). A UAS flying at 50 knots can only travel 3 NM in this time (1 NM in 72 seconds). If the UA is more than 3 NM away from the aerodrome then it will not be able to reach the circuit area before the intruder reaches the aerodrome.

11.5.2.4   The Minimum Altitude of an Aircraft

Aircraft should not generally fly near any city, town or populous area, at a height lower than 1000 ft above ground level (AGL). In other areas an aircraft should not fly lower than 500 ft AGL. Thus, the region between the ground and 500 ft AGL may provide a safe place for a UA to loiter when intruder aircraft approach the UAS operational area.

Near an aerodrome, by convention, the following circuit (traffic pattern) heights are flown [53]:

- High-performance aircraft, above 150 knots: 1500 ft AGL.

- Medium-performance aircraft, between 55 knots and 150 knots: 1000 ft AGL.

- Low-performance aircraft, maximum 55 knots: 500 ft AGL.

This structure provides an example of a procedure that separates aircraft with different performance characteristics. The structure also shows that aircraft are not expected below 500 ft.

11.5.2.5   Unmanned Aircraft Locations and Actions

There are four general geographic areas where the UA may be located. These areas have been defined below, based on the recommended course of action for the UA. These areas are:

1. **Danger** – The UA is between the intruder and the aerodrome but outside the circuit area. This configuration limits the time for the UAS to act.

2. **Transit** – The UA is outside the circuit area but it is in the likely transit area of the intruder.

3. **Hold** – The UA is outside the circuit area and is positioned laterally to the current flight path of the intruder.

4. **Circuit** – The UA is within the boundary where it can reach the circuit (traffic pattern) area before the intruder.

Figure 11.12 shows the layout of the four geographic areas with respect to where the UA may be located. The idea is to rotate the figure so that the danger area is oriented towards the intruder aircraft.

A 3 NM radius circuit area is shown for the Flamingo. In general, this radius is set by the speed capabilities of the UA. A 6 NM circle is shown for reference purposes. The aim is that this threshold distance and intruder speed will allow a calculation of the time until the aircraft reaches the aerodrome.

The UAS actions, when the UA is located in each area, are as follows:

Area	UA command
Danger	Descend to $\leq$ 400 ft AGL and orbit.
Transit	Track to a hold area and orbit.
Hold	Orbit or maintain speed and heading, whichever is safer.
Circuit	Track to the 'dead' side of the circuit at $\leq$ 400 ft AGL and orbit.

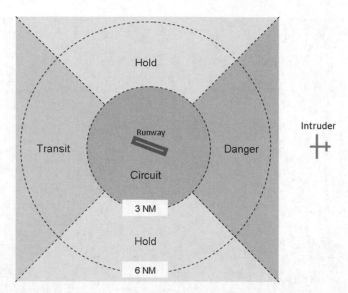

Figure 11.12 The UAS action map. The map shows the layout of the four geographic areas where the UA may be located. The map is rotated so that the danger area is oriented towards the intruder aircraft

The other options considered for the UA were landing or climbing. The general problem with landing the UA is that if there is an accident on landing, the UA will then become a hazard for the intruder aircraft as it lands. The general problem with commanding the UA to climb is that the climb rate of small UA is often too slow to be useful as an avoidance manoeuvre.

Commanding the UA to join the circuit was also considered. This idea was rejected because it would involve a small and difficult-to-see aircraft (the small UA) mixing with manned aircraft where the traffic density may be relatively high.

Zero conflict airspace has been proposed as one method of managing UAS operations in response to other air traffic [20]. The aim of Figure 11.12 and the associated UA actions is to provide a more flexible concept for managing operations.

11.5.2.6 Results

The Smart Skies project provided an opportunity to examine the operation of a UAS when an intruder aircraft approaches the operational area. In these experiments the ASL played the role of the intruder.

For these experiments the Flamingo unmanned aircraft was only permitted to operate within a 2 NM radius of the local airfield and from the ground to 2700 ft AMSL. This restricted the UA options that could be tested in the following experiments.

Figure 11.13 shows one example of the intruder experiment, which was performed on 5 August 2010. The radar and GPS-based 'truth' tracks for the ASL are shown in the figure.

The Flamingo UAS was being operated from its GCS, which was located on the airfield. The radar and GPS tracks for the Flamingo are also shown in the figure.

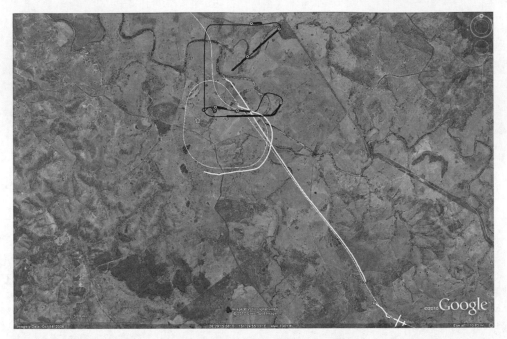

Figure 11.13 The radar tracks of the ASL (white) and the Flamingo UA (black). The thick lines show the radar tracks whilst the fine lines show GPS-based 'truth' tracks. The star symbol shows the location of the MATS. This experiment was performed on 5 August 2010

The MATS was located approximately 0.38 NM to the south-west of the Flamingo's GCS. This means that position reports from the MATS were provided to the UAS pilot using VHF voice radio calls.

Figure 11.14 shows a second example of the intruder experiment. The MATS was moved to a different location, approximately 0.6 NM to the north-east of the airfield. In this case, the MATS initially tracked the aircraft moving away from the airfield towards the north-east. The aircraft then turned towards the airfield.

Position reports from the MATS were again provided to the UAS pilot using VHF voice radio calls.

11.5.2.7 Discussion

Figure 11.13 and Figure 11.14 provide examples of the ASL acting as an intruder. The MATS tracked the intruder and radar-based position reports about the intruder were provided to the Flamingo UAS pilot using VHF voice radio calls.

The UAS pilot, located in the GCS, was responsible for operating the UAS. The UAS pilot was also responsible for acting on the position, speed and heading reports provided by the MATS operator.

The experiments confirmed that it was difficult to control the UAS and mentally track the location of the intruder from the radio reports. Receiving and comprehending the

Figure 11.14 The radar tracks of the ASL (white) and the Flamingo UA (black). The thick lines show the radar tracks whilst the fine lines show GPS-based 'truth' tracks. The star symbol shows the location of the MATS. This experiment was performed on 10 December 2010

radar-derived reports was found to lengthen the SAA timeline [20, 21]. As one report was being interpreted by the UAS pilot another radio broadcast would begin. Training and practice may help with the management of the workload. There is a suspicion, however, that if the number of aircraft in the vicinity increases then the UAS pilot will again be overwhelmed.

A common approach is to operate the UAS with two people inside the GCS: one operating the UAS and the other managing airspace-related issues. In this configuration the aim of the MATS is to provide information to the airspace manager. This method has been used successfully in other flight trials.

The next technological step is for the radar to display the real-time airspace picture inside the GCS. The radar's remote display, the TVW, provides this capability [46]. The TVW will also help to prioritise the UAS manoeuvres for multiple intruder scenarios. Some initial trials of using the TVW have also been successfully completed.

11.5.2.8   Summary

The aim of this section was to consider the practical problems associated with operating a UAS near a non-towered aerodrome. A strategy for manoeuvring the UAS when the MATS detects intruder aircraft was presented. The strategy is particularly relevant for small UAS where a GBSAA system is supporting their operations.

The examples demonstrated the ability of the MATS to detect and track the intruder aircraft. In these scenarios the radar and the UAS pilot form a pilot-in-the-loop SAA system: the MATS provides the 'sense' function and the UAS pilot provides the 'avoid' function by manoeuvring the UA.

The airspace picture provided by the MATS is frequently updated. This must be contrasted with the current situation where intermittent position and intention reports are only provided by radio-equipped aircraft.

GBSAA systems are able to detect and track aircraft at greater ranges than visual observers. This greater range translates into more operational freedom and more time to manoeuvre the UA well clear of any intruder aircraft.

11.5.3   Automated Sense and Avoid

One of the highlights of the Smart Skies flight test programme was the automated SAA experiments. These experiments used the MATS primary radar to provide the ADAC with information about non-cooperative aircraft. The ADAC's role was to control the cooperative aircraft.

The non-cooperative aircraft could not be controlled by the ADAC if a potential conflict was identified. Instead, the cooperative aircraft involved in the conflict could be issued with an updated flight plan to keep all the aircraft separated from each other.

The Smart Skies architecture was shown in Figure 11.2. The automated SAA experiments used this architecture in an integrated manner. In these experiments the MATS and the aircraft were located in Australia while the ADAC was situated in the United States. The 3G public mobile data network was used by the MATS to provide the ADAC with information about the non-cooperative aircraft.

In these experiments the ASL acted as a non-cooperative aircraft, which was tracked by the primary radar of the MATS. The cooperative aircraft used in the experiment were simulated, but this makes no difference to the ADAC as they behave like real cooperative aircraft: they produce periodic position reports and can receive updated flight plans from the ADAC.

Figure 11.15 shows an example of the results from the automated SAA experiments. Initially, the non-cooperative ASL (AID 5) and the cooperative virtual aircraft (AID 1) were following their respective flight plans. The cooperative aricraft (AID 1) was then issued with an updated flight plan, by the ADAC, to avoid the non-cooperative aircraft (AID 5), which was detected by the MATS. After the encounter the cooperative aircraft returned to its original flight plan.

Examples of automated SAA, for cooperative aircraft, were the focus of the early Smart Skies flight trials [47, 51, 52]. These experiments demonstrated the automatic control of aircraft from a remote location using the Internet.

Non-cooperative aircraft are part of today's aviation environment. A capability to detect and track these aircraft was provided by the MATS. The Smart Skies architecture has also provided the opportunity to extend automated SAA to include non-cooperative aircraft. Although only one example is provided here, the Smart Skies flight test programme used both diamond and circular flight plans to test automated SAA.

TCAS and ADS-B provide examples of cooperative technologies that are in use today. The Smart Skies network may be seen as an extension to these cooperative technologies, where additional information can be sent and received over the data links. The ADAC exchanges custom messages with a data-linked pFMS, which is on board each manned and unmanned

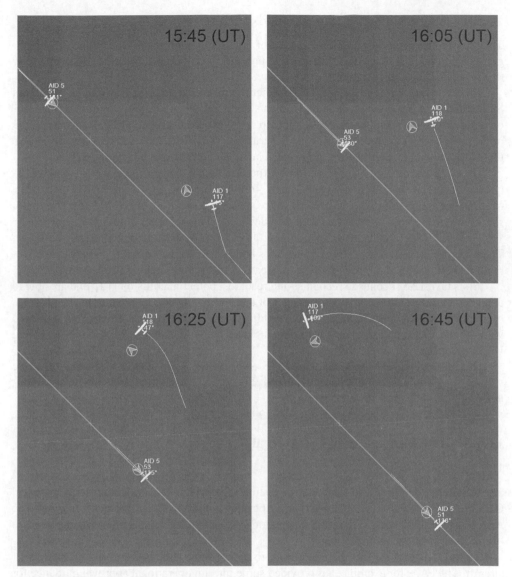

Figure 11.15   A sequence of displays that show the results from an automated SAA experiment. The straight line shows the original flight plan of each aircraft. The cooperative aircraft (AID 1) was then issued with an updated flight plan, by the ADAC, to avoid the non-cooperative aircraft (AID 5), which was detected by the MATS. The cooperative aircraft returned to its original flight plan after the encounter

aircraft [47]. These messages include the desired flight plan of an aircraft and a commanded flight plan, which is issued by the ADAC when a conflict is identified.

Today, small UAS are operating in an environment with both cooperative and non-cooperative aircraft. This section demonstrated the ability of the Smart Skies architecture to automatically manage a variety of aircraft from a remote location, in what is currently non-controlled airspace.

11.5.4   Dynamic Sense and Avoid Experiments

The Smart Skies project developed a number of SAA technologies [3]. One of these technologies is a computer vision system that aims to provide an on-board automated SAA capability that is at least as good as a pilot.

The dynamic sense and avoid (DSA) experiments were performed as part of the Smart Skies flight trial programme. The experiments involved the ASL flying a number of head-on and tail-chase encounters with a target aircraft. The aircraft were always separated by a minimum of 500 ft in altitude during the experiments. The ASL was used as the test platform that carried the computer vision equipment. The target aircraft was a Cessna 182P.

The MATS was used to observe these experiments passively. Figure 11.16 shows a sequence of screen shots from the experiments. Initially, the ASL is shown 5 NM north of the MATS. The target aircraft is located 6 NM to the east south-east of the MATS. The figure shows the two aircraft approaching and then departing the encounter point.

The DSA experiments provided an opportunity to monitor a simulated midair collision. The experiments also demonstrated the radar's ability to track two aircraft and provide information about the local airspace users in an area that is currently non-controlled airspace.

Pilots take 12.5 seconds to recognise and react to a potential collision [12], and are able to detect potential head-on collisions at ranges of approximately one nautical mile. The MATS, by contrast, was able to monitor each aircraft during the entire encounter. This real-time view of the local airspace users also demonstrates the value of the MATS in supporting UAS operations.

The networking capability of the MATS also allows radar-tracking information to be viewed at a remote location. Therefore, the MATS could be used to provide low-cost surveillance at a currently non-towered aerodrome.

11.5.5   Tracking a Variety of Aircraft

11.5.5.1   Introduction

The ASL provided the ideal platform for testing the MATS. The ASL represents a typical GA aircraft, which also represents a typical intruder aircraft from a radar point of view. The ASL was also fitted with a GPS–INS navigation system to provide real-time and independent 'truth' data about the aircraft's state.

For any GBSAA system there is a need to demonstrate the ability to track a variety of aircraft. The FAA, for example, has provided some interim operational approval guidance for UAS flight operations [23], which notes:

> If special types of radar or other sensors are utilized to mitigate risk, the applicant must provide supporting data which demonstrates that:
>
> • both cooperative and non-cooperative aircraft, including targets with low radar reflectivity, such as gliders and balloons, can be consistently identified at all operational altitudes and ranges, and,
> • the proposed system can effectively deconflict a potential collision.

The results shown in this section provide examples of the supporting data that may be included in applications for greater access to airspace for UAS operations.

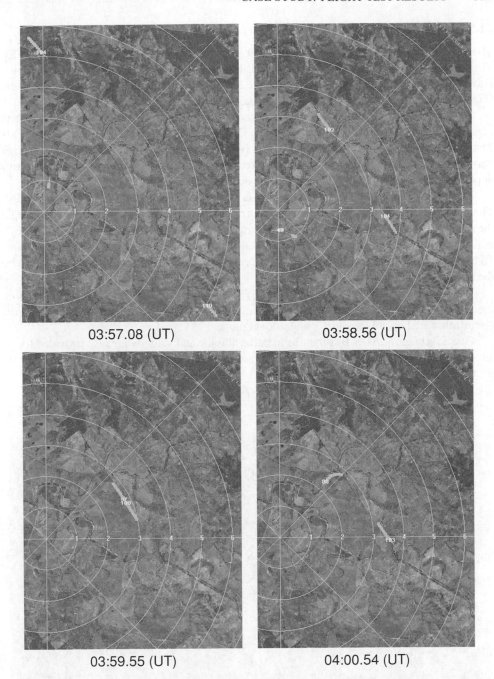

03:57.08 (UT)

03:58.56 (UT)

03:59.55 (UT)

04:00.54 (UT)

Figure 11.16 A sequence of MATS displays from the head-on SAA experiment conducted on 7 December 2010. The images are approximately 1 minute apart. The ASL is initially shown to the north of the MATS. The target aircraft is shown to the east south-east of the MATS. The aircraft speed, in knots, is shown on each track

11.5.5.2 Royal Flying Doctor

The MATS was operated at Kingaroy Airport, Queensland (26° 34′ 48″S, 151° 50′ 30″E). This provided an opportunity to track a Royal Flying Doctor Service aircraft, a twin-engine Hawker Beechcraft B200 King Air. Figure 11.17 shows the radar and ADS-B tracks for the arrival at and subsequent departure from Kingaroy Airport.

The figure shows that the aircraft approached from the south-east. The radar track starts approximately 10 NM away from the airport. The aircraft later departs the airport and heads east. The aircraft is tracked by the radar until it is 11 NM from the MATS. The ADS-B receiver was able to track the approach of the aircraft from 38 NM from the airport. When the aircraft departed it was tracked until 42 NM from the airport.

The correspondence between radar and ADS-B tracks has been reported elsewhere [54]. These tracks had a distinctive 'signature', occurring on predefined flight paths. In one example the received ADS-B information allowed the aircraft to be identified as a Boeing 777-2D7ER, which was on descent to Brisbane International Airport. At the closest point the aircraft was 16 NM from the MATS. A large variety of aircraft have since been simultaneously tracked by the radar and ADS-B systems.

The aim of an ADS-B receiver for the MATS is to: (a) provide detailed information about the aircraft that are also detected by the radar, and (b) provide information about equipped aircraft that are beyond the radar's operational range.

One advantage of ADS-B is that the identity, altitude and velocity are also provided for an aircraft. Therefore, ADS-B-equipped aircraft are able to act as alternative calibration sources for the primary radar.

Figure 11.17 Radar (thick lines) and ADS-B (fine line) tracks of the Royal Flying Doctor Service arriving at and then departing from Kingaroy Airport, Queensland, on 28 September 2010. The inset shows a photograph of the King Air

Figure 11.18   The radar tracks from a microlight flight on 13 July 2010. The inset shows a photograph of the microlight

11.5.5.3   Microlight

A wide variety of GA aircraft uses the Watts Bridge airfield. Figure 11.18 shows the MATS radar tracks from a microlight aircraft. As the figure shows, the microlight initially heads south-east, then flies around Mount Brisbane, over Somerset Dam, before returning to Watts Bridge. The figure shows that, for most of the flight, the aircraft stays within radio line-of-sight of the MATS.

One of the aims of this example is to demonstrate the ability of the MATS to track a small aircraft over an extended flight. This aircraft is also of interest because of its construction; it does not have a traditional aluminium-clad airframe.

The construction of an aircraft has a direct influence on its RCS. Obtaining a theoretical model of the RCS for an aircraft is a complex process. Field measurements, however, provide a practical way of understanding the ability of a radar to detect such aircraft.

11.5.5.4   Tracking Unmanned Aircraft

Section 11.4.3.1 noted that tracking the operational UAS is not the main mission of the MATS because the location of the UA is often known. The GCS typically receives real-time information from the UA under its control.

An example of tracking the ScanEagle® unmanned aircraft with the MATS was presented in [54]. The ScanEagle has a wing span of 3.11 m and a length of 1.37 m. Small UA will generally have a RCS that is smaller than GA aircraft. This means that the detection and tracking of small UA will occur at shorter ranges than for GA aircraft.

The Smart Skies flight test programme provided the opportunity to track the Flamingo unmanned aircraft. Details of the Flamingo were provided in Section 11.4.5. Figure 11.13 and Figure 11.14 provided examples of the MATS tracking the Flamingo. Figure 11.19 provides an example with a longer flight time. The figure shows both the radar tracks and the GPS-based 'truth' tracks.

The figure shows that tracking tends not to occur near the radar. At short ranges the position of the aircraft with respect to the radar is important. If the altitude of the UA is too high, for example, then it may not be within the radar's field of view. Operating the radar at a different location may be considered if tracking the UA throughout its flight is important.

A small systematic offset between the UA track and the radar tracks may be observed in Figure 11.19. The data shown is as recorded, but this offset could be calibrated out of the system if required. The offset is small and is approximately one long-pulse range resolution cell.

Figure 11.19 demonstrates the ability of the MATS to track a low RCS aircraft when it is within the radar's field of view. Why track an unmanned aircraft? One option is to use UAS to test and calibrate other systems, such as the MATS. Many UAS already log position, speed and other information as part of their operations. As this section has demonstrated, this independent 'truth' information is valuable for GBSAA systems.

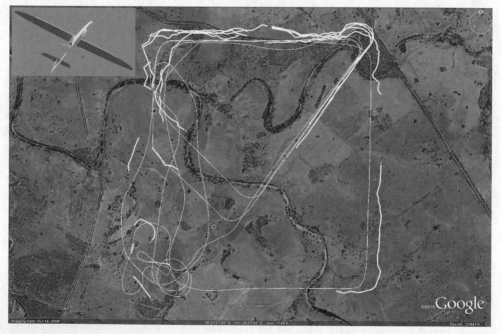

Figure 11.19 The MATS radar tracks of the Flamingo unmanned aircraft from the Smart Skies flight trial on 10 December 2010. The thick lines show the radar tracks whilst the fine line show the GPS-based 'truth' tracks. The star symbol shows the location of the MATS. The inset shows a photograph of the Flamingo

11.5.5.5   Summary

The MATS has demonstrated its ability to track a wide variety of aircraft, from large passenger aircraft to a small unmanned aircraft. The ability to track non-cooperative and cooperative targets was demonstrated. One example also showed the close correspondence between radar and ADS-B tracks. A microlight and an unmanned aircraft demonstrated the ability of the MATS to track low-RCS targets.

Targets of opportunity are valuable for demonstrating a radar's ability to track a variety of aircraft. Limited tracking opportunities and other unknowns, such as altitude, can make the radar performance information from these aircraft difficult to generalise.

On-board GPS logging and ADS-B systems can provide independent information about an aircraft. This information may be used to calibrate and test GBSAA systems.

Targets of opportunity also demonstrate the ability of the MATS to provide information about the local airspace users to the UAS pilot. These aircraft provide realistic examples of unknown aircraft approaching and departing an aerodrome at unpredictable times and directions. Other studies have also shown the ability of radars to improve the situation awareness of UAS pilots [55].

11.5.6   Weather Monitoring

Weather information has always been important to aviation. With advances in technology, more accurate information is available to assess the current conditions and to make forecasts.

Many countries now have dedicated weather radar networks. The individual radars are often located to provide coverage over major population centres. From a UAS perspective these radars will not always be able to provide coverage in the required operational area. This section shows that the MATS may be used by the UAS pilot to sense and avoid weather.

Figure 11.20 shows a sequence of four clutter maps from the MATS radar taken at 30-minute intervals. Within 5 NM the clutter is relatively static, from image to image, as this is ground clutter from the local terrain.

One of the largest features is the storm that is initially located to the south-west of the MATS. The images show that the storm moves towards the radar over the next hour and a half.

The images also show other rain areas. One less-intense rain area is initially observed towards the north-west of the MATS. The rain is seen to move south-east until it is located to the north-east of the radar in the last image.

Other smaller rain areas may be seen to move between each image. These other areas can be tracked by using the higher time resolution data that is available – the radar has an update rate of 2.5 seconds.

This example demonstrates the ability of the MATS to provide information about precipitation within the radar's coverage. This near real-time weather information is useful because it provides a picture of the conditions that may affect UAS operations.

The calibration and use of an X-band marine radar as a cost-effective weather radar has been studied [56]. The cost per installation was estimated to be less than one-sixth of that of a state-of-the-art X-band weather radar system. The use of low-cost and low-infrastructure X-band radars for meteorology has also been investigated [57]. The study compared the results of the X-band radar with an S-band weather radar and reported 'outstanding' results. Therefore, the MATS is able to provide valuable weather information that can support UAS operations.

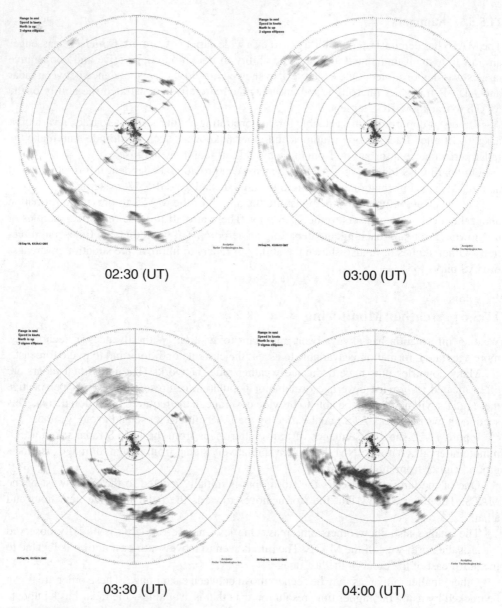

02:30 (UT)   03:00 (UT)

03:30 (UT)   04:00 (UT)

Figure 11.20 The radar clutter maps show the approach of a storm on 28 September 2010. The clutter signal strength is displayed using a greyscale, where black indicates the strongest clutter. The time between each image is 30 minutes. The range rings are in 5 NM increments

11.5.7 The Future

The next phase of development of the MATS is likely to be an upgrade of the standard antenna system. The aim is to reduce the effects of ground clutter and increase the radar's vertical coverage.

The data from the MATS sensors are not currently fused in real time. Work is underway to present a fused common operating picture (COP) to the UAS pilot, which is similar to what is provided to air traffic controllers [58]. This fused picture will make it easier to identify non-cooperative aircraft – an important aspect of operating in the NAS.

11.6   Conclusion

Unmanned aircraft require a capability to sense and avoid in order to gain greater operational freedom within the NAS. Many solutions to this problem are being developed. In this chapter we have demonstrated the ability of a low-cost mobile radar system and an ADS-B receiver to support UAS operations.

GBSAA systems are particularly relevant to small unmanned aircraft as they often have a very limited capacity to carry additional sensors or systems. GBSAA systems can, of course, support a wide variety of unmanned aircraft. One advantage of GBSAA systems is that they are able to support UAS operations without any modifications to the unmanned aircraft.

The MATS was tested as part of the Smart Skies project using a specially equipped aircraft, which flew predetermined flight plans, and with other aircraft of varying shapes and sizes. The flight trial results show that the MATS is able to support UAS operations in non-segregated airspace within the NAS.

The flight trial results show that the local terrain can have an influence on the performance of GBSAA systems. As a result, some consideration must be given to the positioning of the system to achieve the desired surveillance coverage. Another aspect of the operating environment is the weather. The flight test results also demonstrated the ability of radars to monitor rainfall, which is useful information for UAS operations.

The military also has an interest in flying UAS in the NAS. This interest includes training, research and development and testing. The US Department of Defense has provided an incremental NAS Access Strategy that starts with line-of-sight operations [41]. The next step uses a GBSAA system to support terminal area operations. The subsequent steps all involve the use of GBSAA systems to support the UA transiting airspace. Thus, GBSAA systems are likely to be a key component of a layered approach to the integration of UAS into the NAS.

GBSAA systems can help to make a case for 'file and fly' UAS operations. These routine and regular operations will provide important experience with operating UAS in the NAS, a key ingredient for conducting civilian and commercial operations using unmanned aircraft.

Acknowledgements

I would like to thank the Smart Skies team for all their efforts during the project. In particular I would like to thank Mr Duncan Greer, Mr Rhys Mudford, Mr Scott NcNamara, Mr Ryan Fechney, Mr Chris Turner and Professor Rod Walker, who were the pilots and crew of the ASL and made an important contribution to the MATS. The constructive comments of all the reviewers were greatly appreciated. I would also like to thank Insitu Pacific Limited for making the MATS available during the Smart Skies project.

The expert advice of Mr Peter Rogers was also greatly appreciated.

This research is part of the Smart Skies Project and is supported, in part, by the Queensland State Government Smart State Funding Scheme.

References

1. 'Comprehensive set of recommendations for sUAS regulatory development', Small Unmanned Aircraft System Aviation Rulemaking Committee, April 2009.

2. 'Civil aviation safety regulations 1998 (CASR) part 101, unmanned aircraft and rocket operations', Civil Aviation Safety Authority (CASA), 2010.

3. R. Clothier, R. Walker, R. Baumeister, M. Brünig, J. Roberts, A. Duggan, and M. Wilson, 'The Smart Skies Project', *IEEE Aerospace and Electronic Systems Magazine*, 26, 14–23, 2011.

4. 'Unmanned aerial vehicles roadmap: 2000–2025', Office of the Secretary of Defense, April 2001.

5. M. DeGarmo, 'Issues concerning integration of unmanned aerial vehicles in civil airspace', Tech. Rep. MP 04W0000323, The MITRE Corporation, 2004.

6. 'Annex 11 to the convention on civil aviation: Air traffic services', International Civil Aviation Organization (ICAO), 2001.

7. 'Limitations of the see-and-avoid principle', Research Report ISBN 0 642 16089 9, Australian Transport Safety Bureau (ATSB), 1991.

8. C. Morris, 'Midair collisions: Limitations of the see-and-avoid concept in civil aviation', *Aviation, Space, and Environmental Medicine*, 76(4), 357–365, 2005.

9. J. W. Andrews, 'Unalerted air-to-air visual acquisition', Lincoln Laboratory, Massachusetts Institute of Technology, November 1991.

10. R. Wolfe, 'NASA ERAST Non-Cooperative DSA Flight Test', *AUVSI Conference Proceedings*, pp. 1–15, July 2003.

11. R. J. Kephart and M. S. Braasch, 'Comparison of see-and-avoid performance in manned and remotely piloted aircraft', *Digital Avionics Systems Conference*, vol. 25, pp. 4.D.2-1–4.D.2-8, May 2008.

12. 'Pilots' role in collision avoidance', Advisory Circular 90-48C, Federal Aviation Administration, 1983.

13. K. Colvin, R. Dodhia, and R. Dismukes, 'Is pilots' visual scanning adequate to avoid mid-air collisions?', *Proceedings of the 13th International Symposium on Aviation Psychology*, pp. 104–109, 2005.

14. 'Introduction to TCAS Version 7', Federal Aviation Administration, November 2000.

15. R. C. Matthews, 'Characteristics of U.S. midairs', *FAA Aviation News*, 40, 1–3, 2001.

16. 'Review of midair collisions involving general aviation aircraft in Australia between 1961 and 2003', Research Report B2004/0114, Australian Transport Safety Bureau (ATSB), May 2004.

17. J. N. Simon and M. S. Braasch, 'Deriving sensible requirements for UAV sense-and-avoid systems', *Digital Avionics Systems Conference, DASC*, pp. 6.C.4-1–6.C.4-12, October 2009.

18. A. Lacher, D. Maroney, and A. Zeitlin, 'Unmanned aircraft collision avoidance – technology assessment and evaluation methods', *The 7th Air Traffic Management Research & Development Seminar*, pp. 1–10, 2007.

19. D. Seagle, 'NATO developments in UAS airworthiness and sense & avoid functional requirements', *International Council of the Aeronautical Sciences (ICAS)*, pp. 1–22, 2007.

20. 'Sense and avoid (SAA) for unmanned aircraft systems (UAS)', Federal Aviation Administration, October 2009.

21. A. Zeitlin, 'Issues and tradeoffs in sense & avoid for unmanned aircraft', *IEEE Systems Conference*, pp. 61–65, April 2010.

22. 'Airspace integration plan for unmanned aviation', Office of the Secretary of Defense, November 2004.

23. 'Interim Operational Approval Guidance, 08-01, Unmanned Aircraft Systems Operations in the U.S. National Airspace System', Federal Aviation Administration, AIR-160, 2008.

24. M. Contarino, 'All weather sense and avoid system for UASs', Report to the Office of Naval Research for R3 Engineering, 2009.

25. A. Lacher, A. Zeitlin, D. Maroney, K. Markin, D. Ludwig, and J. Boyd, 'Airspace integration alternatives for unmanned aircraft', *AUVSI's Unmanned Systems Asia-Pacific*, pp. 1–19, February 2010.

26. S. Hottman, K. Hansen, and M. Berry, 'Literature review on detect, sense, and avoid technology for unmanned aircraft systems', Tech. Rep. DOT/FAA/AR-08/41, Federal Aviation Administration, 2009.

27. Y. K. Kwag and C. H. Chung, 'UAV based collision avoidance radar sensor', *International Geoscience and Remote Sensing Symposium, IGARSS*, pp. 639–642, 2007.

28. B. Korn and C. Edinger, 'UAS in civil airspace: Demonstrating "sense and avoid" capabilities in flight trials', *Digital Avionics Systems Conference, DASC*, pp. 4.D.1-1–4.D.1-7, October 2008.

29. D. Longstaff, M. AbuShaaban, and S. Lehmann, 'Design studies for an airborne collision avoidance radar', *Proceedings of the 5th EMRS DTC Technical Conference*, pp. 1–9, 2008.

30. S. Kemkemian, M. Nouvel-Fiani, P. Cornic, and P. Garrec, 'MIMO radar for sense and avoid for UAV', *Phased Array Systems and Technology (ARRAY), 2010 IEEE International Symposium*, pp. 573–580, October 2010.

31. R. Carnie, R. Walker, and P. Corke, 'Image processing algorithms for UAV "sense and avoid"', *Proceedings of the 2006 IEEE International Conference on Robotics and Automation*, pp. 2848–2853, 2006.

32. J. Lai, L. Mejias, and J. Ford, 'Airborne vision-based collision-detection system', *Journal of Field Robotics*, 28(2), 137–157, 2011.

33. 'Standard Specification for Design and Performance of an Airborne Sense-and-Avoid System', F2411-07, ASTM International, 2007.

34. B. Karhoff, J. Limb, S. Oravsky, and A. Shephard, 'Eyes in the domestic sky: An assessment of sense and avoid technology for the army's "Warrior" unmanned aerial vehicle', *Proceedings of the 2006 Systems and Information Engineering Design Symposium*, pp. 36–42, April 2006.

35. 'Guidance Material on Comparison of Surveillance Technologies (GMST)', Tech. Rep. Edition 1.0, International Civil Aviation Organization (ICAO) – Asia and Pacific, September 2007.

36. P. Howland and M. R. Walbridge, '"What is the range of your radar?" and other questions not to ask the radar engineer', *IEE Colloquium on Specifying and Measuring Performance of Modern Radar Systems*, pp. 3/1–3/9, March 1998.

37. P. Weber and T. Nohara, 'Device and method for 3D height-finding avian radar', US Patent 7,864,103 B2, 2011.

38. T. McGeer, 'Safety, economy, reliability, and regulatory policy for unmanned aircraft', *Aerovel Corporation*, pp. 1–9, March 2007.

39. M. R. Endsley, 'Toward a theory of situation awareness in dynamic systems', *Human Factors*, 37(1), 32–64, 1995.

40. M. R. Endsley, B. Bolte, and D. G. Jones, *Designing for Situation Awareness: An approach to human-centered design*. Taylor & Francis, London, 2003.

41. 'Final Report to Congress on Access to National Airspace for UAS', US Department of Defense, 2010.

42. C. E. Fossa, R. A. Raines, G. H. Gunsch, and M. A. Temple, 'An overview of the IRIDIUM low Earth orbit (LEO) satellite system', *IEEE National Aerospace and Electronics Conference*, pp. 152–159, 1998.

43. S. Pratt, R. Raines, C. Fossa, and M. A. Temple, 'An operational and performance overview of the IRIDIUM low earth orbit satellite system', *IEEE Communications Surveys & Tutorials*, 2(2), 2–10, 1999.

44. F. Gonzalez, R. Walker, N. Rutherford, and C. Turner, 'Assessment of the suitability of public mobile data networks for aircraft telemetry and control purposes', *Progress in Aerospace Sciences*, 47(3), 240–248, 2011.

45. G. Spence and D. Allerton, 'Simulation of an automated separation management communication architecture for uncontrolled airspace', *AIAA Modeling and Simulation Technologies Conference*, pp. 1–8, 2009.

46. T. Nohara, P. Weber, G. Jones, A. Ukrainec, and A. Premji, 'Affordable high-performance radar networks for homeland security applications', *IEEE Radar Conference*, pp. 1–6, 2008.

47. R. Baumeister, R. Estkowski, and G. Spence, 'Automated aircraft tracking and control in Class G airspace', *International Council of the Aeronautical Sciences*, pp. 1–13, 2010.

48. P. Weber, A. Premji, T. Nohara, and C. Krasnor, 'Low-cost radar surveillance of inland waterways for homeland security applications', *IEEE Radar Conference*, pp. 134–139, 2004.

49. 'Minimum operational performance standards for 1090 MHz extended squitter Automatic Dependent Surveillance – Broadcast (ADS-B) and Traffic Information Services – Broadcast (TIS-B)', December 2009.

50. D. Greer, R. Mudford, D. Dusha, and R. Walker, 'Airborne systems laboratory for automation research', *International Council of the Aeronautical Sciences*, pp. 1–9, 2010.

51. R. Baumeister, R. Estkowski, G. Spence, and R. Clothier, 'Evaluation of separation management algorithms in Class G airspace', *AIAA Modeling and Simulation Technologies Conference*, no. AIAA-2009-6126, pp. 1–14, 2009.

52. R. Baumeister, R. Estkowski, G. Spence, and R. Clothier, 'Test architecture for prototyping automated dynamic airspace control', *Council of European Aerospace Societies (CEAS), European Air and Space Conference*, pp. 1–14, 2009.

53. 'Operations in the vicinity of non-towered (non-controlled) aerodromes', CAAP 166-2, Civil Aviation Safety Authority (CASA), 2010.

54. M. Wilson, 'A mobile aircraft tracking system in support of unmanned aircraft operations', *International Council of the Aeronautical Sciences*, pp. 1–11, 2010.

55. J. Denford, J. Steele, R. Roy, and E. Kalantzis, 'Measurement of air traffic control situational awareness enhancement through radar support toward operating envelope expansion of an unmanned aerial vehicle', *Proceedings of the 2004 Winter Simulation Conference*, pp. 1017–1025, 2004.

56. R. Rollenbeck and J. Bendix, 'Experimental calibration of a cost-effective X-band weather radar for climate ecological studies in southern Ecuador', *Atmospheric Research*, 79, 296–316, 2006.

57. G. A. Pablos-Vega, J. G. Colom-Ustáriz, S. Cruz-Pol, J. M. Trabal, V. Chandrasekar, J. George, and F. Junyent, 'Development of an off-the-grid X-band radar for weather applications', *IEEE International Geoscience and Remote Sensing Symposium (IGARSS)*, pp. 1077–1080, July 2010.

58. 'Guidance Material on Issues to be Considered in ATC Multi-Sensor Fusion Processing Including the Integration of ADS-B Data', International Civil Aviation Organization (ICAO), 2008.

59. 'Unmanned Aircraft Systems (UAS). ICAO Circular 328', International Civil Aviation Organization (ICAO), 2011.

Epilogue

This book comes to you as a result of the concerted effort of a team of experts in the area of unmanned aircraft systems (UAS) and sense and avoid (SAA), specifically. They were led and organised by the Editor, who himself had the opportunity and privilege to work in a series of large-scale (multi-million) projects in this area in the UK and EU. He is also a leading authority and one of the pioneers of the autonomous learning and dynamically evolving/adaptive intelligent systems. The other contributors include:

- George Limnaios, Nikos Tsourveloudis, and Kimon Valavanis (Technical University of Crete, Greece and University of Denver, CO, USA), the authors of the first chapter which introduces the topic including its historical prospective.

- Andrew Zeitlin (MITRE, USA), the author of the second chapter which focuses on performance tradeoffs and the development of standards.

- Pablo Royo, Eduard Santamaria, Juan Manuel Lema, Enric Pastor, and Cristina Barrado (Technical University of Catalonia, Spain), the authors of the third chapter which describes the integration of SAA capabilities into a UAS distributed architecture for civil applications; this chapter is very valuable, because it provides a systems point of view and puts SAA into the context of UAS as a whole which is not the main topic of this book.

- Xavier Prats, Jorge Ramirez, Luis Delgado, and Pablo Royo (Technical University of Catalonia, Spain) who authored Chapter 4, on regulations and requirements. This chapter is also very interesting because it starts the topic of human factors, regulations and requirements which (somewhat paradoxically) are a serious (often impeding) element of the implementation of UAS, especially in non-segregated airspace and more routine scenarios.

- Marie Cahhilane, Chris Baber, and Caroline Morin (Cranfield and Birmingham Universities, UK), the authors of Chapter 5 which provides a thorough and professional analysis of the human factors and their role in UAS.

Sense and Avoid in UAS: Research and Applications, First Edition. Edited by Plamen Angelov.
© 2012 John Wiley & Sons, Ltd. Published 2012 by John Wiley & Sons, Ltd.

- Stepan Kopriva, David Sislak, and Michal Pechoucek (Czech Technical University, Prague, Czech Republic), the authors of Chapter 6 on SAA concepts. This chapter starts the more technical part of the book covering the methodology of the SAA problem for the vehicle-to-vehicle case.

- Hyo-Sang Shin, Antonios Tsourdos, and Brian White (Cranfield University, UK); the authors who represent the largest European Defence Academy and have extensive research, development, and educational experience in the area propose in Chapter 7 a thorough examination of the UAS conflict detection and resolution problem from the point of view of differential geometry.

- Richard Baumeister (Boeing, USA) and Graham Spencer (Aerosoft Ltd, UK) authored Chapter 8, which focuses on aircraft separation management using common information network SAA; this approach puts the SAA and UAS in the context of the network-centric warfare concept which is regarded as the future approach.

- David Sislak, Premysl Volf, Stepan Kopriva, and Michal Pechoucek (Czech Technical University, Prague, Czech republic), the authors of Chapter 9 on AgentFly.

- John Lai, Jason Ford, Luis Mejias, Peter O'Shea, and Rodney Walker (Australian Research Centre for Aerospace Automation and Queensland University of Technology, Australia), the authors of Chapter 10, who provide a detailed report of the visual-based SAA solution that is vital for electro-optical and passive approaches. Sadly, Rod Walker passed away on October 2011 while the book was in production. This book is dedicated to his memory.

- Michael Wilson (Boeing R&T-Australia), the author of the last Chapter 11 on the use of low-cost mobile radar systems for small UAS SAA which provides another interesting approach to the problem based on active (radar) but mobile and cheap systems.

The topic of sense and avoid is pivotal for the viability of the UAS, which themselves are key to the future capability of defence and have huge potential for civilian applications with a great impact on society, the economy, and the environment. The problem may look trivial since the skies are not as congested as our roads and highways, but it is clear from the previous chapters that to reach the requirements of the regulatory authorities and/or to use military UAS safely, a level of safety comparable to (or not less than) that achieved by human-piloted vehicles should be obtained. This is no trivial task, and especially when the only information that can be used comes from passive sensors. However, the direction of the technical, scientific, and technological progress points towards a future with UAS and therefore, this book is very important in its pioneering role and will, quite certainly, be followed by others.

Index

Sense and Avoid in UAS: Research and Applications, First Edition. Edited by Plamen Angelov.
© 2012 John Wiley & Sons, Ltd. Published 2012 by John Wiley & Sons, Ltd.